MATHEMATICAL INVESTIGATIONS

AN INTRODUCTION TO ALGEBRAIC THINKING

DeMarois • McGowen • Whitkanack

INSTRUCTOR'S RESOURCE MANUAL

Phil DeMarois
William Rainey Harper College

Mercedes McGowen
William Rainey Harper College

Darlene Whitkanack
Indiana University

ADDISON-WESLEY

An imprint of Addison Wesley Longman, Inc.

Reading, Massachusetts • Menlo Park, California • New York • Harlow, England
Don Mills, Ontario • Sydney • Mexico City • Madrid • Amsterdam

Reproduced by Addison-Wesley Educational Publishers Inc. from camera-ready copy supplied by the authors.

Copyright © 1998 Addison-Wesley Educational Publishers Inc.

All rights reserved. No part of this publication may be reproduced, stored in a retrieval system, or transmitted, in any form or by any means, electronic, mechanical, photocopying, recording, or otherwise, without the prior written permission of the publisher. Printed in the United States of America.

ISBN   0-321-40031-3

1 2 3 4 5 6 7 8 9 10 VG 00999897

CONTENTS

Instructor Notes — 1

    Chapter 1: Instructor Notes — 2
    Chapter 2: Instructor Notes — 14
    Chapter 3: Instructor Notes — 20
    Chapter 4: Instructor notes — 26
    Chapter 5: Instructor Notes — 33
    Chapter 6: Instructor Notes — 38
    Chapter 7: Instructor Notes — 45

Classroom Management — 51

    Creating an Active Learning Environment — 52
    Learning with Others — 54
    Collaborative Groups — 56
    Policy Statement — 59
    Guidelines for Group Behavior — 63
    Team Evaluation Form — 64
    Course Outline (16 week semester) — 65
    Daily Agenda 2 — 66

Section Answers — 67

    Chapter 1 What is Mathematics? — 68
    Chapter 2 Whole Numbers: Introducing a Mathematical System — 77
    Chapter 3 Functional Relationships — 89
    Chapter 4 Integers: Expanding a Mathematical System — 114
    Chapter 5 Rational Numbers: Further Expansion of a Mathematical System — 145
    Chapter 6 Real Numbers: Completing a Mathematical System — 169
    Chapter 7 Answering Questions with Linear and Quadratic Functions — 196

Assessment: Group and Individual Skills Exams 229

    Group Exams   230
    Individual Skill Eams   245
    Final Exam Review   255
    Final Exam (Skills)   257

Alternative Forms of Assessment 261

    Concept Maps   262
    Entrance and Exit Slips   268
    Journals 1   269
    Minute Quizzes   280
    Portfolios   289
    Assessment Rubrics   290

Blackline Masters 295

TI-82 Graphing Calculator Reference Manual 343

TI-83 Graphing Calculator Reference Manual 395

TI-82 Index of Procedures and TI-83 Index of Procedures 449

    Index of TI–82 Procedures   450
    Index of TI–83 Procedures   453

Instructor Notes

CHAPTER 1: INSTRUCTOR NOTES

What is Mathematics?

Rationale

This chapter introduces the essential ideas of algebra that students must learn if they are to be successful in algebra. The book begins with the **Full Parking Lot** problem, a problem that every student can solve, using one of many different strategies. It provides students with an initial successful experience and with the opportunity to observe the variety of ways a given problem can be explored. A second problem, **Sandwich Purchase**, introduces students to a problem that has multiple solutions.

The chapter continues with the SPC investigation, which encapsulates what mathematics and this text are all about. It contains the most important mathematics in the text and is referenced in subsequent investigations throughout the text.

The SPC investigations provide a strong foundation to explicitly deal with

- structure
- notation
- vocabulary
- arbitrary, but reasonable, rules
- justification
- logic (the problem with converses)
- reflection
- problem solving strategies and multiple concepts embedded in a single, rich problem.

The concept of variable, which research has shown is imperfectly understood, is introduced in a nonstandard way, generalizing the rules of the SPC system, so that students can examine their assumptions and previous misconceptions about variables. The chapter concludes with an investigation arising out of the SPC problem that introduces exponential expressions and two basic properties of exponents.

Be aware of administrative details that are a normal part of the first class meeting activities. Provide for those details in a time frame you feel appropriate. We've found it helpful to postpone some, if not all administrative details, until the second class meeting, using that first meeting with students to set the tone for what we expect of students, having them get into the mathematics and work with partners and/or groups. Having students take the syllabus home, along with the first assignment, provides a basis for lively discussion the second class meeting.

Help students become accustomed to the structure of each section including the:

- statement of purpose

- student discourses between Pete and Sandy which are designed to help make connections between sections and to clarify specific mathematical points that often bother students.

- investigations in which students answer questions designed to facilitate their discovery of key mathematical ideas

- discussions which provide some answers to the investigations and provide students with a presentation of the key mathematical ideas

- explorations which allow students to check their understanding of the section and to review previously-learned topics

- reflection which provides the opportunity to reflect and then write about a major mathematical idea from the just completed section.

Students should also be aware that the last section of each chapter provides the opportunity for review. Students are asked to reflect on and write about the key mathematical ideas of the chapter. The set of review questions are provides to help students assess their understanding of the material. The review section often contains questions from all previously-covered chapters, not just the one last completed. Several concept maps are suggested. Use the material on concept maps in this manual to help students understand the assignment that appears at the end of every chapter in this book. Finally, a reflection asks students to write about a key idea of the chapter.

Finally, make sure students notice the blank glossary and the index at the back of the text. The first question of each set of explorations asked them to complete a portion of the glossary by writing definitions of the key words (printed in ***italic boldface***) listed in the section.

Overview

Section 1.1 promotes a classroom environment that encourages active individual and collaborative learning in which success, reflection and collaboration become natural results of the investigations done in class. The numerical approach through the use of tables, especially in the form of Table 2, is helpful for students who are having trouble with algebraic representations and with developing generalizations. While the first problem situation has only one solution, the second, the **Sandwich Purchase** problem, has several solutions.

Section 1.2 explicitly addresses the notion of *metacognition*–reflecting about how one solved the problem. The important ideas of structure and reflecting while working inside a system and recognizing when you are working outside the system are necessary if you are to be successful in mathematics and do more than just find answers. The issues of vocabulary and Standards of Mathematical Notation are explicitly addressed.

Section 1.3 introduces variables in the context of generalizing a rule of the SPC system, rather than as place-holders for missing numbers. Investigations and discussion provide students opportunities to examine their beliefs and misconceptions about variables. The concept of domain is introduced. Assumptions about variables are stated. The table feature on the calculator is introduced as a way of exploring algebraic properties numerically.

Section 1.4 expands the notion of variable and revisits the concept of domain. Students are asked to reflect on their previous learning of mathematics, distinguishing between processes and the results of processes, often expressed using the same notation. In context of the SPC system, nonzero whole number exponents and two exponent properties are introduced.

Section 1.5 is a review section.

Allow approximately 5–7 75-minute class sessions for Chapter 1. Part of this time must be devoted to organizational details and to building groups and a class community. Be careful not to spend too much time on Section 1.2. Keep the class moving through this section. If some students are uncomfortable, they will gain more confidence by seeing where the SPC system is going.

Section 1.1 Learning Mathematics

Pacing   1–2 75-minute classes including administrative details.

Technology

A graphing calculator (TI-82 or TI-83, for example) can be viewed as a tool for students to use to assist their study of mathematics. Introduce the calculator gently on an as-needed basis to help investigate the mathematical ideas. The calculator can be learned incrementally during "teachable moments". Early uses may include how to do calculations and how to use the table feature since many investigations in this book rely on analyzing tables of data. Students already have much that is new to cope with during the first couple of days so the introduction of technology needs to be gentle. On the other hand, introducing the calculator during discussion of the **Full Parking Lot** problem clearly sends a message that the calculator is a necessary part of the course.

The following are some ways the calculator might be employed in Section 1.1:

- It is helpful for students not confident in their arithmetic. If students are having trouble making a table, being able to do the calculation on the calculator may show them what to put in the table.

- Using a calculator with a table feature, Table 1 can be set up with x = number of cars, $y_1 = 4x$, $y_2 = 20 - x$, $y_3 = 2y_2$, and $y_4 = y_1 + y_3$, where the equations give the number of car wheels, the number of motorcycles, the number of motorcycle wheels and the total number of wheels. Use the TblSet feature of the calculator to set the TblStart = 0 and the ΔTbl = 1. (Refer students to the TI-82/83 reference manual for the procedure in entering equations and setting the table minimum and increment values).

Comments

Students initially work alone on the **Full Parking Lot** problem. They should be encouraged to write in the text as they answer the investigations. This problem can be put on an overhead from the blackline masters if students do not yet have a text.

Emphasize to students that they can solve the problem. For some students there is such a fear of failure that failure is guaranteed because the student never gets started. Encourage every student to write down his/her thoughts and questions about the problem and to identify what information is important. After some time to initially think about the problem, the student should be encouraged to work with another student or in a group.

Students then work with a partner, discussing their questions and ideas. Two teams share their collective efforts and compare methods of solution. Students become acquainted with other members of the class and begin the process of collaboratively working together. By first working in pairs, each student is comfortable contributing to the discussion.

Students should explore additional strategies together in Investigations 8–11. A possible assignment: Working together, find as many solutions as possible and be prepared to explain to the class. Students should continue to write in the text as they answer the investigations. Members of the groups share solutions and justification with the class. Many students are reluctant to share their solutions in the class as they have solved the problem by trial and error. Some students believe it is not appropriate to solve a problem using trial and error and "just numbers". They feel that, if they cannot write an algebraic equation, they are not doing mathematics.

Note that when asked for a justification, students are expected to make a statement about why their response makes sense–not a deductive proof. Using the criteria that a good proof is an argument which convinces someone of the correctness of an idea, deductive proofs that make no sense to students are not expected nor encouraged, even though they are mathematically correct.

Class time can be spent discovering and discussing the ways various students found the solution. Validate each of these solutions as a viable way of looking at the problem.

The purpose of listing the problem solving strategies is to put names to the solution processes that students commonly use to solve this problem. They will be using these strategies during the rest of the semester. Do not take time to teach all of the problem-solving heuristics listed in this section. Do take time to:

- Discuss how to create a table.

- Discuss the criteria of a good solution since this is an important reflection about how much work one should do to a problem after finding an answer.

- Ask students to describe which solution they thought was the best solution and why.

Maintain an open-ended discussion and avoid choosing a "best solution" yourself.

- Have students analyze what information can be determined from Table 2 compared with the information provided in Table 1. A discussion about why recording work, not answers, when looking for patterns is helpful as the class analyzes both tables.

- Encourage students **not** to evaluate their calculations in Table 2. Students have been conditioned to "get the answer" and, as a result, ignore the information that is most useful when trying to discover a pattern or write a generalization.

- Questions that help assess the students' understanding of some basics about tables include:

    1. How the column titles were chosen?

    2. Where do the numbers come from?

    3. How do you choose numbers that make sense?

    4. Why are the values for the independent variables chosen sequentially?

    5. What helps you see patterns more clearly?

    6. Why was c used in place of numbers in the last row?

Once the **Full Parking Lot** problem has be fully investigated, let students work in groups on the **Sandwich Purchase** problem. Unlike many of the investigations, answers to **Sandwich Purchase** are not given so that students have the opportunity to fully explore the problem. Emphasize that all solutions should be found and that students should be able to justify that they have found all solutions. Conjecture why multiple solutions occur.

It is important that students feel responsible for their own learning and that they develop responsibility for the learning of members of their group. This development occurs over time and needs to be encouraged. As students move from working with a partner to working collaboratively within a group during the course, this new role for students is demonstrated in class. Working together outside of class should be encouraged. Students need to be given permission to strictly define the responsibilities of each member of the group. Your non-judgemental support is necessary to help groups work through problems and to gain the cooperation of students who may not have developed appropriate study habits supportive of collaborative work prior to enrolling in this course.

Explorations

In general, it is recommended that all explorations be assigned to be completed outside class with discussion during the following class period. The discussion can be done in small groups, with questions and interesting observations shared during whole class discussion.

This set of explorations is particularly short to allow students to reflect on what they have done. Specific comments follow.

- 1–5 allow students to reflect further on various strategies for the **Full Parking Lot** problem.

- 6 is a new problem that has the same structure as the **Full Parking Lot** problem. Do students recognize this similar structure for two very different contexts?

Section 1.2 Thinking Mathematically

Pacing   1–2 75-minute classes

Technology

The calculator is not expressly used in this section. It is possible, depending upon where student questions take the class discussion, that further explorations of the SPC problem using the calculator could occur, given that "teachable moment". Students have occasionally anticipated the discussion about deriving the formula to generate possible paths for a given P-count and the graphing calculator is an appropriate tool for this investigation.

Comments

The **Student Production Corporation (SPC)** problem is the focus of this section. This system has been successfully investigated by students at many different grade levels–middle school, high school, a college algebra course for students with special needs, preservice elementary teachers, and ninth grade algebra students, as well as students enrolled in the developmental studies programs at our colleges and universities.

A need for standard notation becomes apparent as students try to talk about the paths they have found. The whole-class discussion about notation is one that must receive adequate attention and may expose deep-seated confusion about why mathematicians use the conventions they do. It is as appropriate to discuss feelings here as it is to discuss criteria. Ask students if they have additional criteria which would make the listing of the paths or mathematics in general more understandable.

Complete the Investigations 1–7 with the accompanying discussions in class. Assign the rest of the section as homework. Explicitly discuss in the next class the methods used to generate the twenty paths from investigation 8.

Instruction might begin by asking students to come up with a definition of a term from everyday life such as *wheel* from the parking lot problem or *basketball* from Exploration 6 in section 1.1. Working in groups they should record these. During class discussion, record students' definitions on the board or on the overhead. Whenever possible, cite another item which meets their definition.

For example, in one class, students defined "basketball" in such a way that an orange balloon met all the criteria of their definition. These counterexamples help students realize that definitions should make sense and must contain all of the essential information which distinguishes the term from similar but different items.

Because no one in your class is likely to have seen a problem like SPC before, all students begin on a level playing field. Students who ask questions and do the assigned work will find themselves as successful as students who have always done well in math before. The emphasis upon thinking, rather than following a model to do many similar problems, may make this an uncomfortable experience for the latter group of students because what they are asked to do does not coincide with their previous experiences in a mathematics classroom.

History has taught us that students have trouble correctly applying the rules on their own initially. It may be helpful to go through how to apply each of the rules as a whole class. Be sure to make it clear that SP is the initial path and that all valid paths must arise by applying the rules beginning with SP.

If students are to buy into this new view of mathematics, they need opportunities to explicitly discuss what they are being asked to do to and recognize how these activities actually model what it really means to do mathematics.

- The elements of the SPC system are S, P, and C; the elements of geometry include point and line; and the elements of the whole number system are 0, 1, 2, 3,....

- The SPC system only has one *axiom*--SP. Have students recall one or more axioms from geometry or from algebra.

- *Operations* in the SPC system (the legal ways of combining elements) consist of the four rules. What are the operations in the whole number system?

- *Conjectures* in the SPC system relate to stepping outside the system and asking questions about the system. For example, what happens if you apply a rule repeatedly and what generalizations about types of paths result?

- *Theorems* are valid paths that we can prove true by listing the rules needed to derive the paths. This can be compared to the theorems proved in a geometry course.

The formalism should follow the Investigations of the SPC system. Connections can be made by asking students to consider specific questions, not by delivering a lecture on these connections. The repeated use of Rule 1 is a nice diagnostic to see if students can distinguish between doubling and multiples of two (adding two more to the previous answer). Do students list the sequence of paths SP, SPP, SPPPP, **SPPPPPP** (an impossible path) instead of SP, SPP, SPPPP, SPPPPPPPP?

Have students classify their method of generating paths in one of the following ways:

- Randomly use any rule that applies (working inside the system).

- Repeat applications of Rule 1 until a path offers additional possibilities for the application of other rules (working inside the system). Some students have actually written down 128 P's before abandoning this technique.

- Studying the rules, then predicting all of the paths possible based on applying each of the rules to the existing path (working outside the system).

A tree diagram may be helpful for students to reflect on this third method of generating paths, seeing how many different paths are possible when a given rule is applied to an existing path. Have the class make suggestions for the possible choices at each branch.

Begin with the axiom SP and draw an arrow for each possible rule that can be applied. The number tells which rule was applied at each stage.

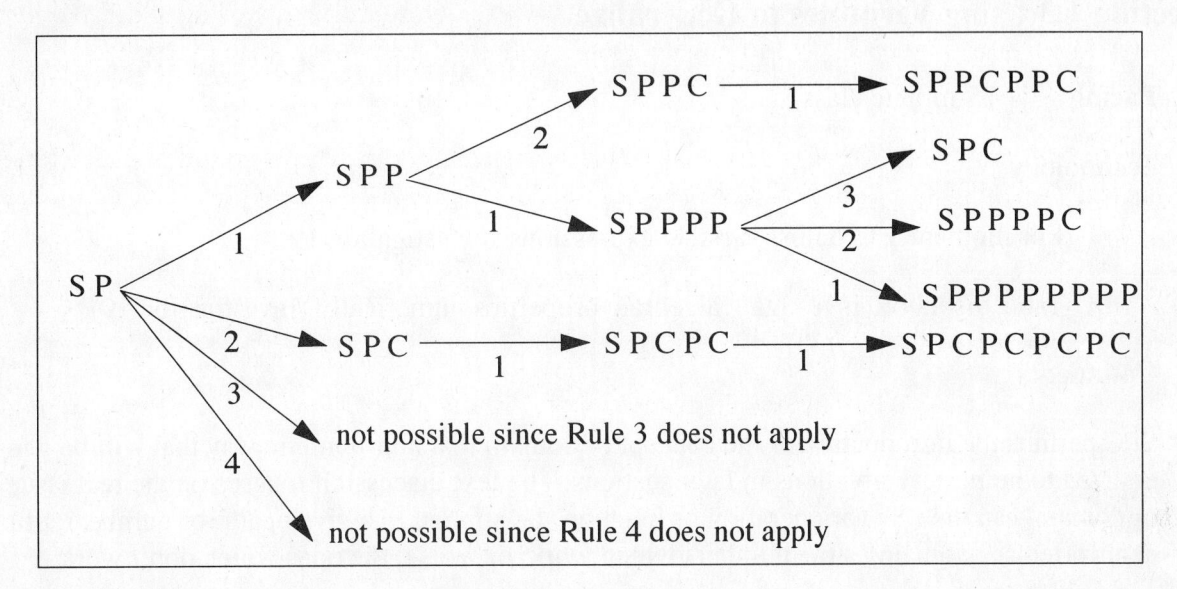

```
                              2    SPPC  ──1──▶  SPPCPPC
                     SPP  ◀
                  1        1                      ▶ SPC
                              ▶  SPPPP  ◀──3──
         SP                                   ──2──▶  SPPPPC
            2                                   1  ▶  SPPPPPPPP
              ▶ SPC  ──1──▶  SPCPC  ──1──▶  SPCPCPCPC
            3
              ▶ not possible since Rule 3 does not apply
            4
              ▶ not possible since Rule 4 does not apply
```

This helps students generalize and also see clearly that multiple paths can lead to the same path. They can also tell that an infinite number of paths exist.

Without the intervention of a teacher who explicitly encourages students to make connections and probes to determine how and with what strength the connections are being made, meaningful, useful connections will probably not be made by many students.

Students will probably ask when they will begin doing "real mathematics" and why they are investigating the SPC system as they work through this section. It is recommended that you help students see the connections between the SPC Investigations and the structure present in any mathematical system. Structure is not a topic appropriate only for higher level mathematics courses. Students who can see how a mathematical system is put together better understand the connections the text materials are trying to help them see.

Throughout the term, it is important to encourage students to work both in and out of the system, and to have them recognize when they are working in one mode or the other.

Explorations

- 1 asks students to begin creating their own glossary. This exploration is the first exploration in every subsequent section of the text. Be sure students realize that the words appear in the back of the book and they need to write definitions of the words as they encounter them.

- 3–9 permits students to test their understanding of the SPC system.

- 10 provides students with an opportunity to work outside the system, using P-counts, to prove that SC is not a valid path.

- 11–16 permit students to exercise metacognition to reflect on their experiences with SPC.

Section 1.3 Using Variables to Generalize

Pacing   1 75-minute class

Technology

- Function entry to define variable expressions (Investigation 19).

- Table display to investigate algebraic properties numerically (Investigation 19).

Comments

SPC permits the introduction of the concept of domain in a non-numeric way that will be generalized to numerical situations in later sections. The text discussion focuses on the real issue of domain–can the rule (or operation or function) be applied to a given path (or number). Students often see domain as the numbers which work, or worse, the ones which don't work.

Research on students' understanding of variable points out that there are at least five uses of variable that are almost never explicitly differentiated for students. In the 1988 NCTM Yearbook, *The Ideas of Algebra, K - 12,* the chapter "Conceptions of School Algebra and the Uses of Variables" by Zalman Usiskin lists these uses. Variables are used in

- a formula such as $P = 2L + 2W$.

- open equations or inequalities to solve such as $2x - 3 = 5$.

- an equation that is a function, showing the relation between two variables and which is not solved such as $f(x) = 2x - 3$ or $y = mx + b$.

- an identity such as $sin^2(x) + cos^2(x) = 1$.

- a generalized statement of a property such as $n + (-n) = 0$.

Note that in these examples, variables are used in differnt ways: 1) to represent numbers; 2) as the argument of a function; and 3) as a pattern generalizer.

Since one's view of algebra is tied to one's understanding of variable, we have attempted to explore the possible uses of variable explicitly in as many ways as possible in this text. Variables are introduced in the context of generalizing a rule as is often done with the properties of arithmetic.The use of the box indicates that a single variable name may represent a whole string of letters, not just a single number which is the usual conception of variable students will bring to your class.

We have attempted to include situations that look at major misconceptions about variables and answer questions such as:

- Why must a variable represent exactly the same thing each time it appears in an expression?

- Since the variable can represent so many different things, how do you know what to choose for the variable to represent?

- How can we use the same variable in different expressions to represent something different – such as in equations $2x = 4$ and $5x + 1 = 1$ or in formulas such as $A = LW$ and $A = P(1 + Prt)$?

- When do we need multiple variables in the same expression?

- Can two variables both represent the same thing some of the time?

- Is it necessary for something to replace the variable?

Look for opportunities to help students reflect on these questions and concepts as students attempt to generalize the rules of the SPC system. All of the assumptions about variables need to be acknowledged in class. Ask students what assumptions they have made about variables and which of those listed they were unaware of before the discussion in this section.

The notion of trade-off should also be discussed. The use of variables makes the mathematics more abstract and so more mental effort is required to remember what is meant. The advantage of having to write less is viewed as an acceptable trade-off. The ease of communicating when symbols replace words works only when everyone uses the symbols in the same way.

One of the important roles of algebra is to generalize arithmetic. We have chosen to follow the generalization of the rules of the SPC system with examples of generalized arithmetic to help students make the connections between the SPC system and the whole number system.

If students have been doing most of their work in class, there may be a tendency for some students to come to class unprepared. It is probably a good idea to hold a class discussion at this time about the responsibility of each person for their own learning and to the members of the group. Students who are absent (for good cause or for no valid reason) should meet with members of their group(s) outside of and prior to class to get assignments and an overview of the class discussion which was missed. Members of the group need to know that it is okay to exert peer pressure on those students who come to class unprepared and that it is okay to refuse to give answers to someone who merely wants to copy them down. Reinforce the idea that class time is to be used to compare answers prepared before class and to discuss, both in groups and as a whole-class, students' discoveries, observations, conjectures and questions. Student-directed regrouping for compatibility is fine.

Explorations

- 3–6 test students understanding of how variable is used in the SPC system.

- 7 allows students to create examples from generalizations.

- 8–10 apply the idea of generalizing using variables to English grammar rules.

- 11 and 12 deal with the problems of over-generalization and should be included in a discussion the following class.

- 13 queries students on the understanding of the assumptions about variables.

Section 1.4 Expanding the Notion of Variable

Pacing   1 75-minute class session

Technology

- Using tables to investigate the difference between powers of two vs. multiples of two, using P-counts addresses a major misconception held by many students.

- Simplifying exponential expressions (investigations 13, 14, 19, 20) allows students to collect data from which to generalize.

- A symbol manipulator, such as the TI-92 (if available), can be used to collect data on variable exponential expressions (investigations 15 and 21) to allow for further generalization.

Comments

David Tall and Eddie Gray, well-known mathematics education researchers from England, have interviewed students and discovered that an implicit idea that mathematicians take for granted confuses students, namely, the ambiguity of mathematical notation. They observed that the ambiguity of notation often causes students great difficulty.

Mathematicians recognize that a process and the result of that process (the answer) can both be represented by the same symbols. They are comfortable thinking of 3×2 as a process to calculate the answer "6" as well as the product 3×2. Students often think of 3×2 as representing the process of multiplying three and two, an equals sign as an indicator to "do something to find the answer," and six as the answer or output of that process. They rarely think of the expression 3×2 as a product and leave the expression as the answer.

Algebraic representations are even more ambiguous. Sometimes it is appropriate to let $2n$ represent the process of multiplying 2 by some number and to find the "answer" when we know the value of n. Sometimes, however, it is appropriate to view the expression $2n$ as the answer.

Many students have great difficulty making the transition to algebraic thinking. Most students are unaware of this ambiguity and have only an inflexible, procedural knowledge. They tend to want to "find the answer" given notation that, to them, indicates only a process by which they find the "answer" and are uncomfortable when they don't have to "do something" to get the answer.

The SPC system is revisited for the purpose of exploring P-counts. The study of P-counts leads to the introduction of exponential expressions and student shave the opportunity to investigate two important exponent properties ($a^m a^n = a^{m+n}$ and $(a^m)^n = a^{mn}$). At this point, exponents are restricted to the domain of natural numbers.

The concept of domain is revisited in light of numbers that make sense in the problem situation versus the algebraic representation.

In the investigations involving tables it is a good idea to revisit table construction by asking students to explain

- the column headings

- why they chose the numbers they did

- why mixed fractions and decimals and large numbers were omitted, if that occurred in your class

- whether fractions could be included

- how they know when they have chosen numbers which disclose untrue generalizations.

Explorations

- 2 asks students to think about the distinction between arithmetic and algebra.

- 4–6 explore the difference between doubling and squaring.

- 8 and 9 make use of the P-count formula.

- 10 asks students to reflect on the ways the word domain is used. 15 also deals with domain issues.

- 11–14 provide practice with exponential expressions.

- 7, 16, and 17 probe student understanding of variable.

Section 1.5 Making the Connections: What Does It Mean to Do Mathematics?

This review section can be assigned as homework, with discussion the following class period.

The students are expected to create their own review by answering and reflecting on investigations 1 and 2.

The problems to be solved in the review section seem to cause a great deal of difficulty. A discussion of why there are three cuts when cutting a log into four pieces emphasizes the value of drawing a picture for example.

The drawing of a mathematician can provide information about what students believe it means to do mathematics. One student covered the space inside the frame with tin foil and informed the instructor: "It's a mirror. Anyone can be a mathematician!"–the goal of this course.

CHAPTER 2: INSTRUCTOR NOTES

Whole Numbers: Introducing a Mathematical System

Rationale

This chapter introduces the properties of closure, applications of commutativity, associativity and the distributive property as applied to the whole number system. Prime numbers and prime factorization are introduced. Students investigate situations that are ambiguous and there is explicit discussion of how to deal with ambiguity. Arithmetic operations and the order in which they are performed are defined as binary or unary operations and investigated using function machines, setting the stage for the introduction of numeric representations of functions in Chapter 3. Polynomials are defined and polynomial addition is introduced as an example of an operation an algebraic expressions. Factoring using the distributive property to extract the common monomial factor concludes the chapter.

Overview

Section 2.1 revisits the parking lot situation. The problem conditions are changed. Students are expected to compare information stated in different problem situations and distinguish relevant information. As a result, the idea of domain is revisited and the whole number system is introduced. Concepts, such as sets, prime numbers, and prime factorization are investigated.

Section 2.2 explicitly addresses the notion of order of operations as a set of arbitrary rules compared with order determined by context. The issue of ambiguity is identified and ways of dealing with ambiguous problems situations is discussed. Function machines are introduced as a method of visually parsing numeric expressions. The meaning of a zero exponent is investigated. This section investigates numeric expressions only. The algebraic extension, using variables, occurs in the next section.

Section 2.3 provides opportunities for students to generalize the order of operations to algebraic expressions containing variables. Evaluation of algebraic expressions is introduced. Polynomials are defined, but investigations on polynomials are postponed until the next section.

Section 2.4 introduces the idea of altering the order of operations using the associative, commutative, and distributive properties of the whole number system. The properties are first investigated on numeric expressions and then extended to variable expressions. Addition and multiplication of polynomials is investigated along with factoring by extracting the common monomial factor. Note that subtraction of polynomials is postponed to Chapter 4 where integers and "taking the opposite" are introduced.

Section 2.5 is a review section.

Allow approximately 5 75-minute classes for this chapter.

Section 2.1 Whole Number Domains

Pacing   1 75-minute class.

If the initial investigations have been assigned previously, class time can be used to discuss the major ideas of the section, both in small groups and with the entire class.

Technology:

- Using a table to test numbers for primality.

- Using a table to determine all factors of a whole number.

- Computations with the P-count.

Comments:

The parking lot situation from Section 1.1 is revisited in this section for several reasons. First, the conditions of the problem have been changed. It is important for students to read the conditions carefully so that they don't assume what applied in Section 1.1 also applies here. Students are expected to compare the information given in the two situations to distinguish when there is a solvable problem and when there is not even a question asked. In relation to the idea of domain, it is important to see that changing the situation to both cars and motorcycles from cars only or motorcycles only changes the domain set.

In terms of the vocabulary chosen for this section, we have used only those terms from the language of sets that will prove useful in exploring and communicating the results of those explorations. Domain, and infinite and finite sets especially, will be referred to continually throughout the book. Frequent questions about these terms in the next few sections will emphasize the importance of these concepts.

The sieve of Erastosthenes for prime numbers has been modified by removing 1, which is not prime. The new arrangement displays some patterns that are not as obvious on a hundreds chart and the concept of closure completes this section. One such pattern is that, except for 2 and 3, every prime number is one less or one more than a multiple of 6. Closer examination of this sieve leads naturally to a discussion about conjectures and generalizations about the form of prime numbers. Closure, or rather the lack of closure, will be the motivation for the development of the other systems of numbers covered in this course.

Prime factorization is introduced as a way of helping students deepen their understanding of prime numbers and to introduce the idea of factored form.

If there are groups of students who do not work well together, the beginning of a chapter is a good time to do some shifting. There are many philosophies about the formation of groups, much of which has been based on work with young children. What we suggest may not be the current wisdom, but it has worked successfully in our college classrooms. Ultimately you will develop techniques you are comfortable with.

One thing to consider is the effect of absenteeism. Sometimes groups take care of their own in a very efficient manner. Other times the frequent absence of a member is a disruption to the group. If the latter is the case, form a group of chronic absentees unless the group they are working with objects.

Consider the role each student plays in the group. If one person is teaching all the others, it may be wise to move this person to a group in which there are others at the same ability level. This provides for greater challenge and personal growth. Group conferences and/or individual conferences to determine how things are going can pay big dividends in terms of student satisfaction, motivation and achievement.

Explorations

- 3–5, 15, and 21 focus on the concept of sets.

- 6 explores closure.

- 7–9 cover prime factorization.

- 10, 11, and 19 explore domain.

- 12–14 apply the ideas of sets to the SPC system.

- 16 and 17 explore primality testing.

- 18 uses variables to generalize.

Section 2.2 Order of Operations with Whole Numbers

Pacing 1 75-minute class.

Technology

All of us make errors in arithmetic, so using a calculator when in doubt provides opportunities to learn to use the calculator in an appropriate manner. Investigating the appropriate use of grouping symbols offer opportunities for diagnostic observations by both students and instructor. Translating from a symbolic form to the equivalent form on the calculator is very challenging for many students. A main reason for using a graphing calculator, rather than a scientific calculator, is that students are able to view both their input and output for a particular problem and are thus able to identify their assumptions about inclusion symbols in discussion with another student. They also learn to use some mental math and estimation skills, as well as calculator entry skills, in these investigations. Finally, students can investigate the meaning of a zero exponent by collecting data with the calculator.

Comments

The interesting feature of the problems in the first set of investigations is that each incorporates the same set of numbers but the context or lack of one produces different answers for each question. As the emphasis on contextual problems grows, it is important for students to not only consider the rules for the order of operations when there is no context, but to also be able to decipher the problem so that a correct order of operations can be determined.

Because of different interpretations of Investigation 1, students have argued justifiably for an answer other than the one discussed in the text. Students need to develop a procedure to deal with ambiguity. An acceptable procedure is to describe in words how the student has interpreted the problem and to solve the problem presented by this interpretation.

Class discussion can help determine the validity of the interpretation (sometimes the grammatical problems of students cause the misunderstanding.) In any event, much of life is ambiguous and it seems reasonable to spend some time addressing the issue.

Grouping symbols are used to change the order of operations. When students are asked what operation should be done first in the expression $3 \langle 2^3 - 4 \times 3 \rangle + 2$, the response "What is in the parentheses" often indicates rote memorization of the rule. Asking an additional question about what operations are used in this problem is worthwhile and frequently illuminating. It is hoped that substituting "8" for "two cubed" as the rest of the problem is copied down would become automatic for students.

There is a concern that students really don't have a clear sense of the meaning of equality when they write the following to evaluate the above the expression:

$$2^3 = 8 - 4 \times 3 = 8 - 12 = 3 \times \langle -4 \rangle = -12 + 2 = -10$$

Many students need help to understand that what occurs on either side of an equal sign must be equivalent. Without that understanding what they have written is grammatically incorrect, even if the final answer is correct.

Function machines are introduced as an aid in parsing numeric expressions. Function machines are used throughout the text so be sure students get a solid start with them in this section.

Investigations 13 and 14 are typical examples of how the calculator can be used to collect data from which students can write generalizations. In this case, the investigate how a zero exponent is defined.

It is worth while it to ask what students now understand about arithmetic which they didn't know before. Emphasize the why, not the how.

Continue to make class risk free by gently dealing with mistakes.

Explorations

- 3, 6, and 10 investigate order of operations.

- 4 and 8 explore function machines.

- 5 reviews and extends exponent properties.

- 7 has students write algebraic expressions from English statements.

- 9 introduces evaluation of expressions.

- 10 and 12 are interesting process problems which are not trivial. A class discussion should explore the multiple ways students approached the problem and the conjectures they made.

Section 2.3 Algebraic Extensions of the Whole Numbers

Pacing: 1 75-minute class.

Time spent addressing student misconceptions about order of operations and developing understanding and appreciation of the properties of whole numbers now provides a solid foundation for later extension of these essential concepts.

Technology

- Investigating order of operations.

- Storing a value for a variable.

- Evaluating algebraic expressions

Comments:

The use of both the graphing calculator and the function machine representation encourage students to use multiple representations to develop deeper understanding of the mathematical concepts of this section.

Revisiting the issue of ambiguous notation and further discussion of the notions of process vs. product (meaning results of the process) are beneficial to many students.

Polynomials are introduced as examples of expressions that contain variables. The terminology is presented, but there are no investigations with polynomials until the next section.

Explorations

- 2 investigates expression evaluation.
- 3, 9–12 explore function machines.
- 5–7 check students understanding of the definition of terms related to polynomials.
- 8 provides more practice with exponential expressions.

Section 2.4 Properties that Change the Order of Operations

Pacing   1 75-minute class.

Technology

- Use table to test various properties.
- Use table to check correctness when performing operations on polynomials.

Comments

The commutative, associative, and distributive properties are investigated using both numeric and algebraic expressions. The properties are used to justify techniques of mental math and algorithms that students learned in arithmetic and the implications for increased understanding of arithmetic as a result of this section are enormous.

Addition and multiplication of polynomials is investigated as an application of the properties. Note how tables are used to help students test whether their simplifications are correct. If a symbol manipulator, such as the TI-92 is available, students can use the symbol manipulator to collect data from which they can generalize. The distributive property is used to introduce factoring by removing common monomial factors.

Explorations

- 2–6, 8, 10–12 all further explore the properties.

- 7–9 provide practice performing operations on polynomials.
- 13 introduces the idea of equations by asking students to find an input given an output and the expression. Reversing the operations and their order provide a nice technique for answering this question. This idea will be used to reverse function machines in Chapter 3.
- 14 relates function machines and the rules of the SPC system.

Section 2.5 Making Connections: What Does It Mean to Generalize?

This review section can be assigned for homework, with discussion the following class period.

CHAPTER 3: INSTRUCTOR NOTES

Functional Relationships

Rationale

This chapter focuses on problem solving and the unifying concept of function. Numeric representations of functions, using tables to organize and investigate data, symbolic representations, using equations, geometric representations, using graphs, and function machines are all used to develop the concept of function. Variables as quantities that change as a situation changes are presented and the behavior of functions is examined. Function notation is introduced. The exploration of whole numbers continues in this chapter as we move from an exploration of order of operations and properties of whole numbers to the use of whole numbers as an appropriate domain for problems–both abstract and contextual. Finite differences are introduced as a way of investigating and classifying functions. This topic foreshadows rate of change and slope.

Overview

Section 3.1 emphasizes the importance of looking for and finding number patterns. Tables and finite differences are used to organize and investigate data and numeric representations of functions are introduced. Function machines and tables are used extensively.

Section 3.2 introduces the notion of a variable as a quantity that changes as a situation changes, and examines the behavior of functions using tables. Students begin with a numeric representation of functions and learn how to build a corresponding symbolic (equation) representation thus improving students' ability to generalize.

Section 3.3 provides students with opportunities to discover relationships between quantities in a problem situation and introduces function notation.

Section 3.4 introduces graphs as a means of communicating information visually and develops a geometric representation of functions. Now students have the capability to express a function using a table, a function machine, and equation, and a graph.

Section 3.5 encourages students to discover relationships between quantities in problem situations, with connections to the triangular numbers. The identity function is introduced.

Section 3.6 continues using finite differences to investigate power functions. The factorial function is introduced.

Section 3.7 is a review of Chapter 3 and the preceding chapters.

Allow approximately 5–7 75-minute class sessions for this chapter. If time is short, Sections 3.5 and 3.6 may be omitted or assigned for extra credit without seriously affecting the flow of the text.

Section 3.1 Investigating Relationships Numerically

Pacing   1–2 75-minute class sessions depending on how much time you want to allow students when investigating **The Nearsighted Professor** problem.

Technology

- Finite differences can be found using the calculator. In particular, the TI–83 has a list operation (ΔList()) that will compute the finite differences in a list automatically.
- Whole number division
- Conversions between decimals and fractions
- Modular operations

Comments

The Nearsighted Professor problem allows students to solve a complex problem step by step, introduces a tremendous amount of mathematics, and provides and introduction to function. It is fun to dramatize this problem as a whole class discussion rather than working in small groups at the beginning of the class. Change the names in the text to five of your students.

Students develop a deeper understanding of the operation of division and the relationships between divisors, quotients and remainders, using the graphing calculator. Using the calculator in class with Investigation 13 affords the instructor opportunities to diagnose student misconceptions about division.When students do the divisions in Investigation 13 with a calculator, they get decimal remainders. It has been productive to have students work in pairs for this and subsequent investigations in this section. One student in each pair should be encouraged to do the division without using a calculator, finding whole number remainders, while the other member of the pair should find and record the decimal form of the remainder using the calculator. The connection between the whole number remainder and the decimal part of the answer illuminates the procedures and thinking of students and provides the instructor with useful diagnostic information about students' understanding of decimal and remainder.

Note that the ultimate solution to **The Nearsighted Professor** problem involves a composite function from any whole number greater than 1 to the column number.

The parking lot problem is revisited with changed conditions and extended to introduce the notion of a relation. Students will need to clarify the difference(s) between the function and the relation machines.

Explorations

- 2–10 explore the concepts of relation, function, domain, and range.
- 11 is a precursor of the investigations in Section 3.5 and should definitely be assigned.
- 12 continues expanding the parking lot problem.
- 13 provides a good opportunity for group problem solving.
- 14 (the locker problem) provides insight into factors, odd numbers, and square numbers.
- 15 and 16 revisit the SPC system.
- 17 is an extension of **The Nearsighted Professor** problem.

Section 3.2 Function: Algebraic Representation

Pacing   1 75-minute class.

Technology

- Using tables to represent relationships.
- Computing finite differences foreshadowing rate of change.
- Entering functions.
- Evaluating functions.

Comments

This section begins with a problem situation (**Concert Receipts**) that is investigated using a table and finite differences. Students are asked to generalize the results of their investigations and to communicate those results algebraically and using a function machine. Functions are investigated using the notions of input and output and real number domains are introduced. Multiple representations of functions are used, including numerical tables, function machines, and algebraic representations.

Explorations

- 2 and 3 involve function machines and reversing function machines.

- 4 investigates evaluating functions and solving equations.

- 5 is the famous "students and professors" problem which research has shown consistently gives students who attempt to solve this problem, based only on their reading of the problem, difficulty.

- 6 and 7 provide students with additional exploration of the important skills needed to successfully interpret and solve proportional problems.

- 8 revisits the parking lot problem, extending the investigation begun in Section 3.2.

- 9 revisits the SPC system.

- 10 requires students to create their own problem situation and function. This can prove difficult for students so class discussion is encouraged.

- 11 focuses on one place from which equations might originate.

Section 3.3 Function: Notation for Algebraic Representations

Pacing   1 75-minute class.

Technology

- Defining functions on the calculator.

- Using tables to explore the behavior of functions.

- Evaluating functions using function notation.

Comments

The general format for function notation is introduced in a manner that hopefully makes sense to students. Explicit class discussion about the benefits and disadvantages of notation is productive.

It is important to answer student questions. However, knowing which student to call on to facilitate a coherent class discussion is difficult and not all questions can be addressed during class time. Using entrance and exit slips is a means of addressing this problem. As students enter or leave the class, they are encouraged to turn in a slip of paper with questions unresolved from the preceding class or which did not get addressed during the present class period. The instructor can then group the questions and address student concerns more efficiently, either during that class period or at the beginning of the next class.

Students will interpret the data in creative ways, some of which are surprising.

Explorations

- 2 asks students to reverse a function defined symbolically.
- 3 and 6 involve evaluating and solving.
- 4 and 5 investigate domain and range.
- 7–9 revisit old problems requiring the use of function notation.
- 10 and 11 require students to create their own problem situation and function.
- 12–14 provide the opportunity to extract mathematical information from text.
- 15–17 provide review of exponent properties and polynomial operations.

Section 3.4 Function: Geometric Interpretations

Pacing   1 75-minute class.

Technology

This section requires no technology since the initial graphing investigations are done by hand.

Comments

This section provides a nice change of pace that students seem to appreciate. They might see the similarity between the Mayan Mix-up and a game many students are familiar with– Battleship. Having students actually play the game during class provides diagnostics both for the students and the instructor. While students, in pairs (if an odd number of students is present that day, the instructor can participate), play the game, the instructor is able to observe the class and quickly determine which students are plotting locations incorrectly. Students frequently diagnose their own errors when, thinking they have found an opponent's treasure, they compare locations already named with their game partner and discover each person has marked different locations.

The **Concert Receipts** problem from Section 3.2 is revisited and investigated graphically.

Scaling is an important and confusing issue for students. It is critical to understand when they must begin to set viewing windows with a graphing calculator. Standards for scale are introduced in this section.

Explorations

- 2, 3, 7–9, and 16 require students to read and interpret graphs.
- 4 and 5 allow students to compare and contrast numeric, symbolic, and geometric representations of function.

- 10–15 proved review of previously-encountered material.

Section 3.5 Triangular Numbers

Pacing   1 75-minute class or assign as extra credit.

Technology

- Calculating cumulative sums.
- Testing conjectures.
- Evaluating formulas.

Comments

The **Triangular Wages** problem concentrates on developing further students' understanding of function notation and their ability to generalize. Two different symbolic representations of the formula for the nth triangular number are used and have been beneficial in challenging students' assumptions about multiplying by one-half vs. dividing by 2. The resulting function is really a composite function that moves from day number as input to day's wages to total wages to date as an indicated sum to total wages to date as the final output.

Explorations

- 2–7 further explore triangular numbers and the derived formula.
- 8 explores the relationship between constant second finite differences in outputs and quadratic functions.
- 9 introduces an investigation of square, rather than triangular, numbers.

Section 3.6 Power and Factorial Functions

Pacing   1 75-minute class or assign as extra credit.

Technology

- Calculating successive finite differences (the ΔList(command on the TI-83 simplifies these calculations immensely).
- Evaluating power functions using the exponentiation key
- Evaluating the factorial function

Comments

Investigating the class of power functions numerically, using finite differences, algebraically by writing equations, and geometrically by constructing graphs, tests students' computational skills and their ability to make conjectures. Discussion of the number of ways five books can be arranged often leads to a discussion of scientific notation. Students enjoy discovering what the largest factorial the calculator can display without using scientific notation.

Explorations

- 2 asks students to look for patterns in data using finite differences in outputs.

- 3 and 4 encourage students to use what they have learned in this section, together with skills and understandings acquired since the start of the term.

- 5 continues the investigation of the factorial function.

Section 3.7 Making Connections: What is a Function?

This review section can be assigned for homework, with discussion the following class period. Note that material from Chapters 1 and 2 is included in the review section.

CHAPTER 4 INSTRUCTOR NOTES

Integers: Expanding a Mathematical System

Rationale

Integers are introduced in order to investigate the opposing function, the absolute value function and the identity function. Graphing is extended from one-quadrant graphing to four quadrant graphing and students use a graphing utility to graph discrete and continuous functions for the first time. The absolute value function is introduced as a piece-wise function whose definition focuses on the sign of the input, not its size, and whose domain is split into two distinct parts.

Overview

Section 4.1 introduces the integers using an investigation Calculator Golf. A new unary operation, opposing, is introduced as a result of the Calculator Golf investigations. In the process students investigate both the random function and the greatest integer function.

Section 4.2 introduces the basic operations on integers. Addition and subtraction are investigated using a number line model and an electron/proton model. Opposites of polynomials are investigated and subtraction of polynomials is introduced.

Section 4.3 begins with a problem situation (**College Tuition**) that is modeled by a piece-wise defined function, followed by another problem situation that introduces the concept of absolute value. Subscript notation is explicitly discussed. The absolute value function is explicitly defined as a piece-wise function.

Section 4.4 revisits the Mayan Mix-up investigations, graphing in four quadrants instead of one quadrant.

Section 4.5 introduces students to the graphing utility. Graphing of order pairs (scatter plots), and finding appropriate viewing windows are initially explored using locations from the Mayan Mix-up game. An appropriate viewing window is determined by looking at a table of values. The transition from rectangular coordinate graphing to parametric graphing in the tee-shirt problem is motivated by considering the domain of a function. The domain of the problem situation is discrete, making parametric plotting of the graph of the function appropriate. Tracing on a connected graph gives nonsensical results, whereas parametric graphing allows us to control the T-values so that T, and automatically x, take on only domain values.

Section 4.6 asks students to investigate the algebraic, numerical and graphical representations of the identity, opposing and absolute value functions.

Section 4.7 reviews material studied to this point.

Allow approximately 7 75-minute classes for Chapter 4. The pace for Sections 4.1 and 4.2 depends on students' previous acquaintance with signed numbers. Section 4.3 needs careful attention due to the difficulty students experience with the absolute value function. Section 4.4 usually causes little difficulty, but offers the opportunity to diagnose problems students have with plotting points, correctly identifying axes, and scaling. Section 4.5 requires much attention as this is the students introduction to graphing with technology. At least 2 class sessions are usually required on this section.

Section 4.1 Integers and the Algebraic Extension

Pacing 1 75-minute class.

Technology

- The INT function (greatest integer function).
- The RAND function (random number generator).
- The "finding the opposite" unary operation (compared to the operation of subtraction).

Comments

This section begins with students playing a game of calculator golf. This is a section in which the initial investigations should be done in class, not assigned as homework preceding the class. The calculator golf game provides an opportunity for students to deal with negative numbers. Par is used to introduce the idea of an arbitrary zero. The opposite, or additive inverse, is then investigated using two models: the electron/proton model and the number line model. Function machines are used to analyze the unary operation of "finding the opposite". Significant time is devoted to asking students to confront their understanding of the symbol "–". Students try to distinguish between the operations of "finding the opposite" and subtraction using the calculator and function machine representations. The three common uses of "–" (negative, opposite, subtraction) are specifically discussed.

Explorations

- 2 and 4 focus on finding the opposite.

- 3 addresses the issue of closure.

- 5 and 6 explore the uses of the "–" symbol.

- 7 tests students understanding of the use of the *int* and the *rand* functions to generate golf scores.

- 8 proves another problem situation involving proportional reasoning.

Section 4.2 Operations on Integers

Pacing   1 75-minute class.

Technology

- Using tables to collect data on signed number operations.

- Computations with signed numbers.

- Using tables to compare a given polynomial expression with one that has been modified using the basic operations to check for the equivalence of the result.

Comments

Students investigate addition and subtraction of signed numbers beginning with golf scores and then with tables on the calculator providing data. An electron-proton model and a number line model are used to help students better understanding addition and subtraction of integers. Investigations involving multiplication and division of signed numbers are saved for the explorations. Expressing subtractions in terms of additions is investigated.

Operations on polynomials are revisited. Students investigate how to find the opposite of a polynomial. Subtraction of polynomials is then introduced. While students can use the table feature on their calculator to check their answers, a symbol manipulator, such as the TI-92, is very helpful in helping students collect data to discover how the operations actually work.

Explorations

- 2–4 ask students to add and subtract using the electron-proton model and the number line model.

- 5–8 review the algorithms for multiplying and dividing signed numbers. Students are asked to generalize the rule based on their numerical investigations.

- 9 provides practice doing signed number computations.

- 10–13 asks students explore common properties with integers.

- 14 presents a model that can be used for multiplication of integers.

- 15 and 16 provide practice in using the "–" symbol in variable expressions.

- 17 and 18 involve evaluating functions.

- 19 involves equation solving with signed numbers.

Section 4.3 The Absolute Value Function

Pacing   1 75-minute class.

Technology

- Defining and analyzing piece-wise defined functions.
- Defining and analyzing the absolute value function.

Comments

This section begins with an investigation of a problem situation that requires a split domain. The function is evaluated for a given input using a function machine representation. This investigation leads to a consideration of another piece-wise defined function, the absolute value function. Ivan's test to determine when to buy or sell a certain stock builds on the ideas of an arbitrary zero and split domain introduced in the previous section. The use of absolute value to guarantee a positive output and the use of delta notation to represent the change are addressed explicitly. Having students use function machines to analyze the behavior of an absolute value function reinforces the idea that the output is determined only after a decision is made about the necessity of changing the output to its opposite (in the case when the input was negative) or of leaving the input alone (do nothing). Solving absolute value equations is introduced at the end of the section.

Explorations

- 2 explores the notation used for indicating change.

- 3 and 5 require students to analyze the meaning of the sign when measuring change.

- 4 provides practice in calculating change from initial and final values.

- 6–9 ask students to measure change in various contexts.

- 10–12 ask students to consider the application of absolute value to find the distance a number on the number line is from zero.

- 13 and 14 involve solving absolute value equations.

Section 4.4 Graphing with Integers

Pacing   1 75-minute class.

Technology

The graphing calculator is not used in this section since graphing is done by hand.

Comments

Plotting ordered pairs is extended from graphing in the first quadrant to graphing in all four quadrants. The Mayan Mix-up game is modified so that the origin is in the middle of the grid, not on the lower left corner. Allowing students to complete at least one game during class provides the instructor with opportunities to observe students who were having difficulty plotting points correctly during the initial Mayan Mix-up investigations to determine whether or not they are still having difficulties plotting ordered pairs correctly.

Explorations

- 2–4 provide the opportunity to read, write, and plot ordered pairs.

- 5 and 6 asks students to create their own problem situation and to represent it numerically, graphically, and using a function machine.

- 7 and 8 ask students to investigate the graphs of the opposing function and the absolute value function.

- 9 provides more practice using signed number operations.

- 10 gives the symbolic form of a function and asks students to create the numeric and geometric representations. Students must also find an input given and output. It is interesting to note which representation they use.

- 11 establishes a connection between linear equations and finding the input to a linear function, given the output. Absolute value equations are included.

Section 4.5 Using a Graphing Utility

Pacing   2 75-minute classes–one class for investigations and one day for small group and whole class discussion of the explorations.

Technology

- Entering sets of ordered pairs in lists.
- Graphing ordered pairs using a scatter plot.
- Setting appropriate viewing windows.
- Graphing functions in rectangular mode.
- Graphing functions in parametric mode.

Comments

The Mayan Mix-up game is revisited in this section, using the graphing calculator screen as the gameboard. Graphs are created from lists of data recorded in tables in previous investigations. Explicit discussion of issues involving appropriate function domains and the relationship of the domain to viewing windows occurs in this section. Choosing appropriate viewing windows proves difficult for students. Sufficient attention at this point in the course may help alleviate continuing difficulties throughout the remainder of the course.

The distinction between discrete and continuous functions becomes apparent in the section by comparing plots in rectangular mode to plots in parametric mode. Parametric equations are introduced as a means of controlling input values, particularly useful when the inputs are whole numbers. The amount of change in the input variable can also be controlled in parametric mode.

Explorations

- 2–4 address setting appropriate viewing windows.

- 5–9 and 11 provide students with the opportunity to further investigate and compare rectangular coordinate graphing and parametric graphing. Class discussion in which various students share their strategies for identifying a particular function domain and viewing window is appropriate.

- 10 focuses on reading and interpreting a graph.

- 12 involves setting an appropriate viewing window and using a graph to read trends concerning numeric data.

- 13 gives students a graph and asks them to create a numeric and symbolic representation of the function.

- 14 and 15 revisit equation solving.

- 16 revisits finding the opposite of a polynomial.

Section 4.6 Functions over the Integers

Pacing   1 75-minute class.

Technology

- Defining and graphing the identity (do nothing) function.
- Defining and graphing the opposing function.
- Defining and graphing the absolute value function.

Comments

Multiple representations are used to formalize the unary operations of opposing and absolute value as functions, comparing and contrasting the opposing and the absolute value functions numerically, algebraically, and graphically.

This section summarizes three functions introduced earlier in the chapter and permits students to compare and contrast their features. If you are short of time, this section could be assigned outside of class or as extra credit.

Explorations

- 2–6 explore the effects of applying an opposite or an absolute value to a given function.
- 7 explicitly focuses on misconceptions centered on interpreting the sign of a variable.
- 8–10 review operations on polynomial functions.
- 11 introduces another problem whose structure is similar to the original **Full Parking Lot** problem.
- 12 and 13 explore the meaning of algebraic expressions.
- 14 provides more practice with equation solving.

Section 4.7 Making Connections: How Do Integers Expand the Mathematics

This review section can be assigned for homework, with discussion the following class period. Material from all preceding chapters is included in the review section.

CHAPTER 5 INSTRUCTOR NOTES

Rational Numbers: Further Expansion of a Mathematical System

Rationale

Rational numbers are introduced in order to study rates of change, reciprocals and negative exponents. The idea of horizontal and vertical change is first introduced graphically, then numerically, measuring the finite differences in output compared with the finite differences in input in a problem situation that involves a variable rate of change. The idea of increasing, constant, and decreasing functions is foreshadowed by investigations involving the relationship between distance, rate, and time. The reciprocal function is introduced and students are provided with opportunities to relate negative exponents to the operation of reciprocaling.

Overview

Section 5.1 investigates rates of change in terms of change in altitude compared with change of distance for various portions of a journey. The problem situation involves a variable rate of change. The second problem situation investigates a constant rate of change by calculating finite differences of inputs and of outputs.

Section 5.2 focuses on the idea of equivalent ratios leading to the study of proportional reasoning. Reciprocals are introduced and the rational number system is formally defined. Numeric and algebraic fractions are investigated side by side. Percent is introduced as a ratio and investigated within problem situations.

Section 5.3 introduces operations on rational numbers and on algebraic fractions. The equivalence of decimal and fractional representations is explored.

Section 5.4 explores reciprocal and power functions. The initial investigation introduces the reciprocal function with time as a function of speed for a fixed distance. A graph is used to analyze the relationship between the input and output variables.

Section 5.5 provides students with opportunities to further investigate the properties of exponents. Negative exponents, using a symbol manipulator to generate data, are investigated, after which students generalize definitions and/or properties of integer exponents.

Section 5.6 is a review section.

Allow approximately 5–6 75-minute classes for Chapter 5.

Section 5.1 Rates of Change

Pacing   1 75-minute class.

Technology

- Computing finite differences in successive inputs and in successive outputs.

- Entering tables as lists of data.

- Calculating rates of change using lists.

- Calculating rates of change using graphs.

Comments

Students are introduced to the idea of horizontal and vertical change graphically in the first set of Investigations. Investigations 8–12 introduce the idea of rate of change by measuring the ratio of the finite differences in output to the finite differences in input. The first problem situation—the monk's climb— involves a variable rate of change. Students study a problem situation involving a constant rate of change in Investigations 17–19. This foreshadows the concept of slope and the study of linear functions later in the text.

Explorations

- 2, 3, 7, and 9 focus on the computation of rate of change in various problem situations.

- 4 and 6 focus on interpreting the meaning of slope in problem situations.

- 5, 17, and 18 ask students to interpret the meaning of slope from a graph.

- 8 and 16 introduce rates of change for quadratic functions.

- 10 focuses on the use of a letter as a variable versus a letter used as a label.

- 11 explores the traditional definition of slope with appropriate practice.

- 12 and 13 provide some linear functions in equation form and asks students to determine the slopes.

- 14 and 15 ask students to compare the rates of change of linear functions with the rates of change of quadratic functions.

Section 5.2 Rational Numbers and Proportional Reasoning

Pacing   1 75-minute class. More time may be needed if the understanding of fractions is weak.

Technology

- Calculating reciprocals.

- Converting between decimals and fractions.

- Entering, evaluating, and graphing power functions.
- Exploring exponential properties graphically in the explorations.
- Graphing functions that arise from proportional reasoning situations.

Comments

Unit cost is used to introduce a rate of change and to introduce the idea of equivalent ratios (Investigations 1–9). Students use finite differences to write equations and to construct graphs in which unit cost acts as the slope. Reducibility of ratios is investigated and the meaning of equivalent fractions is explored graphically. The idea of a reciprocal is also introduced. Investigations 11 and 12 allow students to consider how to reduce algebraic fractions (rational expressions). Investigations 13–15 allow the student to study the relative sizes of numeric and algebraic fractions. Investigations 16–18 focus on graphing rational numbers. Ultimately, the rational number system is defined. Finally, an application involving percent is investigated in Investigations 19–23.

Explorations

- 2 and 9 explore equivalent ratios.
- 3 and 6 focus on reciprocals.
- 4 and 5 focus on the meaning of rational numbers.
- 7 asks students to simplify rational expressions.
- 8 focuses on the domain of rational expressions.
- 10 asks students to compare the relative sizes of ratios.
- 11–14 provide problem situations requiring proportional reasoning.

Section 5.3 Investigating Rational Number Operations

Pacing   1 75-minute class. More time may be needed if the understanding of fractions is weak.

Technology

- Converting between decimal form and fraction form.
- Using tables to check the results of performing operations on rational expressions.

Comments

The section begins with an investigation (Investigations 1–3) requiring addition of both numeric and algebraic fractions. Investigations 4–7 deal with determining the size of a product when multiplying. Too often students who have learned that multiplication means repeated addition assume that the product must always be larger than the factors. Beware of this misconception.

Investigations 8 and 9 focus on multiplication of numeric and simple algebraic fractions. Investigations 10–15 investigated the meaning of reciprocal and introduce both fraction and decimal representations. At this point, order of operations is revisited and the operation of "reciprocaling" is included.

Investigations 16–20 allow students to investigate how divisions can be expressed as multiplications. The remaining investigations focus on solving equations that contain fractions.

Explorations

- 2 focuses on reciprocals.

- 3–8 provide practice with operations on numeric and algebraic fractions.

- 9 provides practice in equation solving.

- 10 explores the denseness of the rational numbers.

- 11 and 12 asks students to find outputs given inputs and vice versa for given functions. The inputs and outputs may involve fractions.

- 13 allows students to numerically investigate the value of $\frac{1}{x}$ as x decreases.

- 14 and 15 involve problem situations that require manipulation of rational numbers.

- 16 revisits the idea of closure.

Section 5.4 Reciprocal and Power Functions

Pacing   1 75-minute class.

Technology

- Entering, evaluating, and graphing the reciprocal function.

- Entering, evaluating, and graphing power functions.

- Exploring of exponent properties graphically.

Comments

A problem situation (**Spring Break Trip**) involving the relationship between distance, rate, and time provides the opportunity to introduce reciprocal functions. The situation involves a trip of fixed distance. Time is defined as a function of speed. Investigations 12 and 13 focus on how the output of a function changes as the input changes. This foreshadows the idea of increasing, constant, and decreasing functions. The reciprocal function is formally introduced in Investigations 14–17. The section concludes with students graphically investigating various power functions. If they studied power functions previously in Section 3.6, this material will be review, for the most part.

Explorations

- 3 and 4 focus on the general shapes of the graphs of polynomial functions.

- 5 asks students to real-world applications for the various types of variation.

- 6 checks students' understanding of the relationship between distance, rate, and time.

- 7–12 provide the opportunity to discover exponent properties by studying graphs. These Explorations serve as advance organizers for the study of exponent properties in the next section.

- 13 focuses on the concept of inverse variation.

- 14 provides additional practice solving linear equations.

- 15 and 16 introduce the idea of a literal equation.

Section 5.5 Integer Exponents

Pacing 1 75-minute class.

Technology

- Use fraction display on graphing calculator to collect data that allows students to generalize the meaning of a negative integer exponent.

- Use a symbolic manipulator, if available, to collect data that allows students to generalize rules for exponents.

Comments

The section extends the meaning of exponents to negative integer exponents. Investigations 1–4 could be done prior to class. A symbol manipulator that can manipulate and simplify exponential expressions is a helpful deal for collecting data concerning exponential properties.

It is important that the students see fractions displayed when investigating negative integer exponents. This allows them to relate negative exponents to the operation of reciprocaling.

Investigations 3–7 focus on the property $\dfrac{a^m}{a^n} = a^{m-n}$ provided a is not zero.

Investigations 8–12 conclude the section by introducing the properties $(a^m b^n)^p = a^{mp} b^{np}$ and $\left(\dfrac{a^m}{b^n}\right)^p = \dfrac{a^{mp}}{b^{np}}$ provided that b is not zero.

Explorations

- 2–5 check students' use of exponent properties.

- 6 and 7 focus on order of operations when opposing and squaring appear in the same expression.

- 8 asks students to explain in writing the meaning of a negative exponent.

- 9 asks students to use patterns in a table to study the reasonableness of the definition of zero and negative integer exponents.

Section 5.6 Making Connections: Is There Anything Rational About Functions?

This review section can be assigned for homework, with discussion the following class period. Material from all preceding chapters is included in the review section.

CHAPTER 6 INSTRUCTOR NOTES

Real Numbers: Completing a Mathematical System

Rationale

This chapter introduces irrational numbers and formally defines real numbers. The square root function is introduced. Classes of basic functions are studied algebraically, numerically, and graphically. Linear and quadratic functions are analyzed in detail to conclude the chapter. This analysis sets the stage for a study of linear equations, systems of linear equations, and factoring in the last chapter.

Overview

Section 6.1 begins with an investigation that requires square roots and motivates the discussion of irrational numbers. Repeating, terminating, nonterminating and nonrepeating decimal representations are investigated to develop a deeper understanding of the real number system. This section concludes with the observation that, with the real number system, it is possible to investigate functions with continuous domains.

Section 6.2 investigates the square root function using multiple representations. The section explicitly deals with the distinction between a square and a square root. In the process, the concept of principal square root is discussed. The opposite of the square root and the square root of the opposite of x are composite functions which are compared and contrasted with the square root function.

Section 6.3 includes investigations of multiple representations of seven basic functions: constant, linear (identity), quadratic, opposite, absolute value, reciprocal and square root. The domain and range for each is studied. The idea of intercepts is introduced. Generalizations for each of the basic functions conclude this section.

Section 6.4 looks at multiple representations of linear functions. The concepts of constant slope and intercepts are explored. Zeros of a function are introduced. Concepts of parallel and perpendicular lines are investigated. Expressing a linear function in slope-intercept form using literal equation solving techniques is included. Discussion of why a vertical line does not represent a function concludes the section.

Section 6.5 defines the general quadratic function by looking at transformations on the basic quadratic function. By looking at the algebraic representation, students predict whether or not the parabola opens upward or downward, where the intercepts are and the x-coordinate of the vertex. The connection between the zeros of a quadratic function and its factors conclude this section.

Section 6.6 is a review section.

Allow approximately 5–6 75-minute classes for Chapter 6.

Section 6.1 Real Numbers and the Algebraic Extension

Pacing   1 75-minute class.

Technology

- Calculating square roots.

- Analyzing decimals as representing rational numbers or irrational numbers.

- Comparing graphs of discrete versus continuous functions.

Comments

A problem situation (**Building a Dog Pen**) leads to the need to define square roots that result in irrational numbers. The need for closure for the operation of square rooting leads to the irrational numbers. Developing an intuitive sense for these numbers is difficult for students. To better grasp the concept of an irrational number, decimal representations of rational numbers are explored. Exact vs. approximate representations of numbers are explicitly discussed.

The structure of a mathematical system is now complete. These investigations lead to an analysis of continuous vs. discrete graphs and students are asked to consider under what circumstances each type of graph is appropriate.

The mathematics in this section can be confusing for many students. Students should be encouraged to express their concerns about the connection between infinite points and the real numbers or their bewilderment over the fact that pi never repeats. Dissonance is a natural occurrence in the process of learning and acquiring understanding. Students should not just accept the words of a textbook at face value, but question and explore the dissonance.

Explorations

- 2 provides the instructor with a quick diagnostic of students' ability to recognize and categorize various numbers.

- 3–6 encourage students to review all of the number systems studied to date.

- 7 provides students with opportunities to investigate rational and irrational numbers graphically.

- 8 deals with approximations of π.

- 9 provides more equation solving practice.

- 10 provides more practice simplifying exponential expressions.

Section 6.2 The Square Root Function

Pacing   1 75-minute class.

Technology

- Defining and analyzing the square root function.

- Numerically and graphically comparing functions that involved both opposites and square roots.

Comments

This section identifies some of the major misconceptions students have about squares and square roots (Investigations 1–8). Students have difficulty deciding what the correct order of operations is for many unary operations (Investigations 9–15). This section encourages students to analyze, compare and contrast $\sqrt{-x}$ and $-\sqrt{x}$, using function machines and a graphing utility. Discussion of issues of domain and range are continued in this section. Provide multiple opportunities for students to explain what is happening when they analyze these functions.

Explorations

- 2 checks students ability to discriminate between rational and irrational numbers.

- 3 asks students to construct function machines for expressions involving several operations, including square root.

- 4 and 5 provide investigations leading to the fact that the $\sqrt{x^2}$ and $|x|$ are equivalent.

- 6 compares on the rates of change in outputs of the squaring function and of the square root function.

- 7 and 9 explore the equivalence of $\sqrt{x}$ and $x^{1/2}$. The use of fraction exponents needs some class discussion so that the instructor can verify that students are making the connection between the square root of a number and that number raised to the one half power. This exploration allows students to expand their understanding of exponents beyond the scope of the course by collecting data and generalizing. Such explorations empower students as mathematicians.

- 8 explores what happens to the output as you iterate $\sqrt{x}$.

- 10 introduces the distance formula for the distance between two points in the plane.

- 11 and 12 present problem situations that involve the square root function.

- 13 reviews operations on polynomials.

- 14 reviews reducing rational expressions. This is in preparation of working with more complex rational expressions in Chapter 7.

Section 6.3 Classes of Basic Functions

Pacing   1 75-minute class.

Technology

- Analyzing the behavior of basic functions using both a table and a graph.

- Identifying intercepts using both tables and graphs.

Comments

Multiple representations are used to investigate some basic classes of functions and students are asked to identify the domain, the range, and the intercepts of these basic functions. They are expected to recognize the multiple representations of these basic functions after completing this section. Students should be encouraged to predict what the output of a particular algebraic representation will look like prior to graphing it and to be able to state why they think their prediction is appropriate. The general form of both linear and quadratic functions is introduced.

Explorations

- 2 deals with the rate of change of each basic function.

- 3 asks students to use graphs to solve equations.

- 4 asks students to relate the basic functions to some practical situation.

- 5 focuses on the meaning of intercepts in a problem situation.

- 6–9 asks students to investigate linear and quadratic functions numerically and graphically. The explorations of this section provide opportunities for students to analyze the graphs of basic functions and to reflect on what information each different representation of the same function provides and what information that representation does not provide.

- 10 allows students to investigate the value of $\frac{1}{x}$ as x increases without bound. Note the representation they choose to use for this exploration.

- 11 asks students to create and analyze a linear function from a diagram.

Section 6.4 Linear Functions

Pacing   1–2 75-minute classes. The material in this section is very important. Take the time to make sure students have made the connections between the various representations of linear functions before going on.

Technology

- Using tables and graphs to determine intercepts and slopes of linear functions
- Using graphs to investigate the role of each parameter in the general linear function.

Comments

Students investigate a problem situation (**Campus Cookies**) modeled by a linear function. Numeric, algebraic, and geometric representations are studied, followed by a discussion of basic features of linear functions. The role of the parameter b and the role of the parameter m in the function $y(x) = mx + b$ are studied graphically. The concept of a zero of a function is defined in the context of the horizontal intercept of a linear function. The process of finding a zero of a linear function serves as an advance organizer for solving linear equations.

Beware: the concept of a zero of a function causes students much difficulty. The relationship between x-intercepts and zeros must be emphasized. These concepts occur in Section 6.5 also and become pivotal when discussing factoring in Chapter 7.

The ramp investigations (15–25) provide the opportunity to investigate the meaning of slope, writing equations of lines, the meaning of parallel lines in terms of slope, the meaning of perpendicular lines in terms of slope, and the use of the distance formula.

Explorations

- 2–7 provide practice in identifying the slope and the intercepts, given the equation, for a linear function.

- 8–12 ask students to write the equation for a linear function given the slope and vertical intercept.

- 13 asks students to write equations for the axes.

- 14 asks students to find the slope of a line given two points.

- 15 asks students to interpret the meaning of the horizontal intercept in a problem situation.

- 16–18 asks to create an equation for a linear function from the graph.

- 19 tests students understanding of linear functions in another problem situation.

- 20 and 21 focus on creating and solving linear equations by asking questions about the output of linear functions.

- 22 focuses on the domain and range of linear functions.

Section 6.5 Quadratic Functions

Pacing   1 75-minute class.

Technology

- Graphically analyzing the effect of parameters a and c on the graph of $y = ax^2 + bx + c$.
- Comparing factored forms and expanded forms of quadratics using tables and graphs.
- Finding zeros of quadratic functions using tables or graphs.

Comments

The general form of the quadratic function is defined. Students are asked to discover the relationships between the quadratic function and its graphic representation, the parabola. The role of the parameters a and c are studied, followed by investigations on transformations. Investigations of the intercepts and vertex of a quadratic function serve as advance organizers for the study of factoring in Sections 7.3 and 7.4. The relationship between the zeros of the function and the factors of the function, using multiple representations, concludes the section.

Explorations

- 2 involves recognizing the effects of parameters on the graph of a quadratic function.

- 3–5 explore the relationships between zeros, factors, and the vertex of the graph of a quadratic function.

- 6 requires students to create the symbolic form of a quadratic function given the graph.

- 7 provides practice with writing a general algebraic expression given a diagram.

- 8 introduces the geometric interpretation of constant, linear, and quadratic terms in terms of lengths and areas of rectangles and squares. This exercise foreshadows the geometric model for completing the square.

- 9 requires equation solving.

- 10 is a review of important ideas about linear functions.

Section 6.6 Making Connections: Is There REALly a Completion to the Number System?

This review section can be assigned for homework, with discussion the following class period. Material from all preceding chapters is included in the review section.

CHAPTER 7 INSTRUCTOR NOTES

Answering Questions with Linear and Quadratic Functions

Rationale

Students continue their study of linear and quadratic functions. In the process, equations arise from the need to find an input for a given output. This allows for a discussion of equation and inequality solving. Technology makes finding zeros relatively easy to find and allows students to make connections between the factors and the zeros of a function. Equation-solving makes sense rather than rote manipulation of symbols.

Linear systems of equations are briefly introduced. Graphical interpretation is investigated first. The substitution approach is used to solve the system by manipulating symbols.

Finally, quadratic functions are revisited with a focus on using the zeros to write the quadratic function in factored form. In the process, students learn to solve quadratic equations by factoring. Simplifying, adding, and subtracting rational expressions is briefly investigated making use of factoring to perform the operations.

While we believe that factoring is an overemphasized skill, we realize some syllabi still require a solid treatment. Sections 7.3 and 7.4 approach factoring by first identifying zeros, but also discuss some of the more standard approaches to factoring. It is hoped that instructors will use as little or as much of these two sections as they believe is needed to meet the needs of their students.

Overview

Section 7.1 Linear equations and inequalities are introduced by asking students to find inputs so that the outputs meet specified conditions. Discussion of what the equal symbol means is included in this section. The section concludes with a discussion of linear equations of the form $mx + b = c$ and three methods for solving this form of a linear equation for x: a numerical method using a table, an algebraic method and a geometric method that involves treating each side as a function and determining the point of intersection of the two sides.

Section 7.2 starts by looking at a graphical solution to a system of relationships that has no algebraic representation. A problem situation gives rise to a system of linear equations which is solved graphically as well as algebraically via substitution. The section concludes by looking at and considering the general linear equation $ax + b = cx + d$ as resulting from the system $y = ax + b$ and $y = cx + d$. The related inequalities are also discussed. Dependent equations and inconsistent equations are investigated in the explorations at the end of the section.

Section 7.3 investigates factoring quadratic trinomials in which the leading coefficient is 1. Students can use a calculator to find the zeros and then create the factors. A symbol manipulator, if available, can be used to factor polynomials. Students look for patterns and address their misconceptions explicitly. Factoring is applied when performing operations on rational expressions.

Section 7.4 extends the coverage of factoring to binomials, square trinomials, and quadratic trinomials in which the leading coefficient is not 1. An exploration provides some introduction to factoring the sum or difference of two cubes.

Section 7.5 is a review section.

Allow approximately 4–6 75-minute classes for Chapter 7, depending on how much time you must spend on factoring and on solving systems of equations.

Section 7.1 Linear Equations and Inequalities in One Variable

Pacing   1–2 75-minute classes.

Technology

- Table analysis of the solution to linear equations and inequalities.
- Graphical analysis of the solution to linear equations and inequalities.

Comments

The section begins with a familiar problem situation: how to find the input given the output or range for the output. The algebraic result is the solution of linear equations and inequalities. Students investigate linear equations of the form $mx + b = c$. Linear equations in which the variable appears on both sides of the equation are treated in Section 7.2.

Using the idea of "undoing" operations, students solve linear equations and inequalities. Use of the "=" in the context of equation solving is explicitly discussed and the steps involved in solving linear equations and inequalities of the form $mx + b = c$ are formally presented.

Investigations 18–22 provide the opportunity to practice the equation-solving steps. Numeric and graphical checks are encouraged. The section concludes with a summary of how to solve the equation $mx + b = c$ for x numerically, algebraically, and geometrically. The solution to the equation is then connected to the solutions to the related inequalities.

Investigations 23 and 24 introduce students to solving literal equations by asking students to write linear equations in two variables in slope-intercept form.

Finally, Investigations 25–28 provide experience with writing the equation of a line perpendicular to a given line through a given point.

Explorations

- 2 asks students to find the horizontal intercepts of linear functions.

- 3–6 and 11–14 involve equation and inequality solving. The problems are presented in a variety of formats to increase students' flexibility in solving linear equations and inequalities.

- 7-9 ask students to write equations for linear functions.

- 15 provides the opportunity to use graphs to solve linear equations and inequalities.

Section 7.2 Systems of Equations

Pacing   1 75-minute class.

Technology

- Using tables and graphs to analyze the solution to linear systems.

- Using tables and graphs to analyze the solution to the equation $ax + b = cx + d$ in which x is the only variable. This is accomplished by defining two functions, $y = ax + b$ and $y = cx + d$, and solving the system created by the two functions.

Issues and Considerations

This section is intended as an introduction to systems to equations. Only the substitution method is introduced. Other algebraic techniques are discussed in the next text.

The section begins with a problem situation (**Monk's Journey**) that includes much unnecessary information. By extracting the crucial information and thinking graphically, students can solve the problem. The model for the problem involves two graphs and the intersection point is the solution.

The development and solution of a linear system of equations in two variables uses the substitution approach to solve the next system (**John's Barbecue**). A table and a graph are used as checks of the solution that was found algebraically.

Students revisit the original **Full Parking Lot** problem (Section 1.1), solving the problem algebraically (Investigations 22–25). The solution to equations Investigations 26–28) with the form $ax + b = cx + d$ is introduced using the system $y = ax + b$ and $y = cx + d$. Students also solve the related inequalities (Investigations 29–32). Graphical interpretations of the solutions are provided at the end of the section.

Explorations

- 2 provides practice in solving linear systems using substitution. Dependent equations (part f) and inconsistent equations (part g) are studied.

- 3, 4, and 6 provide additional practice in analyzing dependent equations and inconsistent equations.

- 5 requires students to write their own problem modeled by a system of linear equations. Beware: this can be difficult for many students.

- 7 provides practice solving general linear equations and inequalities.

- 8 focuses on the incorrect interpretation of the meaning of a variable.

- 9 revisits an earlier exploration. In this case, students are asked to represent the problem as a system of equations.

- 10 and 11 ask students to solve equations and inequalities that arise in the comparison of two linear functions.

Section 7.3 Finding Zeros of Quadratic Functions by Factoring

Pacing   1–2 75-minute classes.

Technology

- Using tables and graphs to determine the zeros of quadratic functions. Alternately, a root finder could be used though the zeros in this section are all integers.

- Using a symbol manipulator, if available, to collect data on the factors of quadratic functions.

- Using a table or graph to check the results of performing operations on rational expressions.

Issues and Considerations

This section is intended as an introduction to factoring quadratics of the form $x^2 + bx + c$. Factoring quadratics in which the leading coefficient is not one is postponed until the next section. The relationship between zeros of the function and factors of the function is emphasized. In fact, the zeros are used to write the factors.

We believe that factoring is a skill that has been overemphasized. We prefer to concentrate on basic forms and their relationship to important features of quadratic functions. Investigations 1–5 introduce the idea of a quadratic equation using the area of a rectangular dog pen as a model for the related function. Again, equations arise from the need to find the input for a given output.

Initially, a calculator is used to find the zeros of the quadratic. the zeros are used to predict the factors. Students may use a symbol manipulator to check their factors. After collecting data by factoring functions of a specific form, students attempt to factor by hand expressions of the same form. Investigations 7 and 8 focus on factoring quadratics of the form $x^2 + bx + c$ where b and c are positive. Investigations 9 and 10 focus on factoring quadratics of the form $x^2 - bx + c$ where b and c are positive. Investigations 12–15 focus on factoring quadratics of the form $x^2 + bx - c$ where c is positive.

In all cases, the relationship between the zeros and the factors is stressed. The section concludes by applying factoring to problems that require operations on rational expressions. These may be skipped if they are not part of your syllabus.

Explorations

- 2 and 4 require students to write zeros and factors of a quadratic function given its graph.

- 3 and 6 emphasize the solution to quadratic equations.

- 5 provides practice in finding the zeros and the factors of a quadratic function.

- 7 presents a problem situation that results in a quadratic equation to solve.

- 8 and 9 involve operations with rational expressions.

- 10 provides practice in solving linear equations and inequalities.

- 11–13 provide practice in solving linear systems.

- 14 reviews writing linear functions satisfying certain conditions.

Section 7.4 Additional Factoring Experiences

Pacing   1 75-minute class.

Technology

- Using tables and graphs to determine the zeros of quadratic functions. Alternately, a root finder could be used.

- Converting rational zeros in decimal form to fractional form.

- Using a symbol manipulator, if available, to collect data on the factors of quadratic functions.

Comments

This section continues to exploring factoring quadratics using the zeros. Three other cases, not discussed in the preceding section, are investigated.

Investigations 1–5 focus on factoring binomials such as the difference of two squares. Investigations 6–10 focus on factoring quadratics of the form $ax^2 + bx + c$ where the leading coefficient is not 1. Creating factors from rational number zeros and vice versa is emphasized. The rational zero theorem, usually part of a college algebra course, is used as an aid in the search for zeros. Finally, Investigations 11–14 focus on factoring square trinomials emphasizing that the graphs of such functions are tangent to the x-axis.

Explorations

- 2 provides extensive practice with factoring.

- 3 and 4 present problem situations in which locating zeros of quadratic functions is helpful.

- 5 and 6 require students to collect data that foreshadows the factoring of the sum and the difference of two cubes.

- 7 provides practice in solving linear equations, linear inequalities, and quadratic equations.

- 8 requires students to solve linear systems.

- 9 asks students to reduce rational expressions by first factoring the numerator and the denominator.

- 10, 11, and 13 review various features of linear functions.

- 12 asks students to perform operations on rational expressions.

Section 7.5 Making Connections: Linking Multiple Representations of Functions

This review section can be assigned for homework, with discussion the following class period. Material from all preceding chapters is included in the review section.

Classroom Management

CREATING AN ACTIVE LEARNING ENVIRONMENT

Essential elements of a workable active-learning environment include:

- course objectives and content clearly communicated to students.
- a positive classroom tone conveyed by the physical environment and by the instructor.
- coping with teaching space not necessarily designed for active learning.
- knowing more about our students.
- less time in class spent covering content without decreasing the amount of content assigned. (In practice most instructors do decrease the amount assigned).

Instead of repeating what students should have read, time should be spent on:

- highlighting issues.
- focusing student attention on key subject matter.
- encouraging discussion.

To decrease course content down to essentials ask yourself:

- what do I want students to know?
- what should they be able to do by the end of this course?

Provide:

- a preview of course objectives at first class meeting.
- clear information on how class time will be used.
- clear information on how class activities will be evaluated.

Classroom tone should communicate a clear message that expectations are different from traditional classroom – the old rules are out.

Teaching Space

The design and arrangement of teaching space says a great deal about an institution's philosophy of education. In most cases, active learning is neither encouraged nor supported, given existing classroom facilities.

3 basic teaching situations include:

1. adequate size rooms with flexible seating (no more than 25 students).

2. large lecture halls and inflexible seating.

3. small, crowded classrooms, with no room for movement.

Possible seating arrangements that encourage active learning include:

- a U-shaped arrangement (10–15; 20–25 students).
- circles (20–25).
- experimenting with different seating arrangements and groupings.

Student surveys serve four basic purposes including:

- uncovering student attitudes and predispositions.
- finding out about students general or specific areas of interest.
- getting a reading on students' present skills and knowledge in a given area.
- revelations about student attitudes and/or negative stereotypes.

In an effort to promote an active learning environment, the instructor:

- finds out what students need to do and know.
- writes a syllabus that clearly outlines course objectives, expectations and methods of learning.
- creates a classroom environment that supports active learning and participation.
- deals creatively with space limitations.
- finds out more about students' lives, experiences, expectations, and capabilities.

LEARNING WITH OTHERS

The classroom is a community of learners, actively working together in groups to enhance each person's mathematical knowledge, proficiency, and enjoyment.

Neil Davidson

Learning Modes

Whole class

- All listen when one person speaks.
- Refrain from talking amongst each other.
- Raise hand to ask questions or contribute to discussion.

Small teams

- Talk to each other.
- Contribute to the work of the team.
- All members are committed to problem analysis. Work is not submitted until all team members agree on all solutions.

Guidelines for Teams

- Work together.
- Cooperate.
- Achieve a *team solution* to each problem.
- Verify that everyone understands the solution to a problem before team goes on.
- Listen carefully to others and try, whenever possible, to build on their ideas.
- Share the leadership.
- Make sure everyone participates and no one dominates.
- Proceed at a pace that is comfortable for your own team.

Models

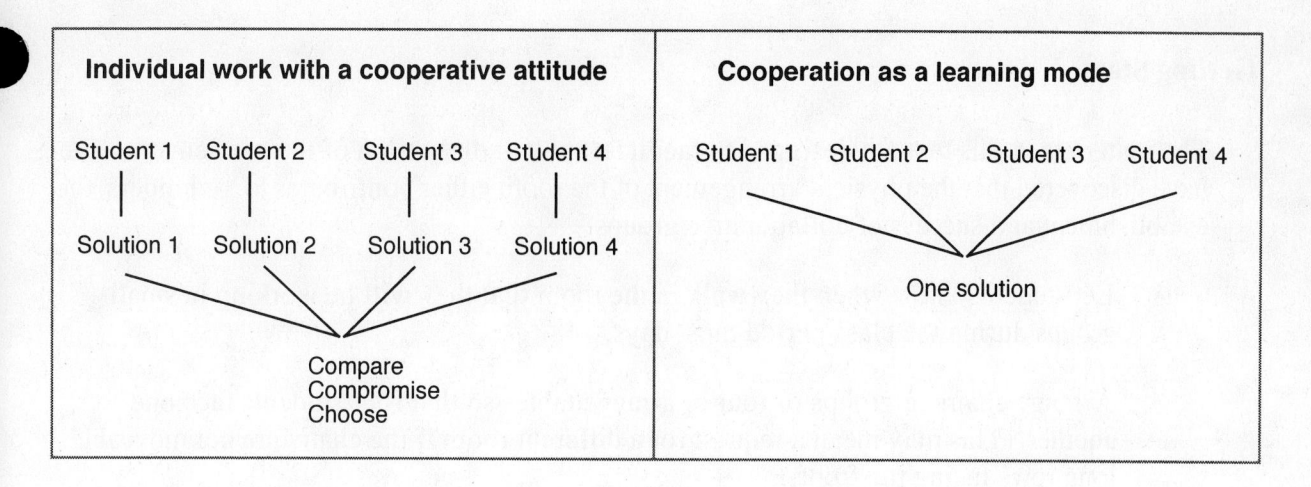

Individual work with a cooperative attitude	Cooperation as a learning mode

Student 1 Student 2 Student 3 Student 4
      | | | |
Solution 1 Solution 2 Solution 3 Solution 4
         ↘   ↘   ↙   ↙
              Compare
              Compromise
              Choose
	Student 1 Student 2 Student 3 Student 4
	         ↘ ↘ ↙ ↙
	                 One solution

Student Responsibility

- Learning material.

- Making sure team members learn the material.

- Helping class members learn material.

Cardinal Rules

- No one is allowed to *interfere* with another student's learning.

- The test for understanding is often the ability to communicate to others; and this act itself is often the final and most crucial step in the learning process.

R. C. Buck

COLLABORATIVE GROUPS

Getting Started

The materials in this text are designed to facilitate student discussion of the mathematics. We have discovered that the physical arrangement of the room either contributes to, or hinders, the establishment and success of collaborative groups.

- Let students know when they walk in the room that they will be working in small groups during the class period most days.

- Arrange chairs in groups of four or arrange tables so that four students face one another. (This may mean a request for a different room if the chairs are not moveable-- long rows facing the front.)

- After each student individually has the opportunity to understand and investigate the parking lot problem, students should be requested to team up with a partner and exchange their solution strategies and results.

This is the first time students will be working in pairs. Working in groups of two, students gain an appreciation of another person's perspective and approach to the problem, which is often different from their own. They discover that their input is of value and that it is difficult not to participate when you are discussing something with only one other individual.

- When the pair has completed their solution of the parking lot problem, ask them to share their work with another pair. As students move from working with a partner to working collaboratively within a group, students experience the new role of collaborative investigator.

Group Size

There are many philosophies about the formation of groups, much of which has been based on work with children. Much has been written about the size of groups, including preferences for groups of four, groups no larger than three members, and pairs. What we suggest may not be the current wisdom, but it has worked in many of our classrooms.

- Students initially uncomfortable working in a group of three or four are more comfortable and willing to work with a single partner.

As you experiment with collaborative groups and the appropriate size of groups for a given class, you will develop techniques you are comfortable with and that work for you most of the time. Often, what works well for students in one class needs to be adjusted for a different mix of students in a different section or course.

We have found that the extent of successful collaboration is as much a function of a particular group of students in a given class and their willingness to put forth effort, as it is a function of the instructor's efforts.

Assignments of Groups

Working together outside of class should be encouraged. Since many of our students commute to class and have outside jobs, we have found that allowing students to find comfortable working relationships with people who have compatible schedules works better than assigning groups arbitrarily.

- Reassure students that the partner and group they work in during the first class session are not necessarily the group and/or partner they will work with throughout the semester.

During the first week or two, students should take the opportunity to work with several other partners and groups, exchanging available times when it is possible to meet outside of class. It is our observation that when each member takes personal responsibility for their group, everyone in the group achieves at a higher level than they would have alone.

During the second or third week, conducting group conferences and/or individual conferences lasting about ten minutes to determine how things are going in the group and in the class may pay big dividends in terms of student satisfaction, motivation and achievement.

Guidelines

At the beginning of the term, we usually ask students how many prefer to work alone. Even when a majority of the class raise their hands, the students are informed that, to succeed in this class, they will have to adjust their preferences to accommodate group work. Effective collaborative learning does not just happen.

- Make the requirement of working in groups explicit.

The amount of time spent working in groups vs.whole class discussion will vary from day to day. There are times when there is a need for the explicit clarification or sharing of information by the instructor. Be flexible. In meeting the needs of your students, work within your own comfort zone.

- Beginning the class with group discussion helps make group work seem the norm. Rather than answering questions from homework, have students compare their results and discuss their different solutions and methods.

There are times when members of a group are unable to come to a consensus.

- Don't fall into the role of referee when the group members disagree.

Ask if every one in the group understands the question. Clarify who doesn't understand the question or who disagrees with the answer other members of the group find acceptable. You might want to turn the question back to the group, redirecting their thinking on it. Encourage students to become independent learners capable of validating their own work and results.

Individual Responsibility

It is important that students feel responsible for their own learning and develop responsibility for the learning of members of their group.

- Students need to be given permission to strictly define the responsibilities of each member of the group.

Your support is necessary to help the groups work through problems and gain the cooperation of students who may not have developed appropriate study habits for collaborative work.

- Consider the role each student plays in the group.

If one person is always teaching all other members of the group, consider moving this person to another group with others at the same ability level. It provides a greater challenge for the student and personal growth for all concerned.

When group activity during class is not on task, it is helpful to hold a class discussion about the responsibility of each person for his/her own learning and to the members of the group. If students have been doing most of their work in class, there may be a tendency for some students to come to class unprepared.

- It must be okay for members of the group to exert peer pressure for students to come to class prepared.

- Students who are absent (for good cause or not) should be expected to meet with a member of their group outside of and prior to the next class to get assignments and notes of the class discussion which was missed.

It is okay to refuse to give answers so someone can merely copy them down. Class time may be used to compare and discuss answers prepared before class, but generally new investigations will be done during this time.

Student directed regrouping for compatibility is okay.

- Sometimes groups take care of excessive absenteeism on their own in a very efficient manner. Other times it is a disruption to the group and to the class. If this is the case, it might be helpful to form a group of chronic absentees unless the group they are presently working with objects.

- If you have groups of students who do not work well together, the beginning of a chapter is a good time to do some adjusting of group assignments.

POLICY STATEMENT

(The next four pages provide an example of a policy statement used at William Rainey Harper Community College. It evolved over several years and is certainly more than a syllabus. The statement prepares students for a new learning environment and addresses the most obvious changes especially in regard to new instruments of assessment. See **Alternative Forms of Assessment**.)

Course format:

Beginning Algebra has a prescribed curriculum required by the state. The course outline includes this curriculum. But studying mathematics is much more than covering a list of concepts or skills. It is an exploration into quantitative methods and how they are useful in life experiences. This course will actively strive to build a community of learners with a commitment to understanding and intellectual growth. In the community of learners our differences are respected and our similarities are celebrated. The community of learners happens through teamwork focusing on the day-to-day challenges of reading, writing, conversation, and problem solving. The course is student-centered, not teacher-centered. Key to the format is the social construction of knowledge. Class time will include the following activities:

- Investigations: student-centered activities from the text designed to promote active involvement with the course material. These activities will usually occur in small groups.

- Whole class discussion: instructor- or student-led discussion of course concepts or problems.

- In-class assessments: individual and group problem sessions designed to assess student understanding of material.

Study groups:

You will do much of the work for this course in cooperative learning groups. It seems to work best if there are three or four students in each group. You will work with your group both inside and outside class. Working well in a group is an important skill that is essential for many of the jobs for which students apply. The objective of group work in this course is two-fold:

- to give you moral and intellectual support while you are working on concepts and problems, and

- to develop skills in working effectively as part of a team.

One of the primary objectives of this course is to help you to learn to think about and solve real-world problems using the tools of mathematics. Working in your group, doing investigation activities, and talking about problems with your group members are all strategies to help you do this.

Assistance:

Numerous resources are available to assist you. These include the required text, your group, other class members, myself, the Tutoring Center, and other students and instructors. This list is not exhaustive. The most important resource will be your study group. If you are unsure of how to get help, please ask me for assistance.

Course objectives:

The primary objective is to build or build upon the intellectual and academic abilities necessary to analyze, understand, and synthesize the course content. The objectives are divided into three main categories:

Attitudes

- Approach the mathematics in this course actively by immersing yourself in the ideas and exploring the connections and limitations of those ideas.
- Be willing to risk making mistakes and realize that learning from mistakes is what learning is all about.
- Take personal responsibility for learning (study resources, use other sources as necessary, devise a time-management scheme, get work in on time).
- Respect the ideas of others (consider multiple ways of attacking problems).

Skills

- Think critically as you analyze problems.
- Communicate effectively (write a clear explanation of your approach and solution to problem situations).
- Work effectively in teams.

Knowledge

- Demonstrate an understanding of the key mathematical concepts of the course.
- Synthesize the key mathematical ideas by applying them to diverse problems and by exploring the interconnections between them.

General requirements and assessment:

A. Investigations

The Investigations are the cornerstone of the course. You work in teams to complete Investigations given in the text. Attendance during the Investigations and completion of the Investigations are critical for success. Reading the Discussion following each Investigation will assist you in assessing your understanding.

B. Explorations

There is a set of Explorations at the end of each section in the text. Completion of all Explorations is expected. Each Exploration begins with a request that you build and maintain a Glossary of key words and phrases. A blank Glossary for you to complete begins on page 425 of the text. Completion of the Glossary is critical since understanding the vocabulary plays a key role in developing mathematical power.

C. Concept Maps

Each section contains a Concept Map to construct. Further instruction will be given in class regarding Concept Maps. Generally, a Concept Map is a visual method of displaying your knowledge of a given concept. Critical components of a Concept Map include a central concept, a set of related concepts, and links between concepts demonstrating relationships. Some, but not all, Concept Maps will be assigned.

D. Reflections

Each section contains a Reflection. These require you to reflect on what you have learned in the given section and preceding sections. Most Reflections require you to write a paragraph or two discussing an important mathematical idea. Often the Concept Map can be used as a brainstorming device prior to the writing of the Reflection. Some, but not all, Reflections will be assigned.

E. Unit Problem Sets

There are three cumulative problem sets. They will be distributed during the fifth, ninth, and fifteenth weeks of the semester. The due date for each will be determined by the class.

E. Journals

Each Monday, you submit a journal entry. The specific format for the journal will be discussed in class.

G. Final Exam

H. Portfolio

You will be required to keep a notebook (three-ring binder) containing your work, aside from the Investigations and the Glossary, for the semester. The Investigations and the Glossary will be completed **in** the text. The binder will include completed Explorations, Concept Maps, Reflections, Journals, Unit Problem Sets, and in-class assessments. I will use this, along with the completed Investigations and the Glossary, as the primary evidence for your grade.

Student self-evaluation

At the end of the semester, you will write a detailed self-evaluation in which you reflect on your performance in light of the objectives of the course. The self-evaluation form will be distributed early in the semester.

Makeups and late submissions:

No make-ups or late submissions are allowed. Exceptions are made only in the event of documented unavoidable circumstances.

Attendance

Full attendance at all class meetings is expected.

Grading: Your semester grade is determined as follows:

Negotiated grading

You regularly will submit assigned written work. About half the work will be evaluated on an individual basis and the other half on a group basis. I will note on your papers what is wonderful and what needs improvement. These will become part of your Portfolio. I will also do observational assessments during class periods. I will meet with each of you at the end of the semester to negotiate your grade. To determine the grade, I will consider the quality of the work in the Investigations, in the Glossary, and in the Portfolio along with your written self-evaluation. From these, we will mutually agree on a grade. Arguments for a certain grade must be backed up by concrete evidence of why that grade is appropriate. We will conduct a practice negotiation approximately halfway through the course. I am willing to conduct an one-on-one grade evaluation with you any time during the semester.

Final exam

- If the final exam grade is above the negotiated grade, you will receive one grade higher than the negotiated grade for the course.

- If your grade on the final is the same as or one grade less than the negotiated grade, you will receive the negotiated grade for the course.

- If the final exam is two or more grades below the negotiated grade, you will receive one grade below the negotiated grade for the course.

Student responsibility

Your cooperative spirit and active participation are what build a strong community of learners. Your responsibility is to others as well as to yourself. Being prepared, helping others when you can, dealing with problems early so they don't interfere with learning–all are responsibilities of learners. A key component of student responsibility will be the study group. Research clearly shows (Astin, Treisman, Bruffee) that **students who form and maintain study groups outside of class more readily succeed in college-level work**. I will encourage you and do my part to help maintain study groups; however, the real responsibility will be with you to make them engaging and productive.

Instructor responsibility

I am here to help you learn and to help you develop as learners. In this role I will provide learning events for your participation. I have high standards but also the commitment to support those standards with appropriate instruction. Please see me to consult about any problems that arise or just to talk about any aspect of the class.

GUIDELINES FOR GROUP BEHAVIOR

Learning mathematics is often perceived as an individualistic, competitive activity that is often lonely and frustrating. Small-group cooperative learning provides an alternative environment in which students work together to maximize their own and each other's learning. In this course, small-group cooperative activities will be used most of the time during class. Students are expected to provide a social support mechanism for each other as they learn mathematics. Small-group cooperative learning offers opportunities for success for all students in the class.

The guidelines under which we will operate are:

1. As you arrive for class, move quickly into your group (four persons per group maximum). If individual desks are in the room, move chairs close together, use first names, and get right to work. If in a room with tables, one group per table.

2. Read all instructions and identify all given information. **Identify assumptions and what the assignment should have as outcome(s)**.

3. Strive for success for all members of the group.

    - Ensure that all members of the group complete the assignment.

    - Achieve a group solution for each problem.

    - Proceed at a pace that is comfortable for all members of your group.

    - Check to see that every group member is able to identify and explain the basic elements of the problem and can justify the solution.

    - Make sure that everyone understands the solution before the group goes on.

4. Listen carefully to others, and try, whenever possible, to build on their ideas.

5. Encourage every member of your group to contribute ideas and participate in the discussion. *Do not allow one individual to take over or to dominate the group!*

6. Foster interdependence among group members. Act responsibly towards your group.

    - Attend class regularly and actively participate in your group.

    - Come to class prepared to discuss assignments and hold fellow group members accountable to that same standard.

    - Help group members who have justifiable absences remain effective, contributing members of the group. Take clear, accurate notes and share them with a group member who is absent *prior* to the next class.

7. Regularly reflect on and assess your own contribution to group activities.

TEAM EVALUATION FORM

Name_____

Assignment_____

1. Overall, how effectively did your team work together on the assignment?

    Poorly   Adequately   Well   Extremely well

2. How many people were in the team?

3. How many team members participated actively most of the time?

4. Give one specific example of **something you learned from the team** that you probably would not have learned working alone.

5. Give one specific example of **something the other team members learned from you** that they probably would not have learned otherwise.

6. Suggest one change the team could make to improve its performance.

COURSE OUTLINE (16 WEEK SEMESTER)

Required Resources and Equipment

- **Text:** DeMarois, P., McGowen, M., & Whitkanack, D., *Mathematical Investigations: An Introduction to Algebraic Thinking* (1st Edition).

- A TI–83, TI–82, or TI–92 graphing calculator.

- A three-ring binder for your portfolio.

Optional Resource

Student Support Manual for DeMarois, P., McGowen, M., & Whitkanack, D., *Mathematical Investigations: An Introduction to Algebraic Thinking* (1st Edition). This manual contains answers to most of the Explorations from the text and information on using the TI–83 and TI-82 graphing calculators.

Prerequisite

Successful completion of prerequisites for this course. It is essential that you see me as soon as possible if you believe that you have been placed in the wrong math class. It is quite easy to make a switch early in the semester, but very hard after a couple of weeks.

Text Content	Time Frame
Chapter 1 What is Mathematics	2 weeks
Chapter 2 Whole Numbers: Introducing a Mathematical System	2 weeks
Chapter 3 Functional Relationships	2 weels
Chapter 4 Integers: Expanding on a Mathematical System	2 weeks
Chapter 5 Rational Numbers: Further Expansion of a Mathematical System	2 weeks
Chapter 6 Real Numbers: Completing a Mathematical System	2-3 weeks
Chapter 7 Answering Questions with Linear and Quadratic Functions	2-3 weeks

Schedule Notes

- Spring vacations is: _____ through _____.

- The last day to withdraw from a class is _____.

- The **final exam** is _____.

DAILY AGENDA 2

Date:_____ **Section in Text:**_____

List the questions you have from the Investigations and from the Explorations

Name	Investigation Questions (write number(s))	Exploration Questions (write number(s))

List 1 or 2 concepts in the section that you found confusing:

Complete the following sentences at the end of class

The most helpful thing we did today was:

The most confusing thing that occurred during class today was:

At the next class meeting, please explain:

Section

Answers

CHAPTER 1

What is Mathematics?

Section 1.1 Learning Mathematics

p. 11

1. Opinion

2. Some patterns include

   a. Column 1 begins at twenty and decreases by ones and, if extended, would end at zero.

   b. Column 2 begins at eighty and decreases by fours and, if extended, would end at zero.

   c. Column 3 begins at zero and increases by ones and, if extended, would end at twenty.

   d. Column 4 begins at zero and increases by twos and, if extended, would end at forty.

   e. Column 5 is a constant of twenty.

   f. Column 6 begins at eighty and decreases by twos and, if extended, would end at forty.

3. In the equation $4c + 2(20 - c) = 66$, the four is the number of wheels on a car, the c is a variable representing the number of cars, so $4c$ is the number of wheels on c cars. The two is the number of wheels on a motorcycle, the $20 - c$ is a variable expression representing the number of motorcycles, so $2(20 - c)$ is the number of wheels on the motorcycles. The 66 represents the total number of wheels in the parking lot.

4. $2m + 4(20 - m) = 66$ where m is the number of motorcycles.

5. $c + m = 20$ represents the fact that the sum of the number of cars and the number of motorcycles equals twenty.
   $4c + 2m = 66$ represents the fact that c cars have $4c$ wheels; m motorcycles have $2m$ wheels, and the total number of wheels is sixty-six.

6. The Bulls made twenty-three two-point shots and four three-point shots. This result could be found by guess-and-test, a table, the equation $2t + 3(27 - t) = 58$, where t represents the number of two-point shots, or the system $\begin{array}{l} t + s = 27 \\ 2t + 3s = 58 \end{array}$ where t is the number of two-points shots and s is the number of three-point shots.

Section 1.2 Thinking Mathematically

p. 25

3. a. $\boxed{\text{S P} \xrightarrow{\text{Rule 1}} \text{S P P} \xrightarrow{\text{Rule 1}} \text{S P P P P} \xrightarrow{\text{Rule 3}} \text{S C P}}$

    b. $\boxed{\begin{array}{l}\text{S P} \xrightarrow{\text{Rule 1}} \text{S P P} \xrightarrow{\text{Rule 1}} \text{S P P P P} \xrightarrow{\text{Rule 1}} \text{S P P P P P P P P} \\ \xrightarrow{\text{Rule 3}} \text{S P P P C P P} \xrightarrow{\text{Rule 2}} \text{S P P P C P P C}\end{array}}$

    c. $\boxed{\begin{array}{l}\text{S P} \xrightarrow{\text{Rule 1}} \text{S P P} \xrightarrow{\text{Rule 1}} \text{S P P P P} \xrightarrow{\text{Rule 1}} \text{S P P P P P P P P} \\ \xrightarrow{\text{Rule 2}} \text{S P P P P P P P P C} \xrightarrow{\text{Rule 3}} \text{S P P P P P C C}\end{array}}$

4. a. S P C S is not a valid path since it contains the letter S in a position other than the first position.

    b. $\boxed{\begin{array}{l}\text{S P} \xrightarrow{\text{Rule 1}} \text{S P P} \xrightarrow{\text{Rule 1}} \text{S P P P P} \xrightarrow{\text{Rule 1}} \text{S P P P P P P P P} \\ \xrightarrow{\text{Rule 1}} \text{S P P P P P P P P P P P P P P P P} \xrightarrow{\text{Rule 2}} \\ \text{S P P P P P P P P P P P P P P P P C} \xrightarrow{\text{Rule 3}} \\ \text{S P P P P P C P P P P P P P C} \xrightarrow{\text{Rule 3}} \text{S P P P P P C P P P P C P C}\end{array}}$

    c. S P P P C is not a valid path since it contains 3 P's. A valid path cannot have a P–count that is a multiple of three.

    d. C S P P is not a valid path since all valid paths must begin with the letter S.

    e. S P P C P P is a valid path. Derive the path. It's challenging but doable.

    f. S C C C C C is not a valid path since it contains 0 P's and 0 is a multiple of three. The P–count cannot be a multiple of 3.

5. A test for non–valid paths might be to check if the path does not begin with the letter S or if it has a P–count that is a multiple of 3.

6. Rules 1 and 2 lengthen valid paths; Rules 3 and 4 shorten valid paths.

7. When working backwards, you think about what rule might have been used to obtain the given valid path. An illegal reversal would mean doing a rule in reverse. For instance, if you did the reverse of what Rule 3 says to do you would replace a C with three P's. This is not a valid procedure, however, and would not be guaranteed to produce a valid path.

8. 0 is a multiple of 3 since there is a whole number, 0, that we can multiply 3 by to obtain 0.

9. The new path is S P P P P P P P P. The number of elements after the S must be doubled. Doubling means multiply the number of elements by 2; counting by 2 means "add 2".

10. SC is not a valid path in the SPC system. Since the number of P's in SC is 0 and 0 is a multiple of 3 it is impossible to arrive at this path by starting with SP and using the rules. This uses the conclusion that a path is not valid if the P-count is a multiple of 3. See the answer to Exploration 5, above.

11. Answers vary. But, in general, when you applied one of the rules to produce a new path you worked within the system.

12. Answers vary.

13. Answer varies. But, drawing general conclusions about the effect of the rules on the P-count of a path as in Exploration 6 above represents working outside the system.

14. Opinion.

15. Opinion.

16. Opinion.

Section 1.3 Using Variables to Generalize

p. 40

3. Using Rule 1, draw a box around the sequence of letters that can be represented by the variable in each of these paths.

   a. S[P]

   b. S[C P]

   c. S[P P P P C]P P P P C

   d. S[C P P C]

4. Using Rule 2, draw a box around the sequence of letters that can be represented by the variable in each of these paths.

   a. S P: empty

   b. S[C]P

   c. S[P P P P C P P P]P

   d. S C P P C: Rule 2 cannot be applied.

70

5. Using Rule 3, draw a box and a circle around the sequence of letters that can be represented by the variables in each of these paths.

   a. S|P C|P P P(P) is one answer.   b. S P P P(P C P) is one answer; box is empty.

   c. S|P|P P P : circle is empty.   d. S|P|P P P(C C P C P P) is one answer.

6. Using Rule 4, draw a box and a circle around the paths that can be represented by the variables in each of these paths.

   a. S|P|C C(P)   b. S C C(P P C): box is empty.

   c. S|P|P P P P P P: Rule 4 doesn't apply.

   d. S P P P C C(P C P P)

7. a. 1 x 7 = 7 for example.

   b. 7 · 7 · 7 = 7³ for example.

   c. $\sqrt{2^2 + 3^2} \leq 2 + 3$ for example.

8. a. Paper is an example of a noun.   b. Boss is a noun that ends in s.

   c. Ax is a noun that ends in x.   d. Fold is a verb.

9. a. Let the variable x represent the singular form of the noun. Then xs is the plural form.

   b. Let the variable x represent the singular form of the noun that ends in s up to the letter s. Then $x$ses is the plural form.

   c. Let the variable y represent the singular form of the noun that ends in x up to the letter x. Then $y$xes is the plural form.

   d. Let the variable x represent the present tense of a verb. Then $x$ed is the past tense.

10. English grammar rules don't work as nicely as the rules in the SPC system because many grammar rules require one to be aware of exceptions.

11. a. $1^2 = 1$; $11^2 = 121$; $111^2 = 12321$.

    b. $1111^2 = 1234321$; $11111^2 = 123454321$.

    c. $111111^2 = 12345654321$; $1111111^2 = 1234567654321$;

    e. 111111111111^2 doesn't follow the previous pattern due to the carry that will occur.

12. John conjectured, based on two examples, that he could park illegally without receiving a ticket. The probable implication of this generalization is he will most likely receive a ticket in the near future.

13. The assumptions about variables with examples are:
    In a problem or statement, every time a variable is used, it represents the same thing.

    Example: In $2x + 3x$, the x represents the same number in both places.

    The variable probably represents something different when we work a new problem.

    Example:

    | Rule 1 | S x $\longrightarrow$ S x x |
    | Rule 2 | S x P $\longrightarrow$ S x P C |
    | Rule 3 | S x P P P y $\longrightarrow$ S x C y |
    | Rule 4 | S x C C y $\longrightarrow$ S x y |

    The variable x means something different in the statement of each rule.

    Variables make it possible to describe a few specific examples as a generalized statement.

    Example: $n + 0 = n$ is an example of generalizing the fact that the sum of any number and zero is the number.

    A generalized statement must always be true and, in order to guarantee this, the use of restrictions, or conditions, may be required.

    Example: $\sqrt{x}$ represents a real number only if x is not negative.

Section 1.4 Expanding the Notion of Variable

p. 57

2. Techniques that are algebraic include writing an equation or writing a system of equations. Both these techniques make use of variables to represent and solve the problem.

3. y is the base and m is the exponent.

4. When you square a number larger than one, you always get a number larger than the original number.

5. When you square a number between zero and one, the square is always smaller than the original number.

6. a. For $2n$, we are computing the product of two and the variable n. For n^2, we are computing the product of n and n.

   b. For 2^n, we are computing the product of n twos.

7. The larger the value of p, the smaller the value of $50 - p$.

8. We will use the formula $P\text{-}count = 2^n - 3x$ where n is the number of times Rule 1 is applied and x is the number of times Rule 3 is applied.

   a. In this case, $n = 7$ and $x = 11$.
   So $P\text{-}count = 2^7 - 3(11) = 128 - 3(11) = 128 - 33 = 95$.
   There are 95 P's in the resulting valid path.

   b. In this case, $n = 4$ and $P\text{-}count = 7$. So, $7 = 2^4 - 3x$.
   $7 = 16 - 3x$. Since $7 - 16 = -9$, Rule 3 must have been applied three times.

   c. In this case, $x = 6$ and $P\text{-}count = 46$. So, $46 = 2^n - 3(6)$.
   $46 = 2^n - 18$. Since $46 = 64 - 18$, we know $2^n = 64$ which leads to $n = 6$. Rule 1 was applied six times.

   d. If Rule 1 is applied four times, then the $P\text{-}count$ is $2^4 = 16$. We could then apply Rule 3 either 0, 1, 2, 3, 4, or 5 times.

   e. By applying Rule 3 successively, we get possible $P\text{-}counts$ of 16, 13, 10, 7, 4, and 1.

9. a. To obtain a $P\text{-}count$ of 13, I must apply Rule 1 four times and Rule 3 once. See the answer to 8d.

   b. To obtain an $P\text{-}count$ of 52, I must apply Rule 1 six times to obtain 64 P's. then eliminate twelve P's. This requires four applications of Rule 3.

   c. Answers may vary. Most likely, the answer was found numerically by an informed trial–and–error process using a calculator.

10. In Section 1.3 the term domain is used in two different ways. On page 30 it refers to the possible paths to which a rule in the SPC system may be applied. SCPPCPP is a path in the domain of Rule 2, for instance. But in using variables to state the rules on page 36 of the text x represents possible sequences of letters that could occur between S and P when applying Rule 2 to a valid path. In SCPPCPP for instance x = CPPCP. The term domain is used to refer to all the possible sequences of letters that might appear between S and P in a valid path where Rule 2 could be applied. In other places in Section 1.3 and Section 1.4 variables represent numbers in algebraic expressions and equations. Depending on the use of the expression in a problem situation the possible numbers which could replace the variable may be restricted. In the P-count equation after Rule 1 has been applied to the path SP six times Rule 3 could be applied 0, 1, 2, 3, 4, or 5 times only. The domain of x in

this situation is this set of numbers. The domain of a variable is the set of all possible replacement values of that variable. The domain of one of the rules in the SPC system is the set of all possible paths to which a rule may be applied.

11. Rewrite each expression using only one exponent, if possible.

   a. $(7^5)(7^{11}) = 7^{16}$
   b. $(2^9)^5 = 2^{45}$

   c. not possible
   d. $(2^9)(2^5) = 2^{14}$

   e. not possible
   f. $(9^{15})(9) = 9^{16}$

   g. $(11^{23})(11^4) = 11^{27}$
   h. $(5)(5^2) = 5^3$

   i. not possible
   j. $(11^{23})^4 = 11^{92}$

12. The difference is in the order of the operations which are indicated. $(2^9)^5$ means raise 2 to the 9th power first and then raise the number you get to the 5th power. $2^9 + 2^5$ means to compute the exponents first and then add the results while $(2^9)(2^5)$ means to compute the exponents first and then multiply the results.

13. $(n)(n)$ means to multiply n by itself and $n + n$ means to add n to itself.

14. Write each as an expression with one exponent.

   a. $(y^6)(y^4) = y^{10}$
   b. $y^5 \cdot y = y^6$

   c. $y^m y^n = y^{m+n}$
   d. $(y^6)^4 = y^{24}$

   e. $(y^m)^n = y^{mn}$
   f. $(y^4)^6 = y^{24}$

15. If Rule 1 is applied six times, the number of P's in the resulting path is $2^6 = 64$. Then reduce the number of P's by three. Recall that x represents the number of times Rule 3 is applied. The domain of the variable is the set of all possible values of that variable. Rule 3 can be applied no times or until less than three P's remain. So, the domain of the variable x in the P–count equation is {0, 1, 2, 3, 4, 5, 6, 7, 8, 9, 10, 11, 12, 13, 14, 15, 16, 17, 18, 19, 20, 21}. If Rule 3 is applied twenty–one times, 63 P's are removed and one P remains.

16. Variables were used in mathematical expressions to represent numbers from the domain of the problem situation. This contrasts with the use of variable in Section 1.3 where variables were used to represent paths that remained constant when a rule was applied. Also, variables were used in this section to generalize a pattern.

17. a. $k + k$ or $2k$ since Rule 1 doubles the number of characters after the S.

b. $k + 1$ since Rule 2 adds one character, C, to the path.

c. $k - 2$ since Rule 3 replaces three P's with C. This is a decrease of 2 characters so we subtract 2 from the original number of characters.

d. $k - 2$ since Rule 4 omits two C's. This is a decrease of 2 characters so we subtract 2 from the original number of characters.

Section 1.5 Review

p. 63

1. a. One could visualize the process with a diagram or picture. Adding or multiplying could be used to come up with the answer of 9 minutes. This, of course, ignores the time required to move from one cut to another.

    b. One must determine the types of coins. Constructing a table with the headings pennies, nickels, dimes, and quarters and filling in possible entries which would total to 43 cents would be one approach. Applying logic one might reason that there could be only 3 pennies. One could soon see that he had no quarters.

    c. Constructing a table to organize the information and search through the possibilities would be a good approach. The table could have headings like number of stools, number of legs on the stools, number of tables, number of legs on the tables, sum of the number of legs. Another approach might be to set up the equation $3s + 4t = 31$ and explore the possible whole number solutions to this equation. In the equation s would stand for the number of stools and t would be the number of tables the carpenter made. There are three possible solutions: 1 stool and 7 tables, 5 stools and 4 tables, or 9 stools and 1 table.

    d. Insight into the problem is the key to this one. Perhaps you should imagine working backwards. Consider when the basket would be full and how much of the basket would be filled 1 second before that time.

    e. A calculator would be useful to produce the number of minutes early you could leave work each day on this one. It would also be great if you could detect a pattern and see how to calculate the solution. You might end up trying to solve the equation $2^{z-1} = 8 \cdot 60$, where the left side represents how many minutes early you would be able to leave on day x. Find x by entering this expression in the Y= menu on your calculator and examining a table that is produced from this expression.

2. a. $a \cdot b = b \cdot a$ b. $\dfrac{0}{a} = 0$ as long as a is not 0.

    $5 \times 10 = 10 \times 5$ $\dfrac{0}{50} = 0$

c. $0 + a = a$ d. $1 \times a = a$

$0 + 23 = 23$ $1 \times 7 = 7$

3. $SCCPPPPC \xrightarrow{\text{Rule 4}} SPPPPC \xrightarrow{\text{Rule 3}} SPCC \xrightarrow{\text{Rule 4}} SP$

4. Rewrite each of the following using only one exponent, if possible.

   a. $(5^{19})(5^8) = 5^{27}$ b. $(3^5)^4 = 3^{20}$

   c. $(3^5)(3^4) = 3^9$ d. not possible

   e. $(17)(17^5) = 17^6$ f. not possible

   g. not possible   h. $(19)(19) = 19^2$

   i. not possible

5. Rewrite using only one exponent, if possible.

   a. $(b^3)(b^{11}) = b^{14}$ b. not possible

   c. not possible   d. $(k^8)(k) = k^9$

   e. $(z^5)^3 = z^{15}$ f. $(b^2)^9 = b^{18}$

   g. $(x)(x) = x^2$ h. not possible

   i. To check part e for instance we replace z with 2 and compute the original expression and our simplification using the specified order of the operations.

   $(z^5)^3 = (2^5)^3 = (32)^3 = 32,768$ and $z^{15} = 2^{15} = 32,768$

6. The order of operations is different in each of the expressions. $(3^5)^4$ indicates raise 3 to the fifth power first then raise the number you get to the fourth power. $(3^5)(3^4)$ would be computed first by doing the two exponents and then multiplying the numbers you get while $3^5 + 3^4$ would be computed by first doing the exponents and then adding the answers.

7. The first indicates multiplication and the second expression indicates addition.

8. a. Negative numbers are not used to measure lengths of sides of rectangles.

   b. No one has probably lived that long.

c. It has been known to get that cold in Minnesota in January.

d. There are teachers who have made that salary.

e. There aren't that many hours in a week.

9. Opinion

10. Opinion

CHAPTER 2

Whole Numbers: Introducing a Mathematical System

Section 2.1 Whole Number Domains

p. 83

3. a. {1, 3, 5, 7, 9, 11, …}

   b. {0, 2, 4, 6, 8, 10, …}

   c. {0, 1, 2, 3, 4, 5, 6, 7, 8, 9}

   d. {0, 1}

   e. {2, 3, 5, 7, 11, 13, 17, 19, 23, 29, 31, 37, …}

   f. { } or ∅

   g. {4, 6, 8, 9, 10, 12, 14, 15, 16, 18, 20, 21, 22, 24, 25, 26, 27, 28, 30, 32, …}

   h. {0, 1}

4. a. The set of odd numbers consists of the whole numbers that are not divisible evenly by 2.

   b. The set of even numbers consists of the whole numbers that are divisible evenly by 2.

   c. The set of digits is the whole numbers less than 10.

   d. The set of whole numbers that are their own squares is the whole numbers that are less than 2.

   e. The set of prime numbers consists of the whole numbers that have no divisors except 1 and themselves.

f. The set of whole numbers between 11 and 12 is empty since there are no whole numbers between 11 and 12.

g. The set of composite numbers consists of the whole numbers larger than 1 that have more than two distinct divisors.

h. The set of whole numbers that are neither prime nor composite consist of the whole numbers less than 2.

5. a, b, e, and g are infinite since they have no greatest element; c, d, f, and h are finite since the number of elements in the set is a finite number.

6. a. Not closed under addition; closed under multiplication.

   b. Closed under addition; closed under multiplication.

   c. Not closed under addition; not closed under multiplication.

   d. Not closed under addition; closed under multiplication.

   e. Not closed under addition; not closed under multiplication.

   f. A moot point!

   g. Not closed under addition; closed under multiplication.

   h. Not closed under addition; closed under multiplication.

7. Prime factorizations

   a. $2^2 \cdot 61$ b. $2 \cdot 3^3$

   c. $3^3 \cdot 2^2$ d. $2 \cdot 5 \cdot 41$

   e. $13 \cdot 11^2$ f. $3 \cdot 7 \cdot 11 \cdot 17$

   g. $2^3 \cdot 5^2 \cdot 7^2$ h. $5^2 \cdot 13^2 \cdot 17$

8. No, because 51 isn't prime.

9. It should be $2^3 \cdot 3 \cdot 5$.

10. The largest element of the set is 80. The next whole number is 81, which is used as the first number above the upper limit of the set. Non–whole numbers between 80 and 81, such as 80.36578, could have been used. The answer also could be stated as the set of whole number multiples of four that are less than **or equal to** 80.

11. 64 represents the starting number of P's. By applying Rule 3, three P's are removed. The division of 64 by three is used to determine how many times three P's can be removed.

12. a. {S, P, C}; finite

    b. {S P}; finite

    c. {Rule 1, Rule 2, Rule 3, Rule 4}; finite

    d. The set of sequences containing only the letters S, P, and C; infinite

    e. The set of strings produced by applying a rule to the initial path or to a previously derived valid path; infinite

    f. The empty set Ø; finite

13. The set of valid paths is a subset of the set of paths. Every valid path is a path but there are many paths that are not valid paths.

14. Yes. If one applies a rule to a valid path in the SPC system you create another valid rule. So one cannot leave the set of valid paths by applying one of the rules.

15. a. This is the set of whole numbers less than 2.

    b. This is the set of one–digit, odd whole numbers.

    c. This is the set of squares of whole numbers.

    d. This is the set of whole number multiples of 7.

16. Zero is not a prime number since it has an infinite number of divisors. One is not a prime number since it has exactly one divisor. Prime numbers have exactly two divisors.

17. a. 287 is not prime since 287 is divisible by 7 and 41.

    b. 283 is prime. It is not divisible by the primes 2, 3, 5, 7, 11, and 13. No larger prime is a divisor of 283 since the square of all other primes exceeds 283.

    c. The largest prime that needs to be tested to determine if 283 or 287 is prime is 13. If a number has a divisor other than itself and one, the divisor must be a prime whose square is less than the given number.

18. a. $m + 9$

    b. $m - 4$

    c. $2m$

    d. $2m + 3$

e. $m+1$ and $m+2$

19. a. The set of all whole numbers that are greater than or equal to 4. $\{4, 5, 6, ...\}$

    b. The set of all whole numbers.

    c. The set of all even whole numbers. $\{0, 2, 4, 6, ...\}$

20. Answers are limited only by your imagination. An example might be to determine how many feet there are in 40 inches.

21. The symbol for the empty set is $\emptyset$. An example of an empty set is the set of labrador retrievers registered for this class.

Section 2.2 Order of Operations with Whole Numbers

p. 101

3. a. $5(7+6) = 65$ b. $18 - 2(3+4) = 4$

   c. $9 - 3(2) = 3$ d. $\dfrac{8}{9-5} = 2$

   e. $28 + 10 \div 5 = 30$ f. $17 - 3 + 5 + 8 - 2 = 25$

   g. $11(5) - 3(4) = 43$ h. $(8+10) \div 4 + 3(2) = 10.5$

   i. $\dfrac{10+8}{5+4} = 2$

4.

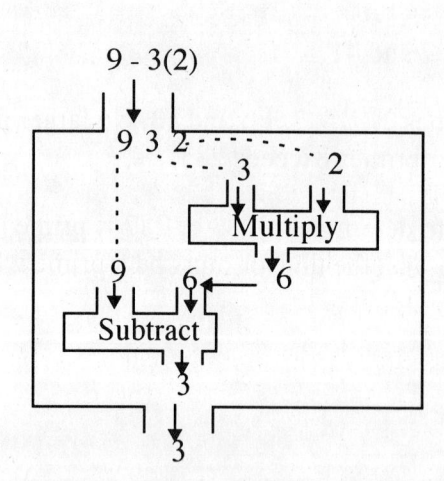

5. Rewrite each of the following using at most one exponent, if possible.

   a. $(7^0)(7^{11}) = 7^{11}$ b. $(2^9)^0 = 1$

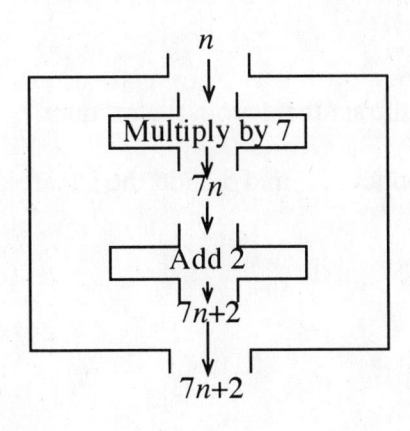

c. $2^9 + 2^0 = 2^9 + 1$
d. $(2^9)(2^0) = 2^9$

e. $(5^3)(0^5) = 0$
f. $(9^0)(9) = 9$

g. $(4^{23})(4^4) = 4^{27}$
h. $(7)(7^9) = 7^{10}$

i. $8 + 8^2 = 8(1+8) = 8(9)$
j. $(4^0)^4 = 1^4 = 1$

6. $\dfrac{5+2(4^3-6)}{\sqrt{8+1}+7} - 2 = \dfrac{5+2(64-6)}{\sqrt{9}+7} - 2 = \dfrac{5+2(58)}{3+7} - 2 = \dfrac{5+116}{10} - 2 = \dfrac{121}{10} - 2$

   $= 12.1 - 2 = 10.1$

7. a. $n+3$
   b. $5n$
   c. $n-8$
   d. $7n+2$

8.

```
         n
         ↓
   ┌───────────┐
   │Multiply by 7│
   └─────┬─────┘
         ↓
        7n
         ↓
   ┌───────────┐
   │   Add 2   │
   └─────┬─────┘
         ↓
       7n+2
         ↓
       7n+2
```

9. a. 9
   b. 30
   c. –2
   d. 44

10. Some examples: $9 - 7 - 1 = 1$; $9 - 7(1) = 2$; $9 - (7-1) = 3$; $(7\sqrt{9}) - 1 = 20$.

11. $\dfrac{20}{2 \cdot 3} = \dfrac{20}{6} = \dfrac{10}{3} = 3\dfrac{1}{3}$. The order of operations is multiplication followed by division.

    $20/2 \cdot 3 = 10(3) = 30$. The order of operations is division followed by multiplication.

12. An example: Start with 11.
    $11 \rightarrow 34 \rightarrow 17 \rightarrow 52 \rightarrow 26 \rightarrow 13 \rightarrow 40 \rightarrow 20 \rightarrow 10$
    $\rightarrow 5 \rightarrow 16 \rightarrow 8 \rightarrow 4 \rightarrow 2 \rightarrow 1$

Section 2.3 Algebraic Extensions of the Whole Numbers

p. 112

2. a. $2(5) = 10$
    b. $5^2 = 25$
    c. $5 + 11 = 16$
    d. $6 - 5 = 1$
    e. $3 + 2(5) = 13$
    f. $4(5) - 7 = 13$
    g. $5^2 + 3 = 28$
    h. $3(5)^2 - 4 = 3(25) - 4 = 75 - 4 = 71$
    i. $(5)^2 + 6(5) = 55$
    j. $6 + 7(5) + (5)^2 = 66$
    k. $9(5) - 2(5)^2 + 8 = 45 - 50 + 8 = 3$

3. a. e. Multiply, add; h. Square, multiply, subtract; j. Square, multiply, add, add.

    b. e. Multiply 2 and 5. Add 3 to the product.

    h. Square 5. Multiply the square of 5 by 3. Subtract the 4 from the product.

    j. Square 5. Multiply 7 by 5. Add 6 to the product of 7 and 5. Add the square of 5 to this sum.

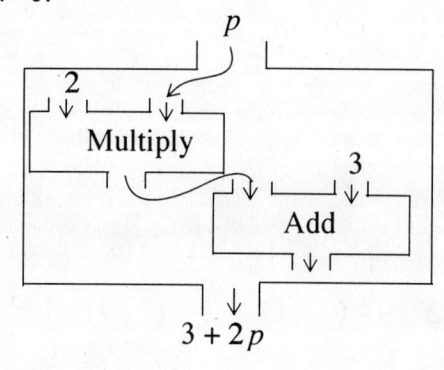

    c. e.

    ```
            p
         ↓
      2 ↓  ↓
      ┌─────────┐
      │ Multiply│
      └─────────┘
              ↓   3
              ↓  ↓
           ┌──────┐
           │ Add  │
           └──────┘
               ↓
             3 + 2p
    ```

h.

[Diagram: p → Square; then combined with p → Multiply by 3; result → Subtract 4 → $3p^2 - 4$]

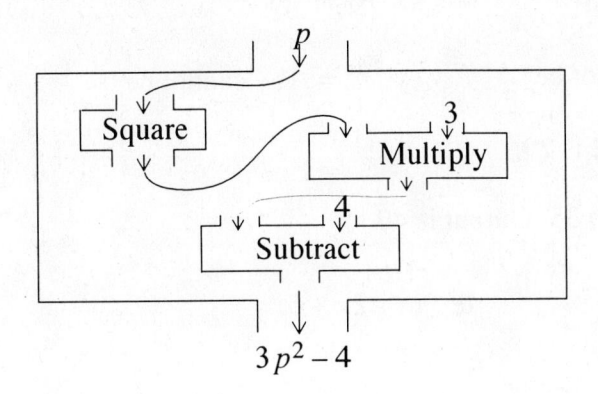

j.

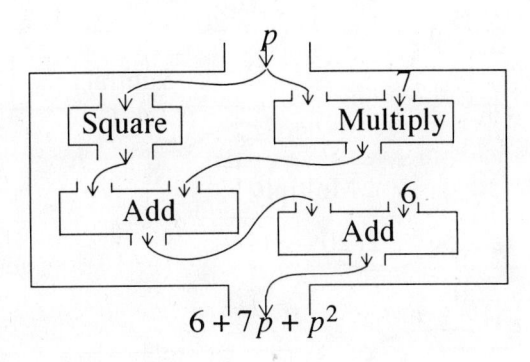

[Diagram: p → Square; p → Multiply by 7; Square result → Add; Multiply result → Add 6; combined → $6 + 7p + p^2$]

4. $\frac{1}{2}n^2 + \frac{1}{2}n$

5. This is not a polynomial since the variable n appears as an exponent.

6. In the linear polynomial $9x + 7$

   a. The terms are $9x$ and 7.

   b. The numerical coefficient of the first term is 9 and the numerical coefficient of the second term is 7.

7. In the quadratic polynomial $4x^2 + x + 3$

   a. the terms are $4x^2$, x, and 3.

   b. the numerical coefficients are 4, 1, and 3.

8. Write an expression with at most one exponent for each of the following expressions. List any restrictions on the variables.

   a. $(y^6)(y) = y^7$

   b. $(y^6)^0 = 1$, y cannot be 0

83

c. $(y^0)^6 = 1$, y cannot be 0   d. $(y^4)(y) = y^5$

e. $(y^m)(y^0) = y^m$, y cannot be 0   f. $(y^m)^0 = 1$, y cannot be 0

9. a. Output 1 is 21; Output 2 is 29.

   b. A1 is 49; A2 is 14; A3 is 35; Output is 40.

10. a. $3x + 8$ b. $x^2 - 2x + 5$

11. a.

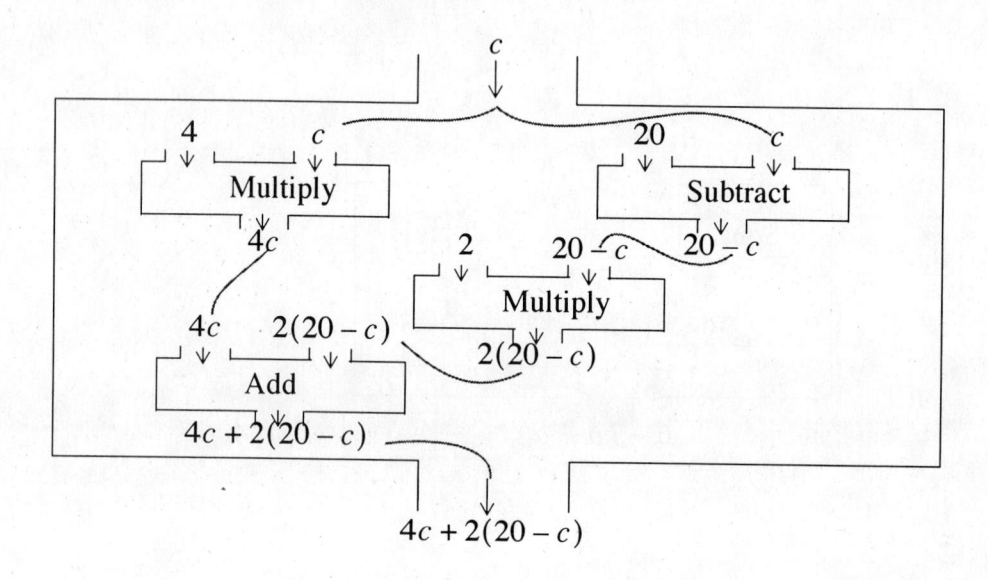

b.

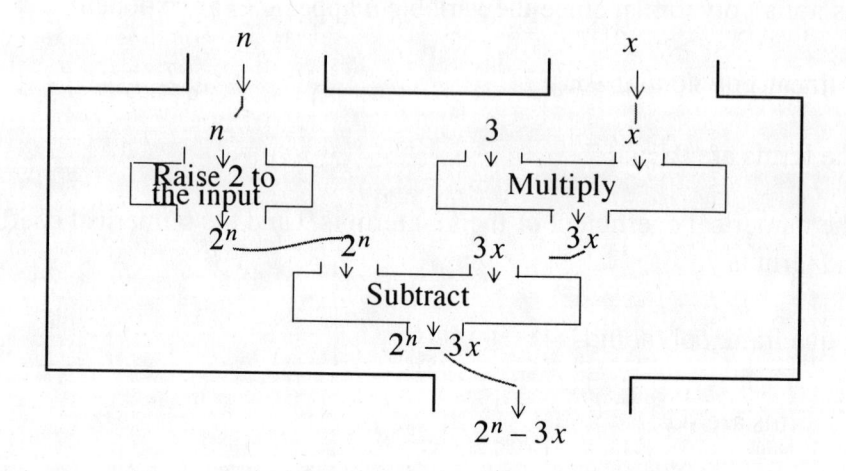

12. a. For example: Input 3; Output 10. Input 15; Output 10. Input 92; Output 10.

    b. $[(2n + 9) + n] \div 3 + 7 - n$

Section 2.4 Properties That Change the Order of Operations

p.132

2.   a. $18 + 9 = 9 + 18$ b. $7(11) = 11(7)$

3.   a. $5 + (3 + 7) = 5 + 3) + 7$ b. $(5(3))7 = 5(3(7))$

4.   a. $7(4 + 9) = 7(4) + 7(9)$ b. $4(8 - 3) = 4(8) - 4(3)$

     c. $5(4) + 5(11) = 5(4 + 11)$ d. $16(3) - 16(2) = 16(3 - 2)$

5.   a. $(8 + 11) + 4 = 8 + (11 + 4)$, for example.

     b. $(16)2 = (2)16$, for example.

     c. $9(6 - 2) = 9(6) - 9(2)$, for example.

6.   Division distributes over addition–an example is $\frac{(8 + 6)}{2} = \frac{8}{2} + \frac{6}{2}$. This is exactly the way the sum of fractions with like denominators is found. Division distributes over subtraction–an example is $\frac{12 - 4}{2} = \frac{12}{2} - \frac{4}{2}$. This is the way the difference of two fractions with like denominators is found.

7.   a. $3x^2 + 4x + 9$ b. $5x + 11$

     c. $5x^2 + 12x + 5$ d. $6x^3 + 15x^2$

     e. $10x^3 + 35x^2 + 45x$ f. $x^2 + 13x + 36$

     g. $12x^2 + 29x + 14$ h. $21x^2 + 34x + 4$

     i. $9x^2 + 30x + 25$ j. $4x^2 + 44x + 121$

     k. $2x^4 + 17x^3 + 29x^2 + 22x + 15$ l. $x^4 + 2x^3 + 2x^2 + 2x + 1$

8.   a. $7(x + 2) = 7x + 14$ b. $5(8 - m) = 40 - 5m$

     c. $6b - 18 = 6(b - 3)$ d. $35 + 7t = 7(5 + t)$

9.   a. $9x^2 + 18x = 9x(x + 2)$ b. $21x^4 + 14x^3 = 7x^3(3x + 2)$

     c. $5x^3 + 25x^2 + 30x = 5x(x^2 + 5x + 6) = 5x(x + 3)(x + 2)$

d. $3x^2 + 6x + 8$ is prime.

10. a. $(3 \cdot 5)^2 = 3^2 \cdot 5^2$ b. $(3+5)^2 = 3^2 + 2 \cdot 3 \cdot 5 + 5^2$

     c. $(ab)^2 = a^2 \cdot b^2$ d. $(a+b)^2 = a^2 + 2ab + b^2$

11. a. The associative property of addition allows you to rewrite the sum.

$$9 + 6 = 9 + (1 + 5) = (9 + 1) + 5 = 10 + 5$$

     b. $9 + 8 = 10 + 7 = 17$

     c. $99 + 57 = 100 + 56 = 156$

12. 9 could be replaced with $10 - 1$ and then the multiplication by 367 could be distributed over this difference. You would need to compute $3670 - 367$ which could be done mentally to get 3303.

13. a. $5a - 8 = 12$ b. $7z + 5 = 12$
              $5a = 20$ $7z = 7$
              $a = 4$ $z = 1$

14. a. [Sx → Rule 1 → Sx x]   b. [Sx P → Rule 2 → Sx P C]

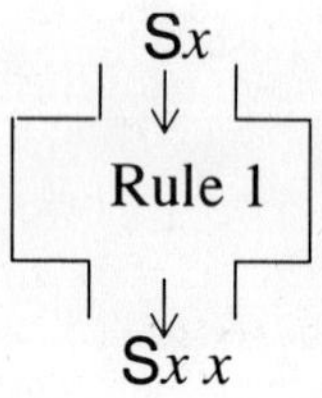

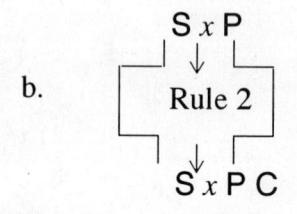

     c. [S x P P Py → Rule 3 → Sx Cy]   d. [S x C C y → Rule 4 → Sx y]

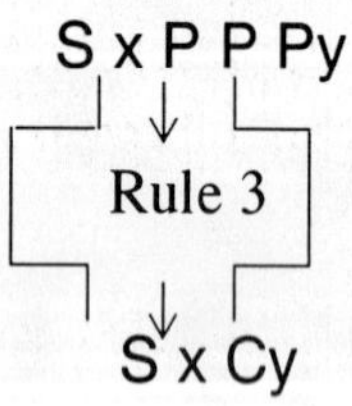

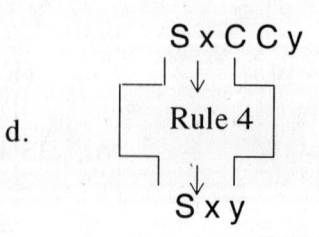

Section 2.5 Review

p. 136

1. Yes.

2. No. 71^2 is the prime factorization.

3. Your response.

4. $\dfrac{7-2(1^3+2)}{\sqrt{3(7)+4}} - 7 = \dfrac{7-2(3)}{\sqrt{21+4}} - 7 = \dfrac{7-6}{\sqrt{25}} - 7 = \dfrac{1}{5} - 7 = -6\dfrac{4}{5}$

5. $5 \cdot 3 - 2 \cdot 3^2 + 3 = 15 - 18 + 3 = 0$

6. 4, 8, 6, 12, 24, 166, 332, 664, 249, and 1992

7. Answers will vary. Check with others in your class.

8.
   a. $3^2 \cdot 7 \cdot 11$
   b. $2^3 \cdot 5 \cdot 13^2$
   c. $29 \cdot 31$
   d. $3^4 \cdot 7 \cdot 11 \cdot 17$
   e. $7^2 \cdot 47$
   f. 8191 since it's prime.

9. To generalize means to draw conclusions about a class of objects or about the way a set of operations works based on observations of only a few (or many) of the objects of a few instances where the property is satisfied. For instance, if we observe that multiplication over division seems to work in the three examples which appear in the following table,

$a(b+c)$	$a \cdot b + a \cdot c$
$5(2+4) = 5 \cdot 6 = 30$	$5 \cdot 2 + 5 \cdot 4 = 10 + 20 = 30$
$4(1+8) = 4 \cdot 9 = 36$	$4 \cdot 1 + 4 \cdot 8 = 4 + 32 = 36$
$8(2+7) = 8 \cdot 9 = 72$	$8 \cdot 2 + 8 \cdot 7 = 16 + 56 = 72$

   The generalization is that $a(b+c) = ab + ac$ for all numbers a, b and c.

10. Note: answers may vary due to interpretation of problem.

    a. Only whole numbers are appropriate since this represents a count.

    b. Whole numbers are not appropriate since measurements involve fractional amounts and rounding.

c. This could involve numbers other than whole numbers since age can be measured in fractions of time units.

d. Only whole numbers are appropriate since this represents a count.

e. This could involve numbers other than whole numbers if the cost of a can of pop is not a multiple of ten cents.

f. Only whole numbers are appropriate since this represents a count.

11. a. $(7^{23})(7^2) = 7^{25}$ b. $(5^7)^2 = 5^{14}$

    c. $(5^7)(5^2) = 5^9$ d. $5^7 + 25$ or $25(5^5 + 1)$

    e. $(11)(11^0) = 11$ f. $(11)(5^0) = 11$

    g. $(2^0)(3^2) = 3^2$ h. $(2^3)(0^2) = 0$

12. a. $(b^5)(b^9) = b^{14}$ b. $s^7 + a^0 = a^7 + 1$ as long as a is not 0

    c. $(z^0)(a^6) = a^6$ d. $(x^4)x = x^5$

    e. $(a^0)^3 = 1$ f. $(b^2)^0 = 1$

    g. $x^0 x^0 = 1$ h. $x^0 + x^0 = 1 + 1$

13. a. $n + 5$ b. $3n$

    c. $4 - n$ d. $3n + 7$

14. a. 14   b. 27

    c. -5   d. 34

15. $3x - 2 = 3 \cdot 5 - 2 = 13$

16. a. $18x + 52$ b. $9x^2 + 4x + 8$

    c. $21x^3 + 24x^2$ d. $10x^2 + 53x + 63$

    e. $64x^2 + 80x + 25$ f. $35x^3 + 66x^2 + 37x + 6$

17. a. $4(4x + 3)$ b. $2x^3(2x^2 + x + 9)$

    c. $14x^2 + 27x + 7$ is prime.   d. $3(5x^2 + 8x + 1)$

CHAPTER 3

Functional Relationships

Section 3.1 Investigating Relationships Numerically

p. 156

2. Every function is a relation. Not all relations are functions. Since every function is a relation and functions have only one output for each input relations don't always have multiple outputs.

3. {0,1,2,3,...}; Infinite; it has no last element.

4. {0, 1, 2, 3, 4, 5, 6, 7}; Finite; it has only 8 elements.

5. {0, 1, 2, 3, 4, 5, 6, 7}; Finite; it has only 8 elements.

6. {1, 2, 3, 4, 5}; Finite; it has only 5 elements.

7. {0, 1, 2, 3, ...}; Infinite; it has no last element.

8. {1, 2, 3, 4, 5}; Finite; it has only 5 elements.

9. {0, 1}

10. A function is a process that receives input and returns an unique value called the output. A relation is a process that receives input and can return more than one unique output for a given input. All functions are relations but not all relations are functions.

11. a. $1 + 3 + 5 + 7 + 9 = 25$

    $1 + 3 + 5 + 7 + 9 + 11 = 36$

    $1 + 3 + 5 + 7 + 9 + 11 + 13 = 49$

    $1 + 3 + 5 + 7 + 9 + 11 + 13 + 15 = 64$

    b. The sum of the first n odd whole numbers is equal to the nth square.

c.

Sum of n odd numbers	Number of odd numbers (n)
1	1
4	2
9	3
16	4
25	5

Table 1: Sum of n odd numbers

d.
```
     1 + 3 + 5 + 7
           ↓
    ┌──────────────┐
    │ Count elements│
    └──────────────┘
           ↓
           4
```

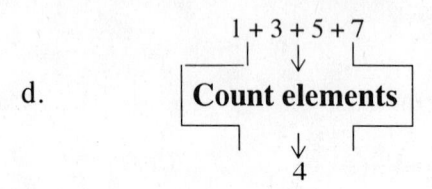

e.

Number of odd numbers (n)	Sum of n odd numbers
1	1
2	4
3	9
4	16
5	25

Table 2: Number of odd numbers (n)

f.
```
           4
           ↓
    ┌──────────────┐
    │ 1 + 3 + 5 + 7│
    └──────────────┘
           ↓
          16
```

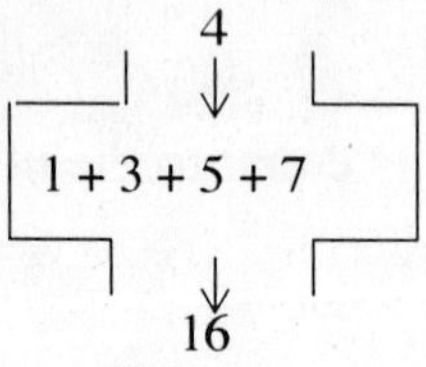

g.
```
     1 + 3 + 5 + 7
           ↓
    ┌──────────────┐
    │ Count elements│
    └──────────────┘
           ↓
           4
           ↓
    ┌──────────────┐
    │ 1 + 3 + 5 + 7│
    └──────────────┘
           ↓
          16
```

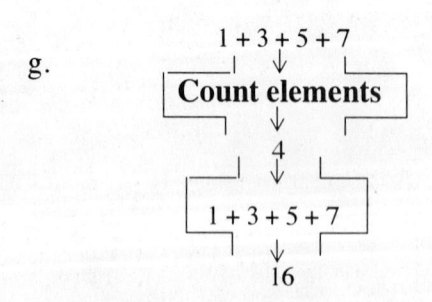

h. The output of the expression $2n - 1$ is the nth odd counting number. So, the sixth odd counting number is $2 \cdot 6 - 1 = 12 - 1 = 11$. The sum of the first six odd counting numbers is $6^2 = 36$. The sum of the first n odd counting numbers is given by the value of n^2.

i. In each problem we determined two functions that were put together so that the output of the first function was the input of the second function. Notice how the function machine diagram in Figure 8 (page 155) and the function machine diagram that appears above in the solution to part g illustrate this connection.

12. a.

Number of cars	Number of cycles
0	20
5	15
10	10
15	5
20	0

Table 3: Cars and cycles

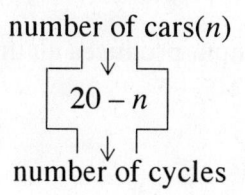

number of cars(n)
↓
$20 - n$
↓
number of cycles

b.

Number of cars	Total wheels
0	40
5	50
10	60
15	70
20	80

Table 4: Cars and wheels

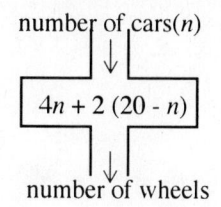

number of cars(n)
↓
$4n + 2(20 - n)$
↓
number of wheels

13. a. Each row begins and ends with a 1.

6	15	20	15	6		
7	21	35	35	21	7	
8	28	56	70	56	28	8

    b. The row sums are 1, 2, 4, 8, 16, 32, 64, 128, 256. The sum of the nth row is the nth power of 2.

    c. The non-zero whole numbers 1, 2, 3, 4, 5, 6, ... appear by reading down two different diagonals; outside diagonals are always ones; triangle is symmetric.

    d. Yes, whole numbers play an important role. The squares seem to be involved as well as adding consecutive whole numbers. Notice the numbers 1, 3, 6, 10, ... are found by adding 1, 1 + 2, 1 + 2 + 3, etc. This sequence of numbers appears in the table too.

14. This problem is so much fun and takes time to investigate. We do not want to spoil your fun by suggesting an answer yet. Keep exploring.

15. Stop when you produce a valid path already produced.

16. a. [Rule 1: input Sx → output Sxx]   a function; a given input produces an unique output.

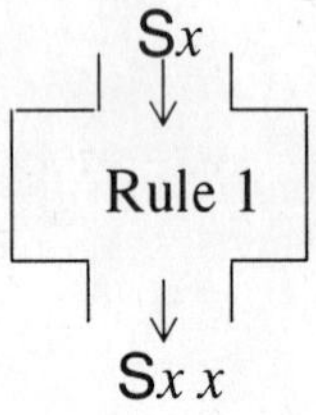

    b. [Rule 2: input SxP → output SxPC]   a function; a given input produces an unique output.

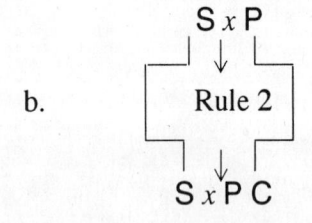

    c. [Rule 3: input S x P P P y → output S x C y]   a relation; more than one output possible for a given input.

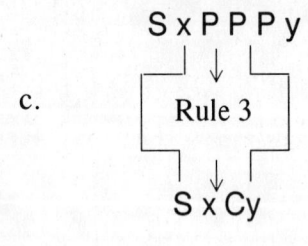

92

d. 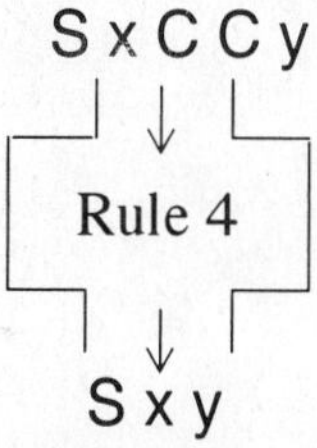
```
    S x C C y
      ↓
   ┌────────┐
   │ Rule 4 │     a relation; more than one output possible for a given input.
   └────────┘
      ↓
    S x y
```

17. We offer a few suggestions for approaching this problem. First, construct a table like the one on page 144 to familiarize yourself with the problem. Second, look at the finite differences to see if any number seems to be important. Does a remainder seem to provide you with a good predictor?

Section 3.2 Function: Algebraic Representation

p. 171

2. a.
    ```
       13      b
        ↓      ↓
      ┌──────────┐
      │ Multiply │
      └──────────┘
            ↓
           13b
    ```

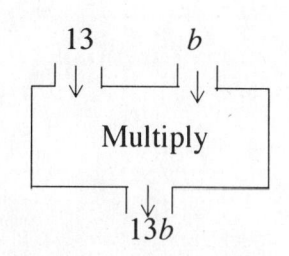

    b.
    ```
               z
               ↓
           ┌──────┐
           │ Cube │
           └──────┘
              ↓ z³
              z³        4
               ↓        ↓
              ┌──────────┐
              │ Multiply │
              └──────────┘
                    ↓
                   4z³
                    ↓
                   4z³
    ```

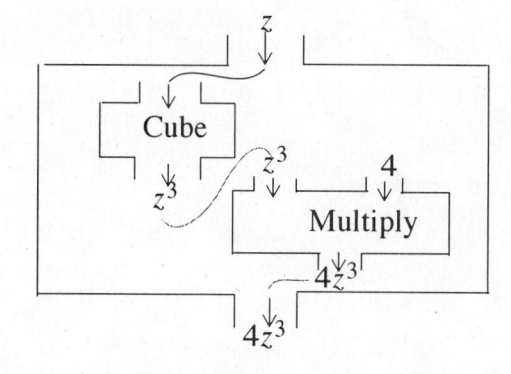

    c.
    ```
                m
                ↓
           ┌────────┐
           │ Square │
           └────────┘
              ↓
              m²       m²      5
                        ↓      ↓
                      ┌──────────┐
                      │ Multiply │
                      └──────────┘
                            ↓
                           5m²
                            ↓
                           5m²
    ```

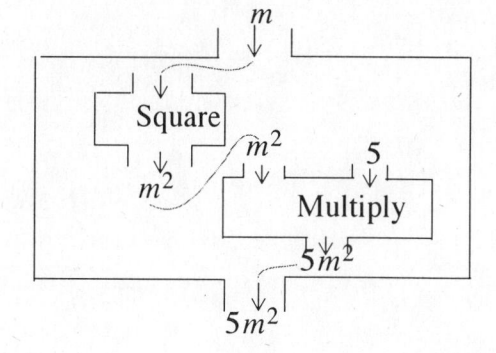

d.

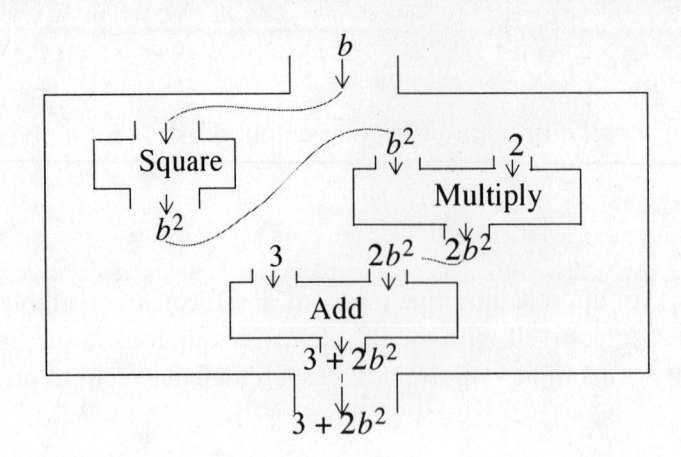

```
         b
         ↓
   ┌─────────────────────────────────┐
   │  ┌────────┐      b²    2        │
   │  │ Square │      ↓     ↓        │
   │  │        │    ┌──────────┐     │
   │  │        │    │ Multiply │     │
   │  │   ↓    │    │    ↓     │     │
   │  │   b²   │    │   2b²    │     │
   │       3    ↓    2b²              │
   │       ↓  ┌──────────┐            │
   │          │   Add    │            │
   │          │    ↓     │            │
   │          │  3 + 2b² │            │
   └─────────────────────────────────┘
              ↓
           3 + 2b²
```

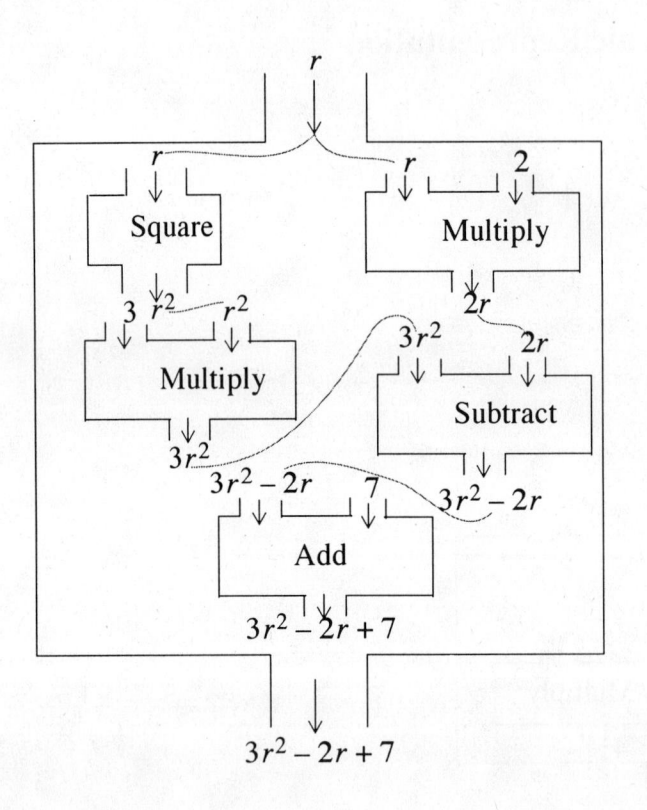

e.

```
                    r
                    ↓
   ┌──────────────────────────────────────────┐
   │    r                 r        2          │
   │    ↓                 ↓        ↓          │
   │ ┌────────┐        ┌──────────┐           │
   │ │ Square │        │ Multiply │           │
   │ │   ↓    │        │    ↓     │           │
   │ │   r²   │  r²        2r                 │
   │  3  ↓    ↓       3r²      2r             │
   │  ↓ ┌──────────┐   ↓       ↓              │
   │    │ Multiply │  ┌──────────┐            │
   │    │    ↓     │  │ Subtract │            │
   │    │   3r²    │  │    ↓     │            │
   │         3r² − 2r    7    3r² − 2r        │
   │            ↓        ↓                    │
   │         ┌──────────────┐                 │
   │         │     Add      │                 │
   │         │      ↓       │                 │
   │         │ 3r² − 2r + 7 │                 │
   └──────────────────────────────────────────┘
                    ↓
              3r² − 2r + 7
```

3. a.

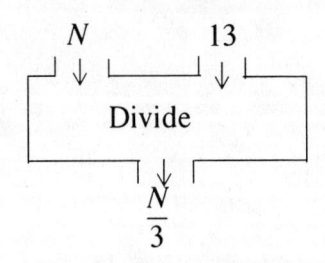

```
       N        13
       ↓        ↓
     ┌────────────┐
     │   Divide   │
     │     ↓      │
     │    N/3     │
```

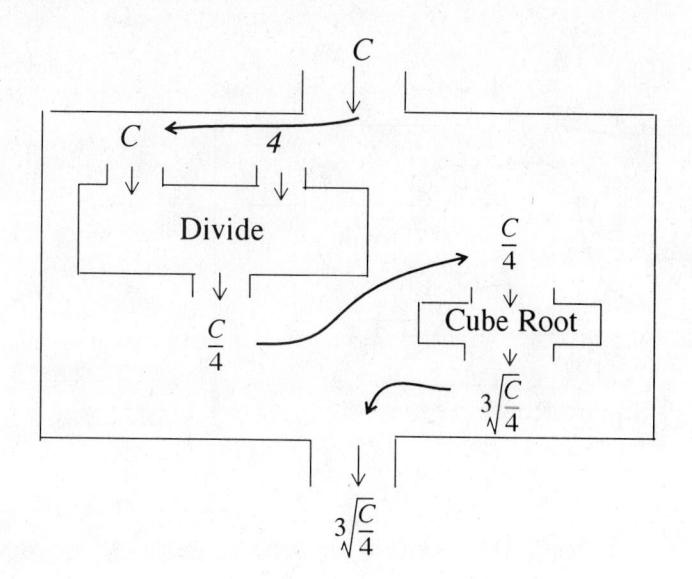

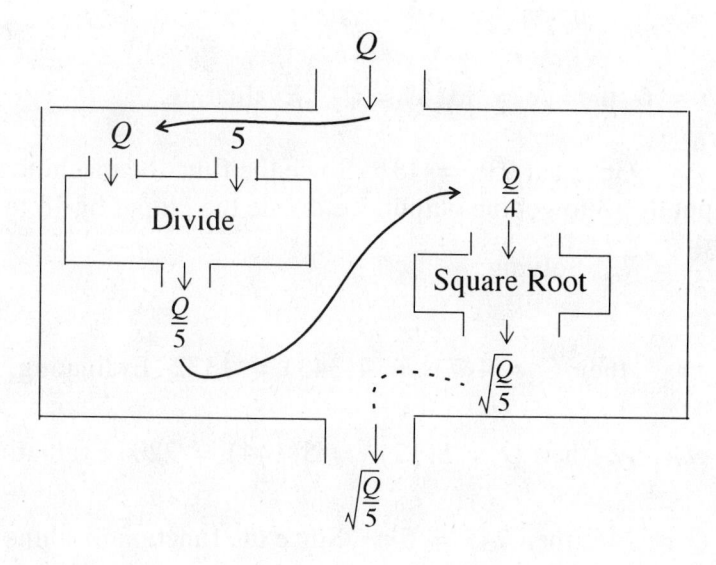

b.

c.

d.

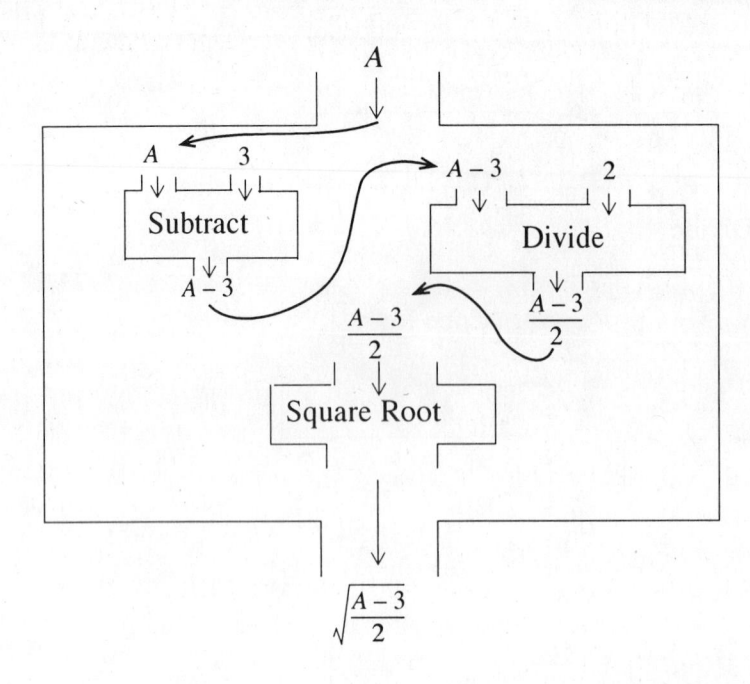

```
         A
         ↓
   ┌─────────────┐
   │ A    3      │  A - 3    2
   │ ↓    ↓      │   ↓       ↓
   │  Subtract   │      Divide
   │    ↓        │        ↓
   │   A - 3     │      A - 3
   │             │       ───
   │         A - 3        2
   │         ─────
   │           2
   │           ↓
   │      Square Root
   │           ↓
   └───────────────┘
              ↓
          √((A-3)/2)
```

4. a. If $N = 13b$ and $b = 6$, then $N = 13(6) = 78$. Evaluating.

   b. If $N = 13b$ and $N = 936$, then $936 = 13b$. Since the function machine requires us to multiply the input by 13 to get the output, we divide the output by 13 to get the input. So, $b = \frac{936}{13} = 72$. Solving.

   c. If $C = 4z^3$ and $z = 7$, then $C = 4(7)^3 = 4(343) = 1372$. Evaluating.

   d. If $Q = 5m^2$ and $m = 12$, then $Q = 5(12)^2 = 5(144) = 720$. Evaluating.

   e. If $Q = 5m^2$ and $Q = 245$, then $245 = 5m^2$. Since the function machine requires us to square the input and multiply the square by five, we must divide by 5 and find the square root of the quotient to get the input when given output. So, $m = \sqrt{\frac{245}{5}} = \sqrt{49} = 7$. Solving.

   f. If $A = 3 + 2b^2$ and $b = 13$, then $A = 3 + 2(13)^2 = 3 + 2(169) = 3 + 338 = 341$. Evaluating.

   g. If $A = 3 + 2b^2$ and $A = 75$, then $75 = 3 + 2b^2$. The function machine process is square, multiply by 2, add 3. To reverse the process, subtract 3, divide by 2, and find the square root. So, $b = \sqrt{\frac{75 - 3}{2}} = \sqrt{\frac{72}{2}} = \sqrt{36} = 6$. We are solving an equation.

   h. If $L = 3r^2 - 2r + 7$ and $r = 6$, then

96

$L = 3(6)^2 - 2(6) + 7 = 3(36) - 2(6) + 7 = 108 - 12 + 7 = 96 + 7 = 103$. Evaluating a function.

5. a. **Table 6: Students Versus Professors**

Professors	Students
50	300
100	600
150	900
200	1200
250	1500
300	1800

b. **Table 7: Students Versus Professors Pattern**

Professors	Students
50	6(50)
100	6(100)
150	6(150)
200	6(200)
250	6(250)
300	6(300)
p	6p

c. $s = 6p$

d. The finite differences between the number of students is 300. They become constant in the first finite difference.

e. As the number of professors increases by 50, the number of students increases by 300. This is verified by the finite differences.

f. The domain and range of the mathematical process is all real numbers.

g. The domain of the problem situation is all whole numbers. The range of the problem situation is all whole numbers that are multiples of 6.

6. a. There are 72 students for each mathematics faculty member.

b. If $M = 12$ then $S = 72(12) = 864$. If there are 12 full–time mathematics faculty, there are 864 students enrolled in mathematics courses.

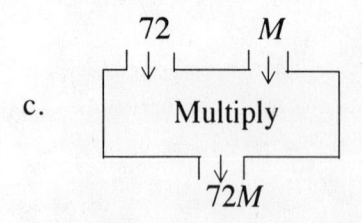

c.

```
    72        M
     ↓         ↓
   ┌─────────────┐
   │   Multiply  │
   └─────────────┘
         ↓
        72M
```

d. The domain and range of the mathematical function are both all real numbers.

e. The domain of the problem situation is the set of all whole numbers. The range of the problem situation is the set of all whole number multiples of 72.

7. $\dfrac{S}{E} = \dfrac{720}{18} \Rightarrow \dfrac{S}{E} = 40 \Rightarrow S = 40E$. So $S = 40E$.

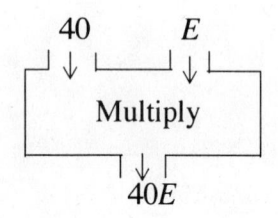

```
    40        E
     ↓         ↓
   ┌─────────────┐
   │   Multiply  │
   └─────────────┘
         ↓
        40E
```

8. Let c represent the number of cars parked in the lot and w represent the number of wheels in the lot.

   a. $w = 4c$

   b. The independent variable is c, the number of cars.

   c. The domain and range of the mathematical process is all real numbers.

   d. The domain of the problem situation is the set of whole numbers less than 21. The range of the problem situation is the set of whole numbers less than 81 that are multiples of 4.

9. a. Input: S P P; Output: S P P P P

   b. For example, apply Rule 2 to input S P to get output S P C.

   c. The domain of Rule 1 is the set of all valid paths.

   d. The domain of Rule 2 is the set of all valid paths that end in a P. This is true since Rule 2 cannot be applied if the path does not end in P.

10. Answers will vary.

11. a. Suppose $s = 5t$ defines the function. When the output $s = 35$ the input will be 7.

    b. Suppose $k = l + 11$ defines the function. When the output k is 24 the input l will be 23.

    c. Suppose $d = 3c - 8$ defines the function. When the output is $d = 13$ the input will be found by adding 8 to 13 to get 21 and then dividing 21 by 3 to get 7. So, if c is 7 then d will be 13.

    d. Suppose $y = 7x + 2$. If $y=58$ the input would be $\frac{58-2}{7} = 8$.

Section 3.3 Function: Notation for Algebraic Representations

p. 183

2. Note the direction of the arrows.

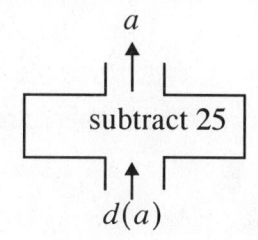

    $$\begin{array}{c} a \\ \uparrow \\ \boxed{\text{subtract 25}} \\ \uparrow \\ d(a) \end{array}$$

3. a. The unknown is the alarm time a. We must solve the equation $10:00 = a + 25$. To do this we reverse adding 25 by subtracting 25 from the time 10:00. This gives us 9:35. The alarm time should be 9:35 A.M.

   b. Here we must determine the value of $d(8:32) = 8:32 + 25 = 9:00$. We will depart at 9:00 A.M.

4. a. The domain of the mathematical process is all real numbers.

   b. The domain of the car rental problem is the set of all possible miles that the car could be driven in a given day. This would be the set of whole numbers less than some maximum number of miles.

5. a. The range of the mathematical process is all real numbers.

   b. The range of the car rental problem is the set of all possible charges for the car. This would be the set of all numbers starting at 35 in increments of 0.17.

6. a. We solve $11 = r - 4$ by adding 11 and 4 to get $r = 15$.

   b. $y(2) = 9(2) = 18$

c. $F(5) = 2(5) - 7 = 3$

d. We solve $14 = 3p + 2$ by subtracting 2 from 14 then dividing the result by 3 to get $p = 4$.

7. a. $R(n) = 19n$

b. $R(2500) = 19(2500) = 47500$. If 2500 tickets are sold, the receipts are $47,500.

c. If the receipts were $59,185 we would have to solve $59185 = 19n$ to determine the number of tickets sold. The solution would be 59,185 divided by 19 or $n = 3115$.

8. a. $s(p) = 6p$

b. $s(75) = 6(75) = 450$. If there are 75 professors, there are 450 students.

c. If there are 486 students there would be 81 students.

9. a. $w(c) = 4c$

b. $w(17) = 4(17) = 68$. If there are 17 cars in the lot, there are 68 wheels in the lot.

10. This is your personal problem.

11. Answers will vary.

12. Some quantities that are constant include $1.36, twenty dollar bill, twenty minutes late, one–half, one–quarter, one–sixteenth, two feet, four feet, $17, six people, three toll booths, one dollar.

13. Some variables include:

    amount of gas pumped (input) and gas charge (output);

    gas charge (input) and change (output);

    number of students enrolled (input) and number of students still remaining (output);

    amount of light two feet away from bulb (input) and amount of light one foot away from bulb (output);

    how well I could see one foot away (input) and how well I could see four feet away (output);

    number of hours I work in the summer (input) and total pay in summer (output);

    amount paid at a toll booth (input) and my change (output).

14. amount of gas pumped (input) and gas charge (output): Increasing the input increases the output.

gas charge (input) and change (output): Increasing the input decreases the output.

number of students enrolled (input) and number of students still remaining (output): Increasing the input increases the output.

amount of light two feet away from bulb (input) and amount of light one foot away from bulb (output): Increasing the input increases the output.

how well I could see one foot away (input) and how well I could see four feet away (output): Increasing the input increases the output.

number of hours I work in the summer (input) and total pay in summer (output): Increasing the input increases the output.

amount paid at a toll booth (input) and my change (output): Increasing the input decreases the output.

15.
   a. 4^{26}
   b. 4^{153}
   c. not possible
   d. x^{15}
   e. 1
   f. not possible
   g. x^8
   h. not possible
   i. not possible
   j. 11^2

16.
   a. $7x + 10$
   b. $5x^2 + 4x + 9$
   c. $15y^2 - 24y$
   d. $16x^2 + 24x + 9$

17.
   a. $3(3b + 2)$
   b. prime
   c. $4y(3y^2 + y + 7)$

Section 3.4 Function: Geometric Interpretations

p. 197

2. Site of Worship: (4, 7), (4, 8), (4, 9), (4, 10).

   Site of Ancient Secrets: (7, 6), (8, 6), (9, 6).

Math Site 1: (5, 1), (6, 1). Math Site 2: (1, 3), (2, 2).

3. Let the points be labelled: $A(30, 2)$; $B(50, 7)$; $C(10, 9)$; $D(60, 0)$; $E(35, 6)$; $F(0, 4)$.

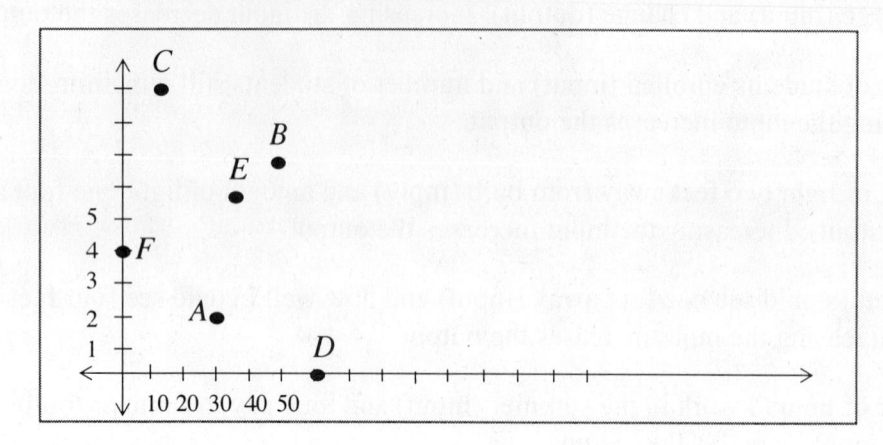

4. a. **Professors and Students**

Professors	Students
10	60
20	120
30	180
40	240
50	300
60	360

b. $S(P) = 6P$

c.

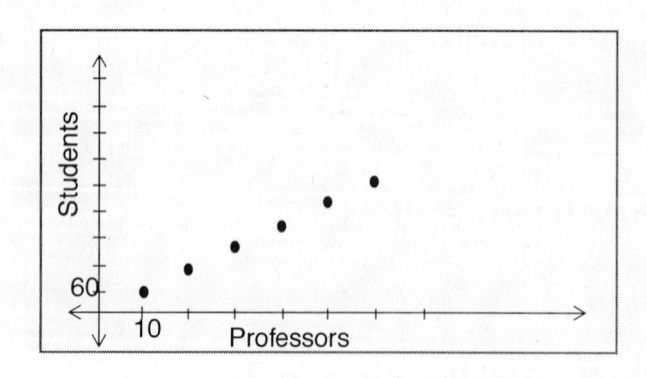

5. a. **Cubes**

Number	Cube
1	1
2	8
3	27
4	64
5	125
6	216

   b. $C(n) = n^3$

   c.

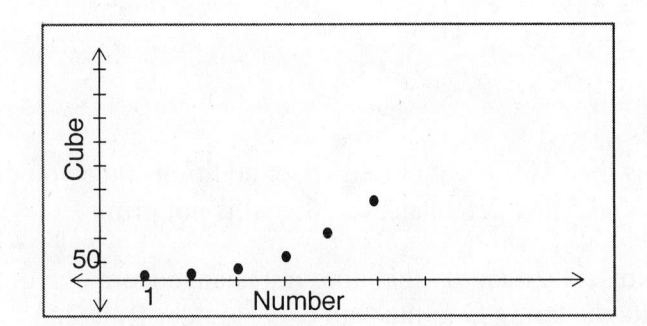

6. a. **Table 3: Rule 2 of SPC**

Input	Output
S P P	S P P C
S C P	S C P C
S P C C P	S P C C P C

   b. S P C is not in the domain of the function defined by Rule 2 since the path does not end in P.

7. Someone turned on the water and made the water level rise at a constant rate until 7:05 when the water was turned off for 5 minutes. At 7:10 the water was turned on stronger until the desired amount of water was attained at about 7:15 or a little before. At 7:20 the drain was opened until about 7:25. There is still water in the tub when our information ends at 7:30.

8. a. *B*: the relatively long time and the steady slope indicate walking.

   b. *A*: the quicker pace and varying speeds suggest a leisurely bike ride.

c. *C*: the rapid change in distance during certain time intervals suggests running.

d. The students stopped for a rest.

9. a. Only Graph *C* could represent a journey.

b. Graph *A* indicates that something moved from the starting point to some positive distance from the starting point instantaneously then stayed there.

Graph *B* may be two graphs which describe two points coming together at a certain time and distance from the beginning. One starting at the starting point the other some distance away from the beginning.

Graph *C* shows the path of someone who starts at the starting point and goes a certain distance away, at the first corner starts going towards the starting point, at the second corner resumes proceeding away from the starting point at a greater speed than on the first or second leg.

10. {97}

11. The set of prime numbers less than 100 is not closed under addition since you can add a pair of prime numbers like 3 and 5 and get an answer, 8, that is not prime.

12. A variable can be used to represent a quantity that does not change from one time to the next. The variable *x* represents the string C in the path S C P under Rule 2.

A variable can be used to generalize such as $a + 0 = a$.

A variable can be used to represent any value in a given replacement set. For example, the variable *n* in the function $R = 19n$ represented the number of concert tickets sold. This could be any whole number less than or equal to some number that represents the maximum seating capacity.

13. a. **Yards to feet**

Yards	Feet
1	3
2	6
3	9
4	12
5	15

b. $F(Y) = 3Y$.

14. $2x - 5 = 17$ for example.

15. a. $x = 7$ b. $y = 12$
    c. $t = 5$ d. $x = 0$

16. a. $(0, 2)$

    b. $(5, 0), (8, 0)$

Section 3.5 Triangular Numbers

p. 216

2. If $n = 7$ then $T = \frac{1}{2}(7)(7 + 1) = 28$. Yes, the answer agrees with the table value.

3. Yes. The generalization does not depend on the number of numbers being even or odd.

4. If $n = 113$ then $T = \frac{1}{2}(113)(113 + 1) = 6441$.

5. If $n = 2173$ then $T = \frac{1}{2}(2173)(2173 + 1) = 2,362,051$.

6. The nth triangular number equals one-half the product of n and one more than n.

7. We must find the value of n that makes $\frac{1}{2}(n)(n + 1) = 465$. If we use the table display on our calculator to examine input-output pairs for the triangular number function we can scroll until we find 465 in the output column. The corresponding input value is 30. In other words we have found the solution to our equation using our calculator's table feature.

8. a. Table 4: Numeric Representation of Triangular Numbers

Finite Differences of Index	Index	Triangular Number	Finite Differences in Triangular Number	Second Finite Differences in Triangular Number
2 – 1 = 1	1	1	3 – 1 = 2	3 – 2 = 1
3 – 2 = 1	2	3	6 – 3 = 3	4 – 3 = 1
1	3	6	4	1
1	4	10	5	1
1	5	15	6	1
1	6	21	7	1
1	7	28	8	1
1	8	36	9	1
1	9	45	10	
	10	55		

b. In the concert receipts problem the first finite differences are all the same.

c. $T(n) = \dfrac{n^2 + n}{2}$ The variable appears to the second power in this function.

d. The functions like $R(n) = 19n$ are defined by multiplying the variable by a constant. The variable appears only raised to the first power. This makes the first finite differences constant. The graphs of such functions follow a straight line. This is an example of a linear function.

Functions like $T(n) = \dfrac{n^2 + n}{2}$ or $S = n^2$ contain the variable raised to the second power. The first finite differences computed in a table are not constant but the second finite differences are. The graph of such functions follow curves called parabolas. These are examples of quadratic functions.

9. a. 1, 4, 9, and 16.

b. 25.

c. 36, 49, 64, 81, 100. Hopefully you recognize the pattern of squares.

d. **The Squares**

Dots on a side	Total number of dots
1	1
2	4
3	9
4	16
5	25
6	36
7	49
8	64
9	81
10	100

e. If $n = 4$ then $S = 16$. If $n = 8$ then $S = 64$

f. $S = n^2$

g. The graphic display produced on a TI-83 together with the viewing window is shown. The table columns are stored in list number 1 and 2 in the calculator. One of the points is displayed using the trace feature of the calculator.

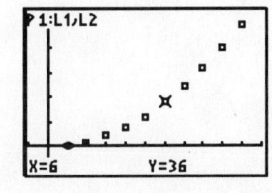

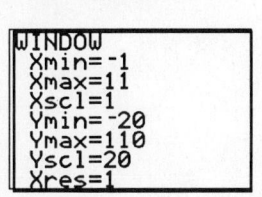

h. The graph is more like the graph of the triangular numbers function. They both follow the shape of a curve rather than a straight line as in the graph of the concert receipts function.

Section 3.6 Power and Factorial Functions

p. 230

2. a. The finite differences become constant after two computations.

b. The model is based on squares since the second finite difference is constant.

c. $D(t) = 16t^2$.

3. a. The labels are a help. It looks like $S(x)$ does not appear in Figure 7. There does seem to be a curve that the three graphs follow. $f(x)$ curves up the steepest.

   b. $f(x)$ increases most rapidly followed by $F(x)$ then by $G(x)$ the next steepest would be $S(x)$ if it were shown.

   c. They would appear to be straight lines because of the scale used on the vertical axis. Neither grow above 1000 when the input is 10.

   d. We would only see the first three points. The other points would be off the top of the screen.

   e. The point (1, 1) is on the graph of all three functions.

4. There are at most two students in the room. Can you figure out why?

5. a. 9! = 362,880. There are 362,880 different arrangements of the players in the nine positions.

   b. 362,880 innings are necessary to try all possible arrangements.

   c. $\dfrac{362,880}{6} = 60,480$. 60,480 six–inning games are required to try all possible arrangements.

   d. $\dfrac{60,480}{365} \approx 165.7$ Approximately 165.7 years are required to try all arrangements.

Section 3.7 Review

p. 233

1. a. $N = 17(4) = 68$; Evaluating.

   b. b is a little bigger than 43.4. To find b we divide 738 by 17. Solving.

   c. $Q = 60.75$; Evaluating.

   d. $m = 9$ or $m = -9$; Solving.

   e. $A = 725$; Evaluating.

f. $b = 7$; Solving.

g. $L = 72$; Evaluating.

2. a. $b, m,$ and r are the independent variables.

   b. $N, Q, A,$ and L are the dependent variables.

3. a. For N multiply by 17. c. For Q first square then multiply by 3.

   e. For A square, multiply by 2 then add 3.

4. a. 33

   b. First cube, multiply by 4, divide the 8 by 4, then add and finally subtract.

   c. $3 + (4(2^3) - 8 \div 4)$

5. a. m the number of miles driven.   b. $C(m)$ the cost.

   c. $C(m) = 0.17m + 27$

   d. $C(137) = 0.17(137) + 27 = 50.29$ The cost is $50.29. We evaluated a function to find this cost.

   e. $m = \dfrac{42.81 - 27}{0.17} = 93$ We solve the equation $0.17m + 27 = 41.81$ to find that we drove 93 miles.

6. a. **Table 1: Foreign Versus U.S.-Made Cars**

Foreign-Made Cars	U.S.-Made Cars
3	12
5	20
7	28
9	36

   b. The input would be the number of foreign-made cars and the output is the number of U.S.-made cars.

   c. $A = 4F$

d. The following graph was produced on a TI-83 graphing calculator.

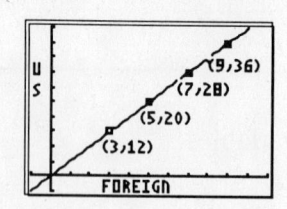

7. a.

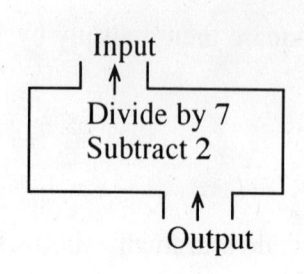

    Input

    Divide by 7
    Subtract 2

    Output

    b. $7(13) + 2 = 93$

    c. $x = 1$

8. a.

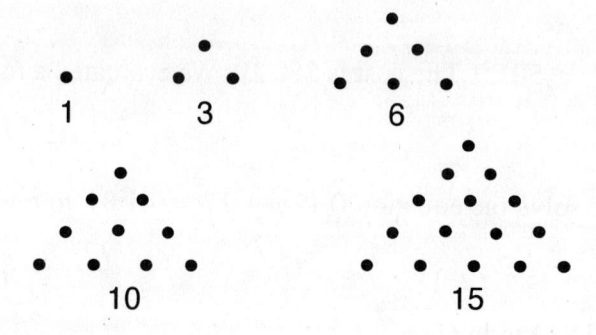

    1   3   6

    10   15

    b. **The Triangular Numbers**

n	T(n)
4	10
5	15
6	21
7	28

n	$T(n)$
8	36
9	45
10	55
11	66
12	78

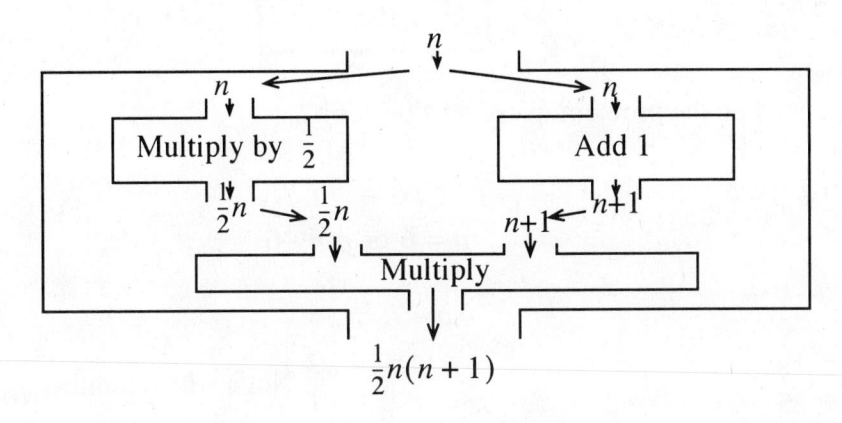

c.

[Function machine diagram: n splits into two paths. Left path: Multiply by $\frac{1}{2}$ giving $\frac{1}{2}n$. Right path: Add 1 giving $n+1$. Both combine via Multiply to produce $\frac{1}{2}n(n+1)$.]

d. $T(n) = \frac{1}{2}n(n+1)$

e. The graph and the viewing window from a TI-83 graphing calculator are shown. The point $(10, 55)$ has been highlighted using the trace feature.

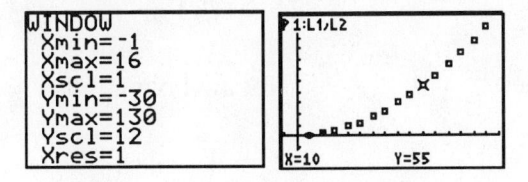

```
WINDOW
Xmin=-1
Xmax=16
Xscl=1
Ymin=-30
Ymax=130
Yscl=12
Xres=1
```

[Graph showing plotted points with X=10, Y=55 highlighted]

9. a. 28 appears in the table.

   b. It might be helpful to refer to the dot pictures of the first five triangular numbers to explain this.

   c. T increases by a greater and greater amount. This can be seen from the graph since the points plotted go up as the first coordinate increases. It is also evident by looking at the table. One might include the first finite differences in another column to show that the outputs increase by a greater and greater amount.

   d. $T(473) = \frac{1}{2}(473)(473 + 1) = 112101$. The algebraic representation using the order of operations indicated in the function machine or a calculator to make the computation seems appropriate here.

e. 153 is the 17th triangular number. Extending the table to an input of 17 reveals this. The display of such a table is shown below together with the Y= menu and the TABLE SETUP screen used to produce the table.

10. It seems to depend on the problem question, doesn't it?

11. a. $2p = 36$
$p = 18$

b. $p^2 = 36$
$p = 6$ or $p = -6$

c. $p + 11 = 36$
$p = 25$

d. $4p - 7 = 36$
$4p = 29$ Not a whole number.
$p = 4\frac{1}{7}$

e. $3 + 2p = 36$
$2p = 33$
$p = 16\frac{1}{2}$

Not a whole number.

f. $p^2 + 3 = 36$
$p = 33$
$p = \sqrt{33}$ or $p = -\sqrt{33}$

Not a whole number.

12. a.

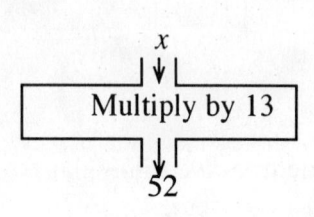

x → Multiply by 13 → 52

$x = 4$

b.

x → Subtract 8 → 14

$x = 22$

c.

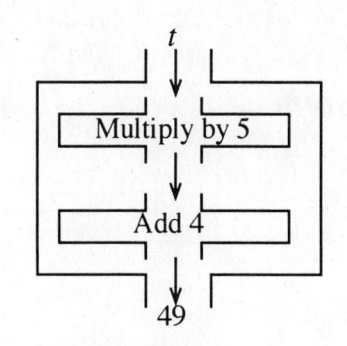

```
    t
    ↓
[Multiply by 5]
    ↓
  [Add 4]
    ↓
   49
```

$5t + 4 = 49$
$5t = 45$
$t = 9$

d.

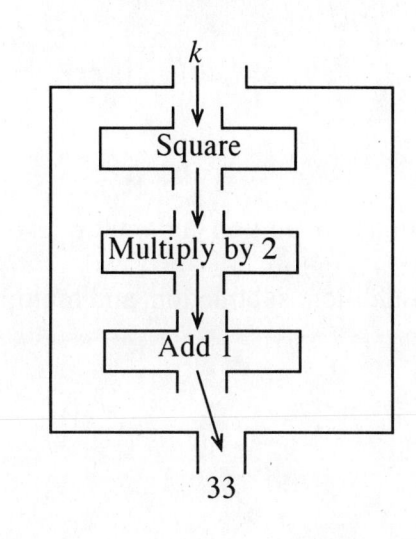

```
    k
    ↓
  [Square]
    ↓
[Multiply by 2]
    ↓
  [Add 1]
    ↓
   33
```

$2k^2 + 1 = 33$
$2k^2 = 32$
$k^2 = 16$
$k = 4$ or $k = -4$

13. a. 3^{12} b. $(5 \cdot 7)^3$

    c. 1   d. 1

    e. x^5 f. z^{99}

14. a. $3x^2 + 7x + 6$ b. $15x^3 + 16x^2 + 24x + 8$

    c. $16y^2 + 56y + 49$ d. $5x + x^2$

15. a. $2 \cdot 3 \cdot 5^2$ b. $3^2 \cdot 7 \cdot 11^3$

    c. $11 \cdot 13 \cdot 17$ d. $2^6 \cdot 13$

    e. $2^4 \cdot 3^2 \cdot 5 \cdot 7$ f. $2^7 \cdot 3^2 \cdot 5 \cdot 7$

16. a. prime   b. $2(a^2 + 2a + 3)$

    c. $x(5 + x)$

CHAPTER 4

Integers: Expanding on a Mathematical System

Section 4.1 Integers and the Algebraic Extension

p. 255

2. a. −6
    b. 5
    c. 0
    d. −5x
    e. 3y
    f. −(x + y) or − x − y

3. a. The whole numbers are closed under the operations of addition and multiplication.
    b. The integers are closed under the operations of addition, subtraction, and multiplication.

4. a. −t = −7
    b. −b = −(−2) = 2
    c. −r = −823

5. a. In − 12 − 5 the first "−" is used to represent negative 12. The second represents subtraction with inputs −12 and 5.

    b. In − 12 − (−5) the first minus sign represents negative 12 and the second represents subtraction with input −12 and −5. The final minus sign is used to indicate negative 5.

    c. In 12 − 5 the minus sign represents subtraction with inputs 12 and 5.

    d. The two minus signs negative 12 and negative 5.

    e. The minus in front of the x represents taking the opposite of the value of x.

    f. This one means subtraction with inputs x and y.

    g. The two minus signs represent taking the opposite of x and y.

    h. The first and the last minus signs represent taking the opposite of x and y. The other one indicates subtraction with inputs −x and −y.

    i. The first represents negative 5, the last indicates taking the opposite of x and the one in the middle means subtract −x from −5.

j. The second minus sign is used to represent negative 5. The first represents subtraction with inputs x and -5.

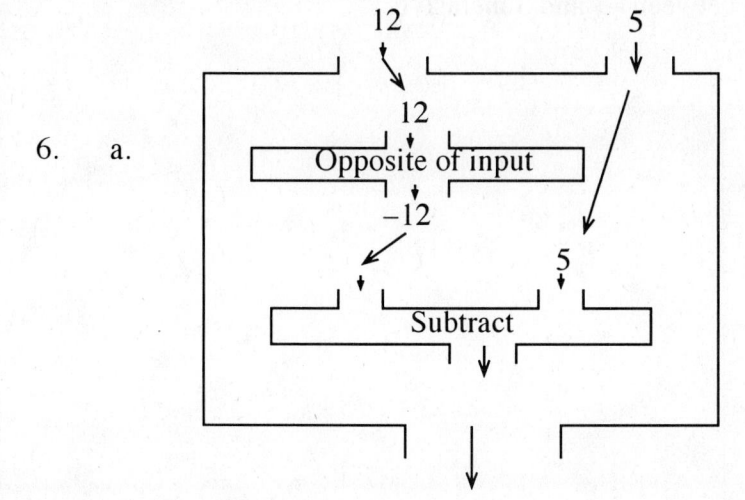

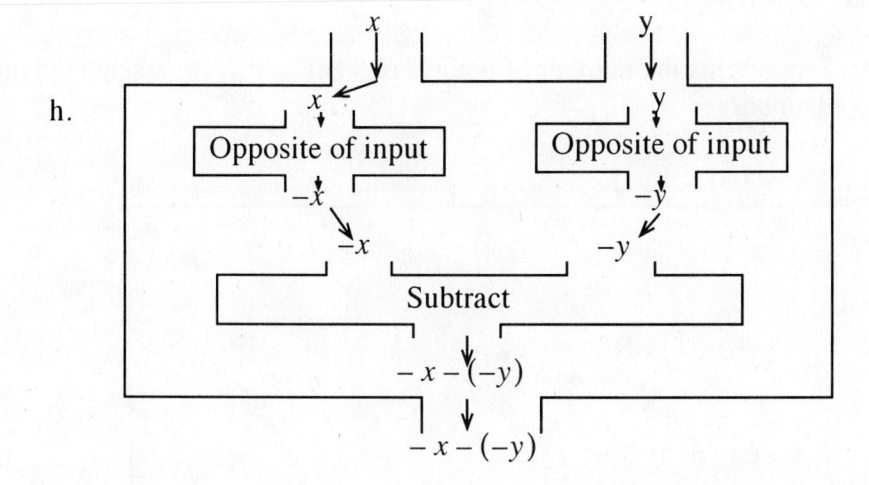

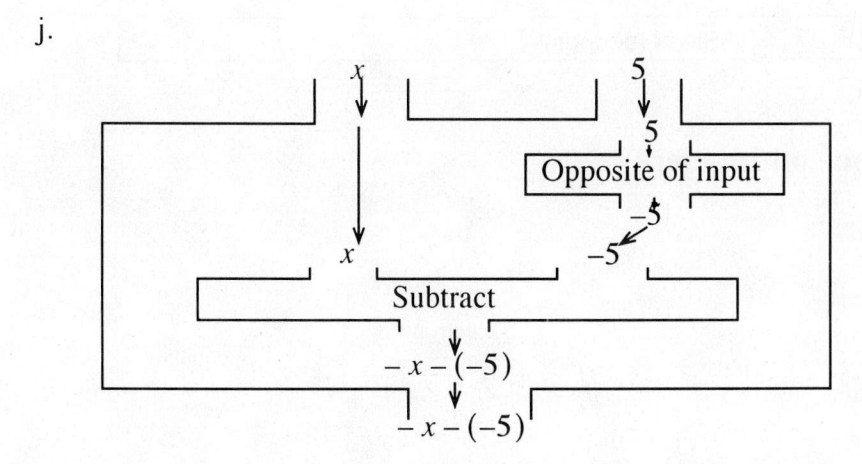

6. a.

    h.

    j.

7. a. $int(rand \cdot 10) + 1$

    b. *rand* is between 0 and 1 exclusive. Seven times *rand* is between 0 and 7 exclusive. $7rand - 4$ is between -4 and 3 exclusive. $int(7rand - 4)$ is between -4 and 2 inclusive. $int(7rand - 4) + 1$ is between -3 and 3 inclusive.

8. a.

Pounds of steak	Pounds of hamburger
10	60
20	120
30	180
40	240
50	300
60	360

    Table 1: Steak and hamburger

    b. $H = 6S$ where S represents the number of pounds of steak and H represents the number of pounds of hamburger.

    c.

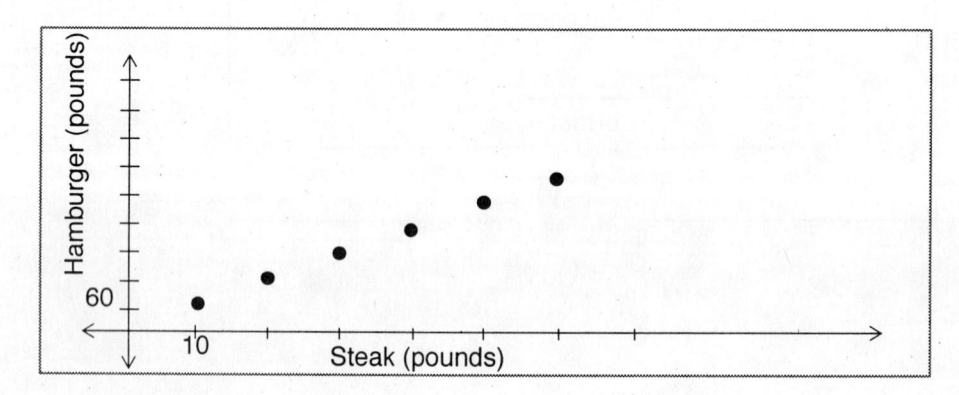

Section 4.2 Operations on Integers

p. 275

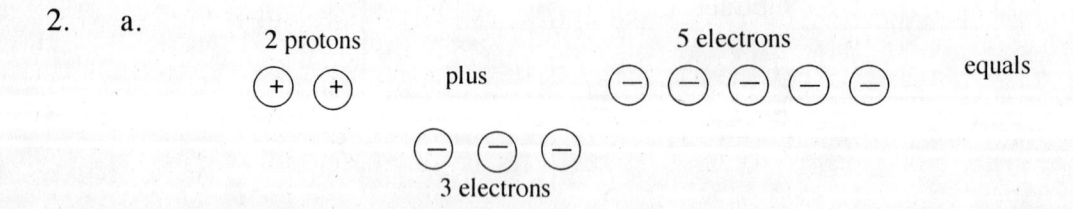

2. a.

116

b.

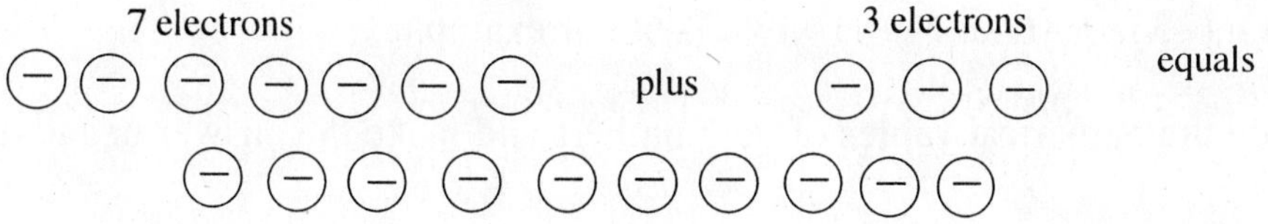
7 electrons
3 electrons
plus
equals

c.

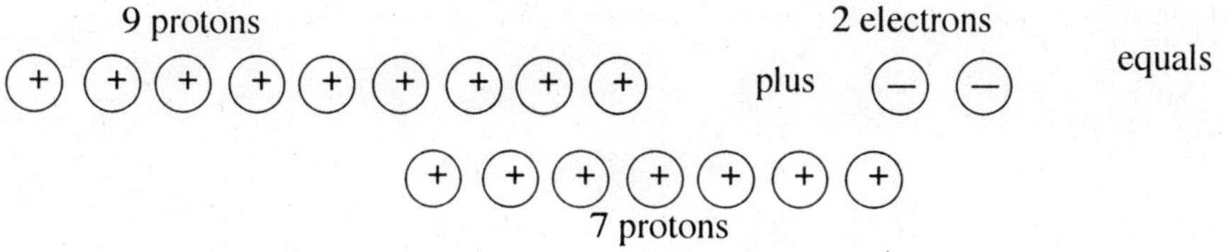
9 protons
2 electrons
plus
equals
7 protons

3. a.

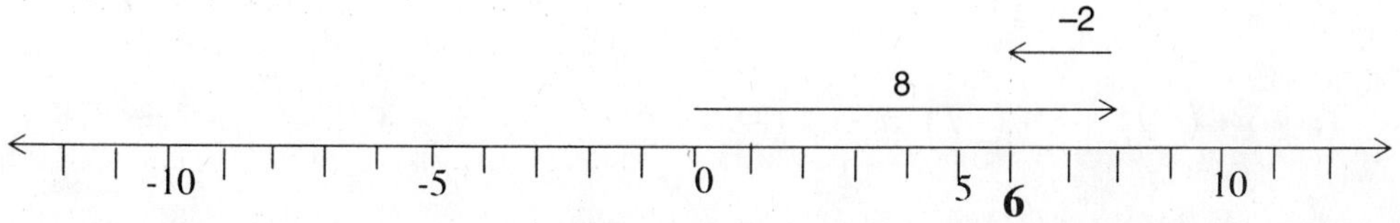
−2
8
-10
-5
0
5
6
10

b.

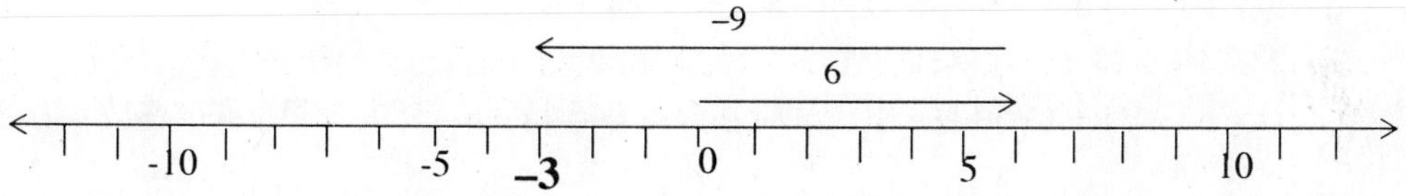
−9
6
-10
-5
−3
0
5
10

c.

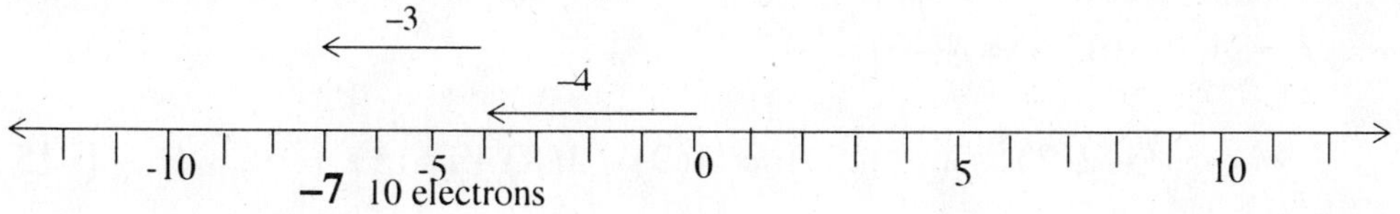
−3
−4
-10
-5
0
5
10
−7
10 electrons

4. a. $3 - 7 = 3 + (-7)$

b. $-2 - 6 = -2 + (-6)$

c. $-4 - (-5) = -4 + 5$

d. $-9 - (-7) = -9 + 7$

5. a. $(-9)(-8) = 72$, for example.

b. Multiply the numerical values of the numbers and make the answer positive.

6. a. $(9)(-8) = -72$, for example.

b. Multiply the numerical values of the numbers and make the answer negative.

7. a. $(-18) \div (-3) = 6$, for example.

b. Divide the numerical values of the numbers and make the answer positive.

8. a. $(18) \div (-3) = -6$ or $(-18) \div 3 = -6$, for example.

   b. Divide the numerical values of the numbers and make the answer negative.

9. a. $-8 - 7 = -8 + (-7) = -15$.

   b. $(-8)(-7) = 56$.

   c. $-(-35) = 35$.

   d. $-15 \div 3 = -5$

   e. $-13 + 2 = -11$

   f. $6 + (-9) - 5 - (-7) = 6 + (-9) + (-5) + 7 = -3 + (-5) + 7 = -8 + 7 = -1$.

   g. $(-3)(8) + (-2)(-6) = -24 + 12 = -12$

   h. $(8 + (-2))(5 - 9) = (6)(-4) = -24$.

   i. $17 - (-9) + (-54) - 38 = 17 + 9 + (-54) + (-38) = 26 + (-54) + (-38) =$
   $-28 + (-38) = -66$.

   j. $-24 \div (-6)(-3) = 4(-3) = -12$.

   k. $4(-3)^2 - 2(2)^3 + (-3)(5)) = 4(9) - 2(8) + (-15) = 36 - 16 + (-15) =$
   $20 + (-15) = 5$.

10. a. Both answers are 3.

    b. Both answers are –4.

    c. Both answers are –10.

    d. The commutative property of addition holds for integers.

11. a. The answers are 2 and –2.

    b. The answers are 8 and –8.

    c. The answers are 3 and –3.

    d. Subtraction of integers is not commutative.

12. a. Both answers are –6.

   b. Both answers are –19.

   c. Both answers are 12.

   d. The associative property of addition holds for integers.

13. a. The answers are –5 and 7.

   b. The answers are –17 and –1.

   c. The answers are 22 and 12.

   d. Subtraction of integers is not associative.

14. a. Case 1: Bathtub filling, play videotape forward.

   Case 2: Bathtub filling, play videotape backward.

   Case 3: Bathtub emptying, play videotape forward.

   Case 4: Bathtub emptying, play videotape backward.

   b. **Table 4 Bathtub Model**

Bathtub filling or emptying	Videotape direction	What you see on the screen
+	+	+
+	–	–
–	+	–
–	–	+

   c. The product of two numbers whose signs are both the same is a positive number and the product of two numbers whose signs are different is a negative number.

15. a. In – 12 – 7 the first "–" is used to represent negative 12. The second represents subtraction with inputs –12 and 7.

   b. In – 12 – (–7) the first minus sign represents negative 12 and the second represents subtraction with input –12 and –7. The final minus sign is used to indicate negative 7.

   c. In 12 – 7 the minus sign represents subtraction with inputs 12 and 7.

d. The two minus signs are used to represent negative 12 and negative 7.

e. The minus in front of the *3x* represents taking the opposite of the value of *3x*.

f. This one means subtraction with inputs *3x* and *y*.

g. The two minus signs represent taking the opposite of *3x* and *y*.

h. The first and the last minus signs represent taking the opposite of *3x* and *y*. The other one indicates subtraction with inputs $-3x$ and $-y$.

i. The first represents negative 9, the last indicates taking the opposite of *x* and the one in the middle means subtract $-x$ from -9.

j. The second minus sign is used to represent negative 9. The first represents subtraction with inputs *x* and -9.

16. a. $-x^2 + 3x - 2$

    b. $(3x - 8) - (2x + 3) = 3x - 8 - 2x - 3 = x - 11$

    c. $(x^2 + 9x - 3) - (4x^2 + 3x + 2) = x^2 + 9x - 3 - 4x^2 - 3x - 2 = -3x^2 + 6x - 5$

    d. $6x^3 - 15x^2$

    e. $-10x^3 + 35x^2 - 45x$ f. $x^2 - 5x - 36$

    g. $12x^2 + 13x - 14$ h. $x^2 - 49$

    i. $16x^2 - 9$ j. $4x^2 - 44x + 121$

    k. $2x^4 - 11x^3 - 29x^2 - 2x + 5$ l. $x^4 - 1$

17. a. $T(-3) = 5(-3) - 2 = -15 - 2 = -17$

    b. $5r - 2 = -17$
       $5r = -15$
       $r = -3$

18. If $y(x) = 3x^2 - x - 4$ then $y(-5) = 3(-5)^2 - (-5) - 4 = 3(25) + 5 - 4 =$
    $75 + 5 - 4 = 80 - 4 = 76$

19. a.
$5x - 2 = -12$
$5x = -10$
$x = -2$

b.
$3a + 7 = -17$
$3a = -24$
$a = -8$

c.
$9q + 1 = 46$
$9q = 45$
$q = 5$

d.
$-2x - 9 = -21$
$-2x = -12$
$x = 6$

e.
$-4c + 5 = 13$
$-4c = 8$
$c = -2$

f.
$-5w + 14 = 14$
$-5w = 0$
$w = 0$

g.
$5 - 8x = -27$
$-8x = -32$
$x = 4$

h.
$-7 + 3y = -28$
$3y = -21$
$y = -7$

i.
$-1 - 6x = 23$
$-6x = 24$
$x = -4$

j.
$-4 - 5p = -69$
$-5p = -65$
$p = 13$

Section 4.3 The Absolute Value Function

p. 295

2. a. x_1 b. x_2

   c. $\Delta x = x_2 - x_1$

3. a. A negative change would imply a decrease in petty cash.

   b. A positive change would imply an increase in petty cash.

4. **Table 7: Daily Change in Petty Cash Account**

Initial Value x_1 in $	Final Value x_2 in $	Change in Petty Cash Δx in $
5	3	–2
3	5	2
3	3	0
0	7	7
7	0	–7
0	–2	–2
–2	0	2
–3	4	7
4	–3	–7
1	–1	–2
–6	–3	3
t	t	0
t	t+4	4
t+4	t	–4
–5	–5	0
anything	anything	0
x	x+5	5
x+5	x	–5

5. a. $\Delta x = 0$ No sign.   b. The change would be positive.

    c. The change would be negative.

6. a. $\Delta t = -12$ At 4:00 P.M. the temperature is 15^0.

   b. $\Delta t = -9$ At 9:00 A.M. the temperature was 36^0, assuming the temperature dropped 3^0 each hour from 9:00 A.M. until noon.

7. At the beginning of the week, I had twenty–seven minutes of leisure time per day. However, I became a member of a committee on Tuesday which will leave me with only 12

minutes of leisure time per day. Let the variable l_1 represent the amount of daily leisure time at the beginning of the week and l_2 be the amount of daily leisure time after Tuesday. Then Δl represents the change in my leisure time where $\Delta l = l_2 - l_1 = 12 - 27 = -15$.

My leisure time per day has decreased by fifteen minutes.

8. The absolute value of a number indicates the magnitude (size) of the number. When we create a number line, we start at the point labelled zero. We then mark off units to the right and to the left of zero. These units represent how far the point is from zero. If we ignore direction, the units represent a magnitude and that is exactly what absolute value represents. Distance is usually used to represent "how far" without considering direction. "How far" is a non–negative idea, while direction is often indicated by a positive or negative sign.

9. $\Delta m = 109 - 143 = -34$. He had driven $|\Delta m| = |-34| = 34$ miles.

10. a. $|6|$ is the distance between 6 and 0 on the number line. $|-43|$ is the distance between -43 and 0 on the number line.

    b. The expression is $|8|$.

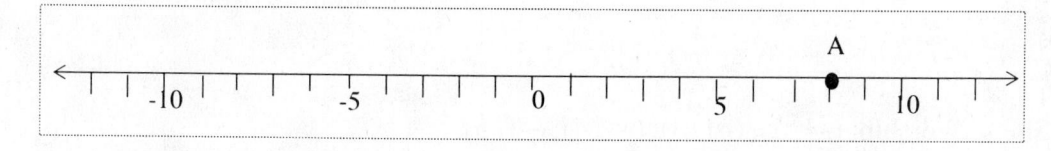

    c. The expression is $|-2|$.

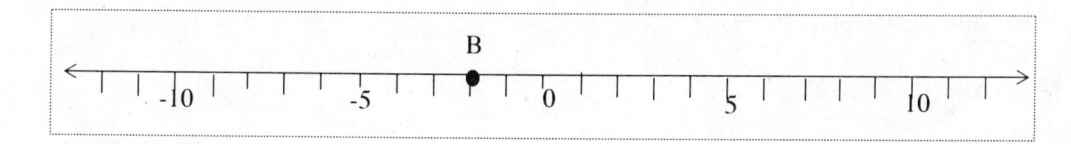

11. a.

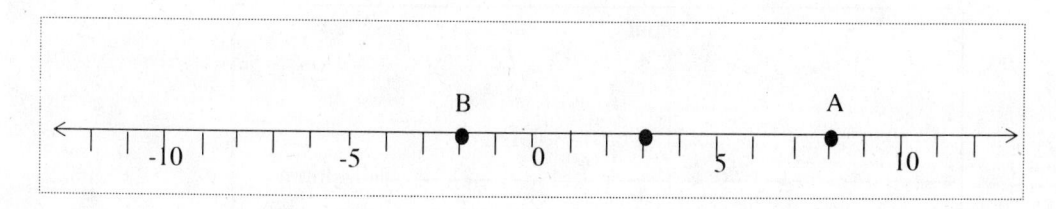

    b. For point A, $|8 - 3| = 5$ For point B, $|-2 - 3| = 5$.

12. a.

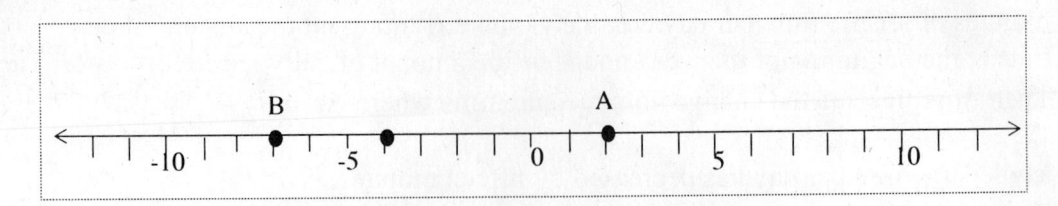

    [number line showing points B between -6 and -5, and A between 1 and 2, from -10 to 10]

    b. For point A, $|2 - -4| = 6$ For point B, $|-7 - -4| = 3$.

13. a. −1 and 2 both produce an output of 3.

    b. −3 and 4 both produce an output of 7.

    c. No number produces and output of -1.

14. a. 12 and -2 are the solutions.   b. -4 is the only solution.

    c. The equation has no solutions.   d. -8 and 14 are the solutions.

Section 4.4 Graphing with Integers

p. 306

2. Site of Worship: $(-1, 2), (-1, 3), (-1, 4), (-1, 5)$.

   Site of Ancient Secrets: $(2, 1), (3, 1), (4, 1)$.

   Math Site 2: $(0, -4), (1, -4)$.

   Math Site 1: $(-4, -2), (-3, -3)$.

3. Label the ordered pairs as follows: A(−5, 40), B(3, −50), C(−2, −10), D(6, 20), E(0, −30), F(−4, 0), G(7, 0), and H(0, 10).

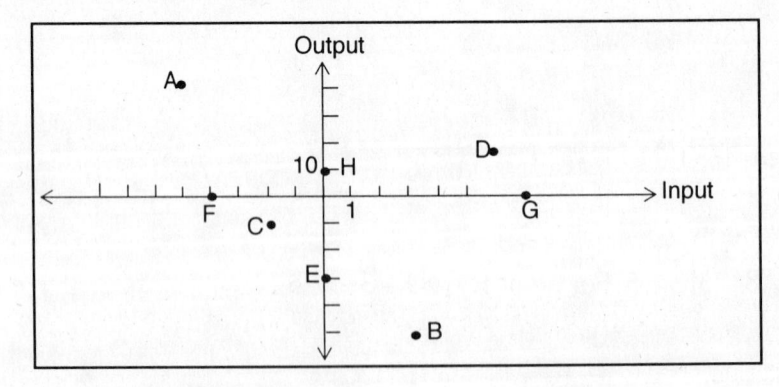

   [coordinate plane with points A, B, C, D, E, F, G, H plotted with axes labeled Output and Input]

124

4. Refer to the T-shirt problem discussed in this section.

   a. To find the profit, you multiply the number of T-shirts by 10 and subtract 1000 from the product.

   b.

   [Diagram: 10 and t inputs into "multiply" box producing 10t; then 10t and 1000 input into "subtract" box producing 10t − 1000. Or: t input into "multiply by 10, subtract 1000" box producing 10t − 1000.]

   c. Let t represent the number of T-shirts sold and P represent the profit. Then $P(t) = 10t - 1000$.

7. **Table 3 Opposites of Integers**

Integer	Opposite of integer
−5	5
−4	4
−3	3
−2	2
−1	1
0	0
1	−1
2	−2
3	−3
4	−4
5	−5

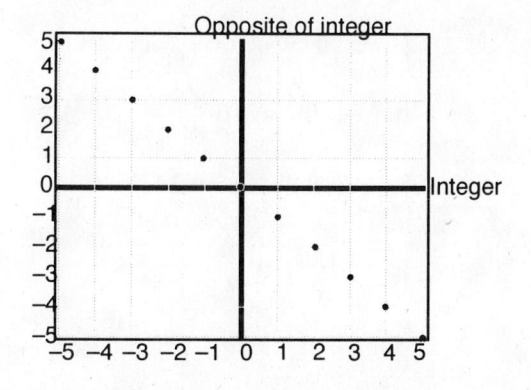

[Graph: Points plotted on coordinate plane with "Integer" on horizontal axis and "Opposite of integer" on vertical axis, showing the line through the points from the table.]

8. **Table 4 Absolute Value of Integers**

Integer	Absolute value of integer
−5	5
−4	4
−3	3
−2	2
−1	1
0	0
1	1
2	2
3	3
4	4
5	5

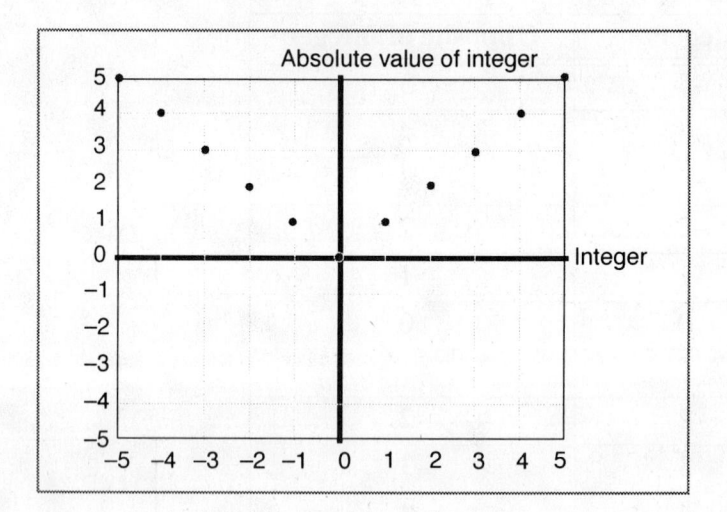

9. a. $-5 + 17 = 12$ b. $2 - 11 = -9$

   c. $13 + (-8) = 5$ d. $-5 - 9 = -14$

   e. $-7 - (-3) = -4$ f. $|-9| = 9$

   g. $-|5| = -5$ h. $-|-6| = -6$

   i. $(-7)(4) = -28$ j. $(-8)(-3) = 24$

   k. $\dfrac{18}{-9} = -2$ l. $-\dfrac{18}{9} = -2$

m. $\dfrac{-18}{9} = -2$ n. $\dfrac{-18}{-9} = 2$

o. $-\dfrac{-18}{-9} = -2$ p. $|-3| - 5 = -2$

q. $-4(|-8|) = -32$ r. $-|-2| + |-9| = 7$

10. The function $P(m) = 3m - 50$ expresses the profit P (in dollars) from the sale of costume jewelry as a function of the number m of pieces sold.

   a. **Profit on sale of costume jewelry**

Pieces sold	Profit ($)
5	−35
10	−20
15	−5
20	10
25	25
30	40

   b. Multiply the number of pieces sold by 3 and subtract 50 from the product.

   c.

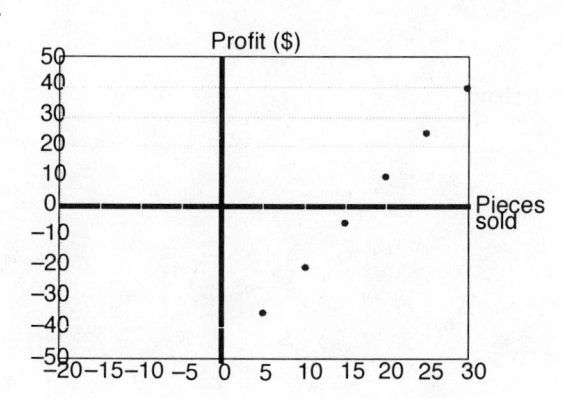

   d. The profit will be a little less than $104 when 51 pieces of jewelry are sold.

11. a. The function is defined by $x - 2$ and the output is -7. $x = -5$

   b. The function is defined by $5y - 3$ and the output is -18. $y = -3$

c. The function is defined by $4 - 3z$ and the output is -17. $z = 7$

d. The function is defined by $-7x - 1$ and the output is 62. $x = -9$

e. The function is defined by $|x - 5|$ and the output is 23. $x = -18$ or $x = 28$

f. The function is defined by $|2x - 1|$ and the output is 5. $x = 3$ or $x = -2$

g. The function is defined by $|4x + 3|$ and the output is -3. There is no input which would produce this output.

h. The function is defined by $|x + 7|$ and the output is 0. $x = -7$

Section 4.5 Using a Graphing Utility

p. 321

2. Xmin = –4, Xmax = 5, Xscl = 1, Ymin = –200, Ymax = 140, Yscl = 20.

3. a. $x = 7.0212766$ and $y = 5.1612903$ on the TI–82 or TI-83.

   b. $(0, 0)$ on the TI–82 and TI-83.

4. The x values are now multiples of five rather than rational numbers. All the values in Quadrants I and IV obtained by TRACE are reasonable values for the problem situation.

5. a. If $P(x) = 10x - 1000$ then $P(132) = 10(132) - 1000 = 320$. This suggests that selling 132 T–shirts results in a profit of $320. The profit indicated in Figure 5 was about $324.

   b. The parametric graph appears below

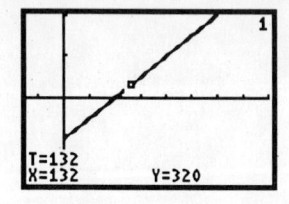

   The result matches the answer to a.

6. a. **Professors and Students**

Professors	Students
100	600
200	1200
300	1800
400	2400
500	3000
600	3600

c.

```
WINDOW FORMAT
Xmin=0
Xmax=700
Xscl=100
Ymin=0
Ymax=4200
Yscl=600
```

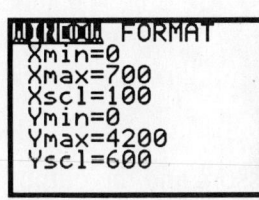

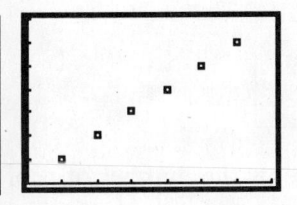

d. The algebraic representation is $S(P) = 6P$.

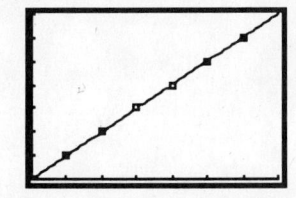

e. **More Professors and Students**

Professors	Students
350	2100
357.44681	2144.6809
364.89362	2189.3617
372.34043	2234.0426

f. The last three values in the table do not make sense in the problem situation. Both the input and the output for this problem must be elements of the whole numbers.

g. Let $x = t$ and $y = 6t$. The graph appears below.

```
WINDOW FORMAT
Tmin=0
Tmax=700
Tstep=25
Xmin=-100
Xmax=700
Xscl=100
↓Ymin=-600
 Ymax=4200
 Yscl=600
```

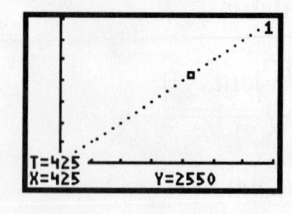

T=425
X=425   Y=2550

Even More Professors and Students

Professors	Students
425	2550
475	2050
525	3150
575	3450

h. The input values (x) represents the number of professors and the output values (y) represent the number of students.

7. a. $C(z) = 2z - 7$

z	$C(z)$
1	–5
2	–3
3	–1
4	1
5	3
6	5

b. and c.

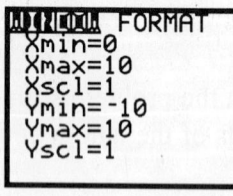

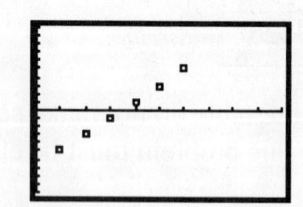

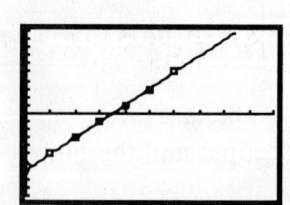

```
WINDOW FORMAT
Xmin=0
Xmax=10
Xscl=1
Ymin=-10
Ymax=10
Yscl=1
```

d. **More of Function C**

z	C
5.212766	3.4255319
5.4255319	3.8510638
2.3404255	−2.319149

e. $x = t$ and $y = 2t - 7$

f. Graph is in the same viewing window as b and c with Tstep equal to 0.5

8. a. $A(b) = 3 + 2b^2$

b	A(b)
−5	53
−2	11
0	3
3	21
5	53
6	75

b. and c.

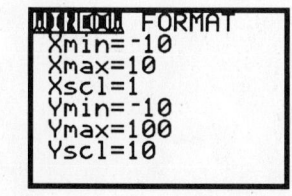

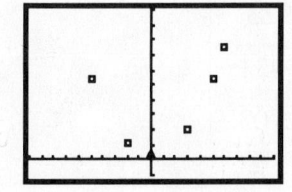

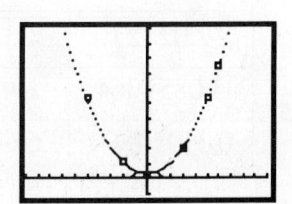

d. **More of Function A**

z	A
−3.829787	32.334541
−5.744681	69.002716
1.2765957	6.2593934

e. $x = t$ and $y = 3 + 2t^2$

f. Graph is in the same viewing window as b and c with Tstep equal to 0.5.

9. a. $L(r) = 3r^2 - 2r + 7$

r	L(r)
−4	63
−3	40
−1	12
0	7
2	15
4	47

b. and c.

WINDOW FORMAT
Xmin=-5
Xmax=5
Xscl=1
Ymin=-10
Ymax=70
Yscl=10

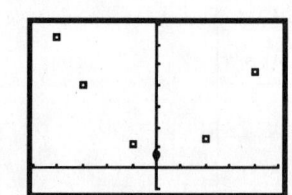

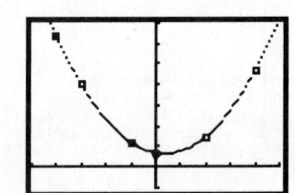

d. **More of Function L**

z	L
−2.021277	23.29923
3.0851064	29.383431
0.31914894	6.6672703

e. $x = t$ and $y = 3t^2 - 2t + 7$

f. Graph is in the same viewing window as b and c with Tstep equal to 0.5.

10. From 0 to 10 seconds the elevator is waiting at Floor 1. From 10 seconds to 20 seconds, the elevator rises from Floor 1 to Floor 2. From 20 to 30 seconds, the elevator is waiting at Floor 2. From 30 to 40 seconds, the elevator has moved up from Floor 2 to just above Floor 3. From 40 to 50 seconds, the elevator has continued to rise to Floor 6. From 50 to 60 seconds and from 60 to 70 seconds, the elevator has remained at Floor 6.

11. a. The graph displays five points that have rapidly increasing *y* values for increasing *x* values.

    b.

    c. The output of the factorial function is undefined when the input is not a whole number. The domain of the factorial function is the set of whole numbers. These are two different ways of indicating what the possible inputs for the factorial function are.

12. a.

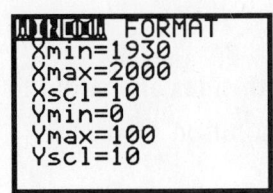

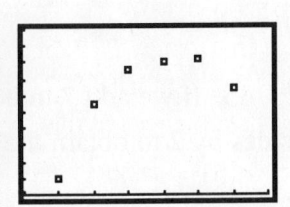

    b. The graph rises rapidly during the first twenty years indicating a rapid increase in the percentage of Americans with private health insurance. The graph levels off for the next twenty years indicating relatively little change in the percentage of Americans with private health insurance. Finally, the percentage covered drops in the last ten years.

    c. The first drop in percentage of Americans covered by private health insurance occurred during the decade from 1980 to 1990

13. a. **Width and Length of Rectangles**

Width (feet)	Length (feet)
0	10
2	8
4	6
6	4
8	2
10	0

b. $l = 10 - w$ where l represents the length and w represents the width.

c. The domain of the length function is the set of all real numbers which are greater than 0 and less than 10.

d. Any number between 0 and 10 would represent the width of one of the rectangles whose perimeter is 20 feet. So, any point on the part of the connected graph which is between the axes would be an ordered pair of values which would make sense in the problem situation.

14. The equation would be $2z - 7 = 35$. If we add 7 to both sides and simplify we get $2z = 42$. Finally, divide both sides by 2 to obtain the solution $z = 21$.

15. $3 + 2b^2 = 35 \qquad 2b^2 = 32 \qquad b^2 = 16 \qquad b = \pm 4$

16. a. $-(-2x + 7)$ \qquad b. $-(x^2 + 5x - 8)$

Section 4.6 Functions over the Integers

p. 334

2. a. **Table 4 $L(x), OppL(x),$ and $AbsL(x)$**

x	$L(x)$	$OppL(x)$	$AbsL(x)$
4	8	−8	8
−3.7	−7.4	7.4	7.4
0	0	0	0
−5	−10	10	10
15	30	−30	30

b. $OppL(x) = -2x$ and $AbsL(x) = |2x|$.

c. All graphs are in the standard viewing window.

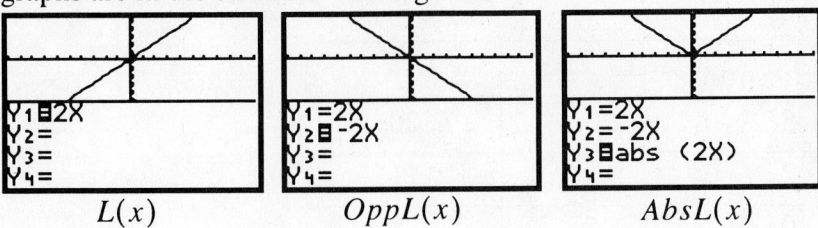

      $L(x)$ $OppL(x)$ $AbsL(x)$

d. The graph of $OppL(x)$ is obtained from the graph of $L(x)$ by reflection about the x-axis.

e. The graph of $AbsL(x)$ is obtained from the graph of $L(x)$ by reflecting all negative outputs about the x-axis. Non-negative outputs are not reflected.

3. Given $Q(x) = x^2 - 5x - 3$.

   a. **Table 5** $Q(x), OppQ(x),$ and $AbsQ(x)$

x	$Q(x)$	$OppQ(x)$	$AbsQ(x)$
–5	47	–47	47
–2	11	–11	11
0	–3	3	3
2	–9	9	9
4	–7	7	7

   b. $OppQ(x) = -(x^2 - 5x - 3) = -x^2 + 5x + 3$ and $AbsQ(x) = |x^2 - 5x - 3|$.

   c. All graphs are in the standard viewing window.

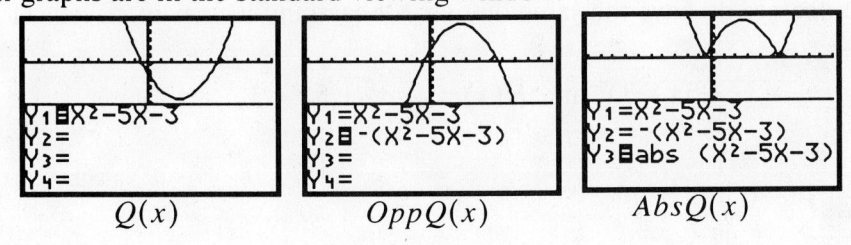

         $Q(x)$ $OppQ(x)$ $AbsQ(x)$

   d. The graph of $OppQ(x)$ is obtained from the graph of $Q(x)$ by reflection about the x-axis.

   e. The graph of $AbsQ(x)$ is obtained from the graph of $Q(x)$ by reflecting all negative outputs about the x-axis. Non-negative outputs are not reflected.

4. Let $G(x) = -(3x-5)$ and $H(x) = -3x+5$.

   a. **Table 6 $G(x)$ Versus $H(x)$**

x	$G(x)$	$H(x)$
−5	20	20
−2	11	11
0	5	5
2	−1	−1
4	−7	7

   b.

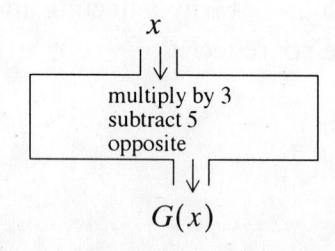

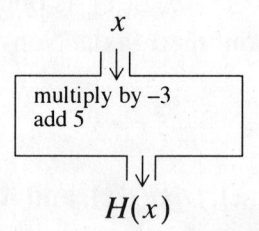

   ```
   x                              x
   ↓↓                             ↓↓
   ┌─────────────┐                ┌─────────────┐
   │ multiply by 3│                │ multiply by −3│
   │ subtract 5  │                │ add 5       │
   │ opposite    │                │             │
   └─────────────┘                └─────────────┘
   ↓↓                             ↓↓
   G(x)                           H(x)
   ```

   c.

   ```
   Y1 ■ -(3X-5)
   Y2 ■ -3X+5
   Y3 =
   Y4 =
   ```

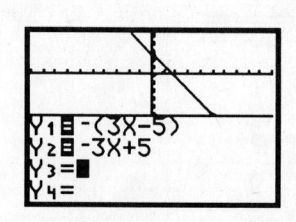

   d. The graphs of $G(x)$ and $H(x)$ are the same.

   e. The process represented by the algebraic representations are different. The function machine diagrams which appear in part b above show the two different processes. The two processes do have the same outputs for each input, however.

5. Let $G(x) = -(x^2 - 5x - 3)$ and $H(x) = -x^2 + 5x + 3$.

a. **Table 7 G(x) Versus H(x)**

x	G(x)	H(x)
−5	−47	−47
−2	−11	−11
0	3	3
2	9	9
4	7	7

b.

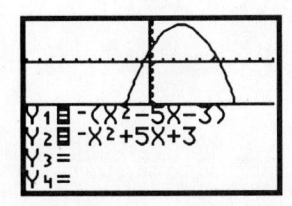

```
Y1■-(X²-5X-3)
Y2■-X²+5X+3
Y3=
Y4=
```

c. The graphs of $G(x)$ and $H(x)$ are the same.

d. In $G(x) = -(x^2 - 5x - 3)$ the polynomial part is evaluated first then the opposite of that number is determined. In $H(x) = -x^2 + 5x + 3$ the polynomial part is the same as the polynomial part in $G(x)$ except all the signs have been changed to their opposites.

6. Let $G(x) = -(3x^2 + 5x - 1)$ and $H(x) = -3x^2 - 5x + 1$.

a. **Table 8 G(x) Versus H(x)**

x	G(x)	H(x)
−5	−49	−49
−2	−1	−1
0	1	1
2	−21	−21
4	−67	−67

b.

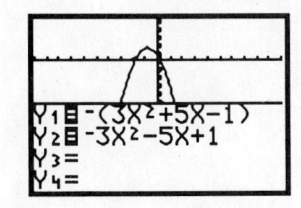

```
Y1■-(3X²+5X-1)
Y2■-3X²-5X+1
Y3=
Y4=
```

c. The graphs of $G(x)$ and $H(x)$ are the same.

d. The minus sign in front of the polynomial in $G(x) = -(3x^2 + 5x - 1)$ indicates taking the opposite of the polynomial last. In the expression which defines H, $H(x) = -3x^2 - 5x + 1$ each of the signs of the polynomial in $G(x)$ have been changed to their opposite.

7. a. The sign of x is negative.

    b. The sign of $-x$ is positive.

    c. The sign of $|x|$ is positive.

    d. Let $x = -43$ then $-x = -(-43) = 43$ and $|x| = -(-43) = 43$.

8. a. $f(x) = -(9 + 2x) = -9 - 2x$.

X	Y1	Y2
-5	1	1
-4	-1	-1

    Y1 = -(9+2X)

    b. $f(x) = -(3x^2 - 4x + 7) = -3x^2 + 4x - 7$

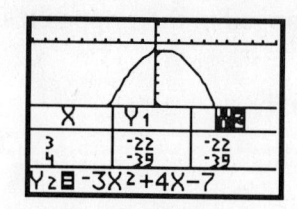

X	Y1	Y2
3	-22	-22
4	-39	-39

    Y2 = -3X2+4X-7

9. For $f(x) = -(9 + 2x)$, the output is negative for some values of x. For example, if $x = -4$, then the output is -1.

    For $f(x) = -(3x^2 - 4x + 7)$, the output is always negative. This is proven by noting the graph is always in Quadrants III or IV. Either input in the given table produces a negative output.

10. a. $fun(x) = -(-13x + 4)$

    b. $joy(x) = -(x^2 + 2x - 3)$

11. A table is an excellent way to investigate this problem.

2–point shots	3–point shots
15	0
12	2
9	4
6	6
3	8
0	10

This problem could be investigated using an equation. If a represents the number of two–point shots made and b represents the number of three–point shots made, then the relationship between a and b is

$$2a + 3b = 30$$

where a and b are whole numbers.

12. Are the expressions $3 + 2m$ and $5m$ equivalent? They are not equivalent. If we substitute 0 for m then $3 + 2m = 3$ and $5m = 0$. Since the expressions have different values for a given value of the variable, they are not equivalent.

13. a. First look at a picture.

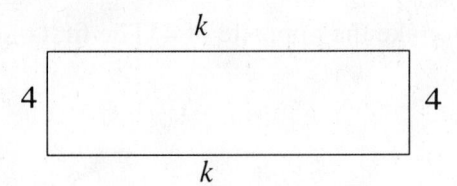

    A rectangle with sides labeled k (top and bottom) and 4 (left and right).

    The perimeter is the sum of the four sides which is $4 + 4 + k + k$ or $8 + 2k$.

    b. The area is the product of the length and the width. So an expression for the area is $4k$.

14. a. $x = -6$ b. $x = -5$

    c. $x = \dfrac{4}{3}$ d. $x = 23$ or $x = -5$

    e. No solutions.   f. $x = 7$

Section 4.7 Review

p. 340

1. a. 8
    b. a
    c. $7b$
    d. $-3y + 2x$
    e. $-5x - 7$
    f. $-x^2 + 3x - 11$

2. a. The first one is used to represent negative 2 the second one is subtraction with inputs −2 and 8.

    b. The second minus sign indicates the operation of taking the opposite of b. The first one represents subtraction with inputs a and $-b$.

    c. Both minus signs represent the operation of taking the opposite of the values of the variables.

    d. The first one is used to indicate the opposite of a. The second one means subtract with inputs $-a$ and b.

    e. The first and last minus signs are used to indicate taking the opposite the second one represents subtraction with inputs $-5a$ and $-2c$.

    f. The second minus sign means to take the opposite of a. The first one indicates subtraction with inputs 4 and $-a$.

    g. The second minus sign means to take the opposite of 4. The first one indicates subtraction with inputs a and -4.

3. b. For $a - (-b)$

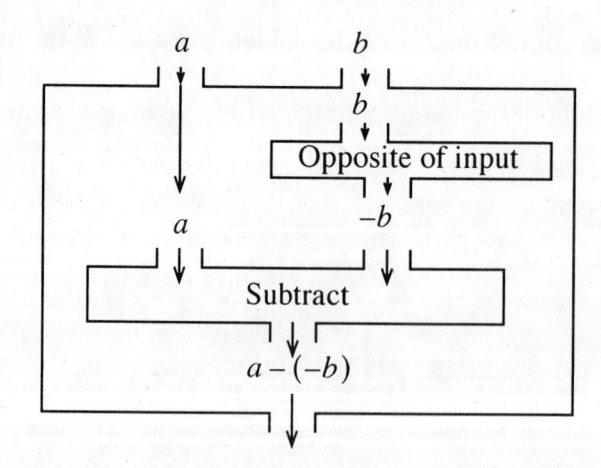

140

c. For $-a + (-b)$

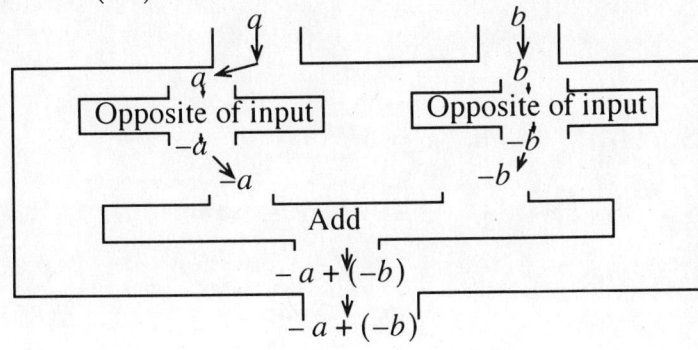

d. For $-a - b$

4. a. $-3 + (-7)$ b. $5 + (-(8)) = 5 + 8$

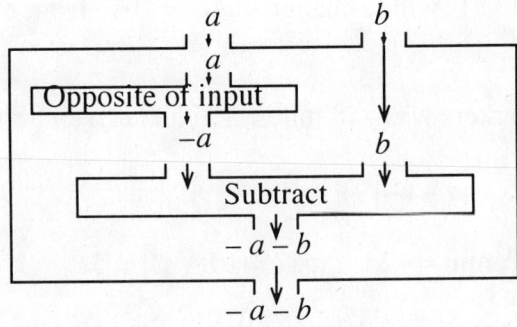

   c. $-a + (-b)$ d. $-x + (-(y)) = -x + y$

5. a. -72 b. 1007

   c. 4 d. 4

   e. -19 f. -17

   g. -38 h. 20

   i. -66 j. 8

   k. 15 l. -15

   m. -11 n. -4

   o. -27 p. 6

6. a. $R(-2) = 7(-2) - 9 = -23$ b. $x = -5$

7. $y(-5) = -5^2 - 4 \cdot 5 + 3 = -42$

8. a. $x = -7$ b. $b = 6$

141

c. $q = -26$ d. $x = 6$

e. $y = 0$ f. $x = 1$

g. $x = 5$ h. $x = 5$

i. $x = 9$ or $x = -5$ j. $x = 3$ or $x = -4$

k. $k = -2$

9. a. The temperature at 5:00 P.M. was 25^0 and the change in temperature since noon was $\Delta t = 20^0$

   b. The temperature at 8:00 A.M. was -11^0 with a change of $\Delta t = 16^0$ from 8:00 A.M. until noon.

10. The change in miles using the mile markers was -88 miles. She had driven 88 miles.

11. a. $|a - 7|$ b. $|a - (-19)| = |a + 19|$

12. Xmin = -160, Xmax = 55, Xscl = 20, Ymin = -25, Ymax = 10, Yscl = 5.

13. Xmin = -12, Xmax = 15, Xscl = 3, Ymin = -150, Ymax = 105, Yscl = 15.

14. Xmin = -15, Xmax = 5, Xscl = 1, Ymin = -1097, Ymax = 343, Yscl = 100 would be one possible window. However, it would be a good practice to provide a larger viewing rectangle so the points at the ends of the graph are visible. The following choices would accomplish that: Xmin = -20, Xmax = 10, Xscl = 1, Ymin = -1300, Ymax = 350, Yscl = 100.

15. a. **Tutors and Students**

Tutors	Students
100	400
200	800
300	1200
400	1600
500	2000
600	2400

c.
```
WINDOW
Xmin=0
Xmax=700
Xscl=100
Ymin=-500
Ymax=3000
Yscl=500
Xres=1
```

1:L1,L2

X=600   Y=2400

d.
```
Plot1  Plot2  Plot3
\Y1=X
\Y2=
\Y3=
\Y4=
\Y5=
\Y6=
\Y7=
```

e. **More Tutors and Students**

Tutors	Students
223.40426	893.61702
327.65957	1310.6383
424.46809	1697.8723
625.53191	2502.1277

f. The ordered pairs found by using the trace feature do not make sense in the problem situation since the numbers must be whole numbers.

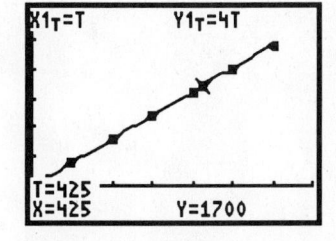

g.
```
X1T=T        Y1T=4T
```
T=425
X=425   Y=1700

Pairs That Make Sense

Tutors	Students
425	1700
450	1800
275	1100
150	600

h. The x-coordinates are the number of tutors and the y-coordinates are the corresponding number of students.

16. a. $-(4x - 7) = -4x + 7$ b. $-(2x^2 + 5x - 11) = -2x^2 - 5x + 11$

These equations are true because of the general principle that the operation of taking opposites distributes over addition and subtraction.

To check our answers numerically we enter the left side of one of the equations as Y1 and the right side of the equation as Y2 in our Y= menu. Then produce a table where the calculated values are produced from both expressions. The outputs should be the same for each input.

17. a. $5x + 9 = -(-5x - 9)$ b. $3t^2 - 5t + 7 = -(-3t^2 + 5t - 7)$

18. a. positive   b. negative

     c. positive   d. If $p = 3$ then $-p = -3$ and $|p| = 3$

19. a. negative   b. positive

     c. positive   d. If $n = -5$ then $-n = 5$ and $|n| = 5$.

20. a. x^{12} b. x^{35}

     c. $6x^{13}y^7$ d. -1

     e. This is simplified.   f. 0

     g. 1 h. c^{x+y}

     i. c^{xy} j. There is no property that can be used to simplify this expression.

21. a. $(5x^3 + 7x^2 - 8x + 2) - (5x - 4x^3 + 7 + x^2) =$

$5x^3 + 7x^2 - 8x + 2 - 5x + 4x^3 - 7 - x^2 = 9x^3 + 6x^2 - 13x - 5$

b. $35x^4 - 26x^3 - 42x^2 + 37x - 6$

c. $16x^2 - 9y^2$ d. $16x^2 - 24xy + 9y^2$

e. $x^3 - 12x^2 + 48x - 64$

CHAPTER 5

Rational Numbers: Further Expansion of a Mathematical System

Section 5.1 Rates of Change

p. 358

2. a. **Table 4 Car Rental Data**

Finite Difference in Input Δm	Miles Driven m	Total Cost C ($)	Finite Difference in Output ΔC	$\dfrac{\Delta C}{\Delta m}$
96	15	37.55	16.32	.17
−19	111	53.87	−3.23	.17
200	92	50.64	34	.17
−142	292	84.64	−24.14	.17
57	150	60.5	9.69	.17
−164	207	70.19	−27.88	.17
	43	42.31		

b. They are the same.

c. The ratio of the change in the outputs to the change in the inputs is the same even if the inputs do not change by a constant amount. This is because in our function for determining the cost when the number of miles is changed the cost is increased by 0.17 times that change or the cost per mile.

3. a. **Concert Receipts**

Number of tickets sold (n)	Total receipts in dollars (R)
200	3800
400	7600
600	11400
800	15200
1000	19000

b. $R(n) = 19n$

c. $slope = \dfrac{\Delta R}{\Delta n} = 19$.

d. The slope represents the rate of change of the receipts per each ticket sold. In this case, for each ticket sold, the total receipts increase by $19.

e. The finite differences of the outputs will not be constant but the ratio of the change in output to the change in input between any two input-output pairs will be 19.

4. All would be labeled significant barriers except DE, which is fully accessible.

5. a. Graph (a) Shows their distance they traveled at each time during the first trip. Graph (b) shows the speed they were traveling at each time during the second trip.

   b. In graph (a) point C shows that they were at the same distance from the starting point at the end of the trip. In graph (b) point C shows that at the time which corresponds to point C they were traveling at the same speed.

   c. The total distance each traveled is the second coordinate of the point C. They both traveled the same distance on this trip. Bev covered more of the total distance in the beginning of the trip than Al did.

   d. Bev was going faster than Al for the entire trip. So, Al's total distance would be less than Bev's.

   e. Bev continued to increase her speed during the entire trip. This is because the graph continues to increase as a point on the graph moves from left to right.

6. a. The second lot is longer. One uses an additive comparison between the lenghts to answer this question.

   b. The second lot has more area. After multiplying the length and the width of each lot to get the areas we use and additive comparison to tell which is bigger.

   c. The 80 feet by 55 feet lot would be closer to a square because $\dfrac{55}{80}$ is slightly closer to one than $\dfrac{75}{110}$. The closer the rectangle is to a square the closer the ratio of the length to the width would be to 1. Here we use a multiplicative comparison.

7. a. $\dfrac{\Delta(mileage)}{\Delta(speed)} = \dfrac{14-17}{60-45} = -\dfrac{1}{5}$. The average rate of change of gas mileage with respect to speed is $-\dfrac{1}{5}$. This means that for every five m.p.h. increase in speed, the gas mileage decreases by one mile per gallon.

b. At 50 m.p.h., the gas mileage is 16 miles per gallon. At 45 m.p.h., the mileage was 17 miles per gallon. By increasing the speed by five m.p.h., we must decrease the mileage by 1 mile per gallon.

At 70 m.p.h., the mileage is 12 miles per gallon. At 60 m.p.h., the mileage was 14 miles per gallon. By increasing the speed by 10 m.p.h., we must decrease the mileage by 2 miles per gallon.

c. A negative rate of change means that as the input increases the output decreases. In this problem, as the speed (input) increases, the gas mileage (output) decreases.

8. a. **Table 6 Change in Distance Fallen of the Stale Corned Beef Sandwich**

Change in time (seconds)	Change in distance fallen (feet)
1 to 2	48
2 to 3	80
3 to 4	112
4 to 5	144
5 to 6	176
6 to 7	208
7 to 8	240
8 to 9	272
9 to 10	304

b. The rate of change in distance fallen (height) with respect to time is varying. As time changes by one second, the distance fallen each second increases by 32 feet.

c. The average rate of change in distance fallen with respect to time represents the ratio of the total distance fallen to the total elapsed time.

d. $\frac{\Delta d}{\Delta t} = \frac{1600 - 16}{10 - 1} = \frac{1584}{8} = 176$ feet per second. The poor sandwich has fallen at an average rate of 176 feet per second. Here we actually used Table 5 instead of Table 6.

e. The sandwich moves the fastest over the time interval 9–10. In this interval, it has fallen 304 feet. The sandwich moves the slowest over the time interval 0–1. In this interval, it has fallen 16 feet.

9. Let d represent distance in miles and t represent time in hours.

   a. 10–12: $\dfrac{\Delta d}{\Delta t} = \dfrac{78334 - 78239}{12 - 10} = \dfrac{95}{2} = 47.5$ m.p.h.

   12–1:15: $\dfrac{\Delta d}{\Delta t} = \dfrac{78334 - 78334}{1\frac{1}{4}} = 0$ m.p.h.

   1:15–2:30: $\dfrac{\Delta d}{\Delta t} = \dfrac{78383 - 78334}{1\frac{1}{4}} = \dfrac{49}{\frac{5}{4}} = 49\left(\dfrac{4}{5}\right) = 39.2$ m.p.h.

   2:35–5:10: $\dfrac{\Delta d}{\Delta t} = \dfrac{78489 - 78383}{2\frac{7}{12}} = \dfrac{106}{\frac{29}{12}} = 106\left(\dfrac{12}{29}\right) \approx 44$ m.p.h.

   b. $\dfrac{\Delta d}{\Delta t} = \dfrac{78489 - 78239}{7\frac{1}{6}} = \dfrac{250}{\frac{43}{6}} = 50\left(\dfrac{6}{43}\right) \approx 34.9$ m.p.h. The average speed over the entire trip was about 35 m.p.h.

10. A is used as a label while a is used as a variable representing altitude.

11. a. $slope = \dfrac{4 - 11}{6 - 3} = -\dfrac{7}{3}$.

    b. $slope = \dfrac{-6 - 3}{4 - (-5)} = \dfrac{-9}{9} = -1$.

    c. $slope = \dfrac{-4 - (-8)}{-7 - 0} = -\dfrac{4}{7}$.

    d. $slope = \dfrac{-3 - (-3)}{15 - 2} = 0$.

    e. $slope = \dfrac{-3 - 2}{4 - 4} =$ undefined.

12. a. 3   b. –2

    c. 5   d. –4

13. a is the rate of change of the outputs to inputs.

14. a. $\dfrac{y(4)-y(1)}{4-1} = \dfrac{(2 \cdot 4 + 3)-(2 \cdot 1 + 3)}{3} = \dfrac{11-5}{3} = \dfrac{6}{3} = 2$ for $y(x) = 2x + 3$.

$\dfrac{y(4)-y(1)}{4-1} = \dfrac{(4^2 + 5)-(1^2 + 5)}{3} = \dfrac{21-6}{3} = \dfrac{15}{3} = 5$ for $y(x) = x^2 + 5$.

b. $\dfrac{y(5)-y(3)}{5-3} = \dfrac{(2 \cdot 5 + 3)-(2 \cdot 3 + 3)}{2} = \dfrac{13-9}{2} = \dfrac{4}{2} = 2$ for $y(x) = 2x + 3$.

$\dfrac{y(5)-y(3)}{5-3} = \dfrac{(5^2 + 5)-(3^2 + 5)}{2} = \dfrac{30-14}{2} = \dfrac{16}{2} = 8$ for $y(x) = x^2 + 5$.

c. $\dfrac{y(9)-y(4)}{9-4} = \dfrac{(2 \cdot 9 + 3)-(2 \cdot 4 + 3)}{5} = \dfrac{21-11}{5} = \dfrac{10}{5} = 2$ for $y(x) = 2x + 3$.

$\dfrac{y(9)-y(4)}{9-4} = \dfrac{(9^2 + 5)-(4^2 + 5)}{5} = \dfrac{86-21}{5} = \dfrac{65}{5} = 13$ for $y(x) = x^2 + 5$.

15. The linear function $y(x) = 2x + 3$ has a constant rate of change of outputs to inputs. In the linear function the variable is raised to the first power only. In the quadratic function, $y(x) = x^2 + 5$, the variable is squared.

16. a.

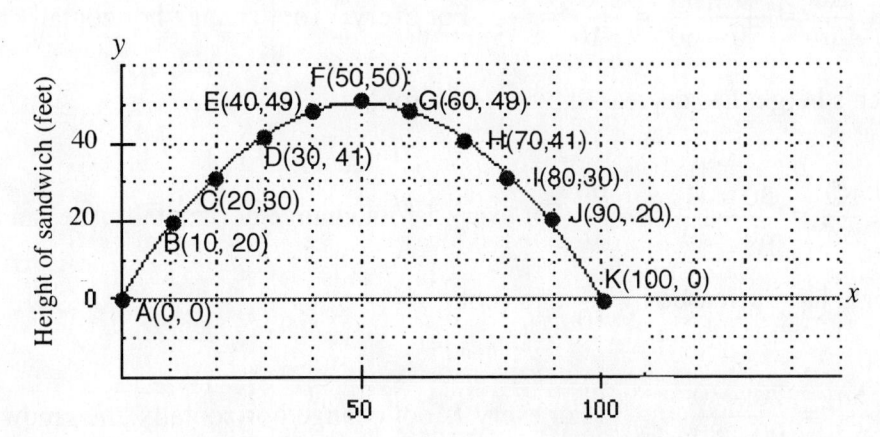

Horizontal distance traveled by sandwich (feet)

b. Let a represent altitude in feet and d represent horizontal distance traveled in feet.

AB: $\dfrac{\Delta a}{\Delta d} = \dfrac{20-0}{10-0} = \dfrac{2}{1} = 2$. For every 1 foot change horizontally, the sandwich's height has increased by 2 feet.

BC: $\dfrac{\Delta a}{\Delta d} = \dfrac{30-20}{20-10} = 1$. For every 1 foot change horizontally, the sandwich's height has increased by 1 foot.

CD: $\frac{\Delta a}{\Delta d} = \frac{41-30}{30-20} = \frac{9}{10}$. For every 1 foot change horizontally, the sandwich's height has increased by $\frac{9}{10}$ of a foot.

DE: $\frac{\Delta a}{\Delta d} = \frac{49-41}{40-30} = \frac{8}{10} = \frac{4}{5}$. For every 1 foot change horizontally, the sandwich's height has increased by $\frac{4}{5}$ of a foot.

EF: $\frac{\Delta a}{\Delta d} = \frac{50-49}{50-40} = \frac{1}{10}$. For every 1 foot change horizontally, the sandwich's height has increased by $\frac{1}{10}$ of a foot.

FG: $\frac{\Delta a}{\Delta d} = \frac{49-50}{60-50} = -\frac{1}{10}$. For every 1 foot change horizontally, the sandwich's height has decreased by $\frac{1}{10}$ of a foot.

GH: $\frac{\Delta a}{\Delta d} = \frac{41-49}{70-60} = -\frac{8}{10} = -\frac{4}{5}$. For every 1 foot change horizontally, the sandwich's height has decreased by $\frac{4}{5}$ of a foot.

HI: $\frac{\Delta a}{\Delta d} = \frac{30-41}{80-70} = -\frac{9}{10}$. For every 1 foot change horizontally, the sandwich's height has decreased by $\frac{9}{10}$ of a foot.

IJ: $\frac{\Delta a}{\Delta d} = \frac{20-30}{90-80} = -1$. For every 1 foot change horizontally, the sandwich's height has decreased by 1 foot.

JK: $\frac{\Delta a}{\Delta d} = \frac{0-20}{100-90} = -\frac{20}{10} = -2$. For every 1 foot change horizontally, the sandwich's height has decreased by 2 feet.

c. The sandwich is rising when the rate of change is positive and the sandwich is falling when the rate of change is negative.

d. The sandwich is at its highest point at about (50, 51).

e. The rate of change of altitude with respect to the horizontal distance traveled is about 0 near the highest point.

f. Before the sandwich reaches its highest point, the rates of change are positive. After the sandwich reaches its highest point, the rate of change are negative.

g. The sandwich is at its highest point and is, for an instant, neither gaining nor losing altitude.

17. The traveler whose motion is graphed started 1000 feet from some fixed reference point. In the first 3 hours he traveled another 1000 feet. From hour 3 to hour 4 he traveled 2000 feet and is 4000 feet from the fixed reference point. From point E to point F the traveler moved closer to the fixed reference point by 1000 feet in two hours. The motions between the other points are similar.

It is legitimate to use horizontal distance traveled as the label on the vertical axis.

18. a. After five years the profits for CDs-R-Us is at $60,000 while the profits for Super Dog Burger Bits is at $300,000.

   b. The rate of change of profits per year for CDS-R-Us is $20,000 per year computed using the following ratio: $\frac{60-20}{5} = \frac{40}{5} = 20$. The rate of change in profits per year for Super Dog Burger Bits is $40,000 per year computed using the ratio $\frac{300-100}{5} = \frac{200}{5} = 40$.

Section 5.2 Rational Numbers and Proportional Reasoning

p.380

2. a. $\frac{32}{20} = \frac{8}{5}$ apples per oranges b. $\frac{33}{51} = \frac{11}{17}$ men per women

   c. $\frac{12}{8} = \frac{3}{2}$ dogs per cats d. $\frac{190}{4} = \frac{95}{2}$ miles per hour

   e. $\frac{36}{8} = \frac{9}{2}$ pounds per square inch f. $\frac{16}{9}$ soda per ice tea; is reduced.

3. a. $\frac{1}{6}$ b. $-\frac{1}{5}$

4. No since there are an infinite number of rational numbers between every pair of rational numbers.

5. $\frac{a}{b}$ can be interpreted as a rational number, as division of a by b, as the answer to the division problem a divided by b, and as the ratio of a to b.

The reason that b cannot be zero in $\frac{a}{b}$ is that the product $b\left(\frac{a}{b}\right)$ must be a. But if b was zero this product would always be zero, since multiplying by zero yields zero.

6. a. **Table 5: More Reciprocal Exploration**

Number	Reciprocal
8	.125
7	.142857...
6	.1666666...
5	.2
4	.25
3	.333333...
2	.5

b. It gets closer to the reciprocal of 1 which is 1.

7. a. Cannot be simplified since the numerator and denominator have no common factor other than 1.

b. $\frac{3x}{2x} = \frac{3}{2}$

c. $\frac{(x+2)(x-3)}{(x-3)(x+4)} = \frac{x+2}{x+4}$

d. $\frac{2x(x+3)}{(x-3)4x} = \frac{2(x+3)}{4(x-3)} = \frac{x+3}{2(x-3)}$

8. a. All numbers except −2.   b. All numbers except 0.

c. All numbers except 3 and −4.   d. All numbers except 3 and −4.

e. All numbers except 3 and 0.

9. a. $\frac{F}{P} = \frac{2}{7}$

b. $\frac{2}{14}, \frac{6}{21}, \frac{8}{28}, \frac{10}{35}, \frac{12}{42}$

c. **Table 6: Full-time versus part-time students**

P	F
7	2
14	4
21	6
28	8
35	10

d. $F(P) = \frac{2}{7}P$

e.

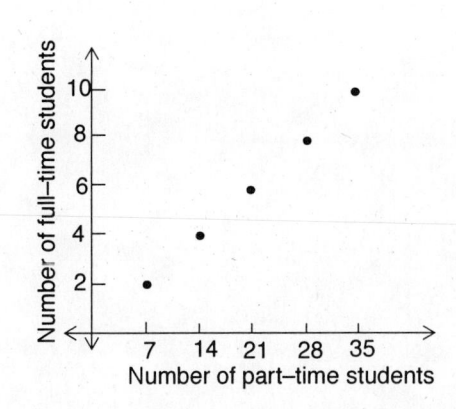

10. Tom is getting the flour at $0.75 per pound, found by changing $\frac{3}{4}$ to a decimal, while Sue is able to buy the flour at little more than $0.71 per pound, found by changing $\frac{5}{7}$ to a decimal. Sue would be getting a better deal.

11. a. If we let w stand for the number of women in the town and m stand for the number of men in the town then since two-thirds of the women are married to three-fifths of the men the equation $\frac{2}{3}w = \frac{3}{5}m$ must be satisfied. If we multiply both sides of this equation by $\frac{5}{3}$ we convert the equation into the form $m = \frac{5}{3} \cdot \frac{2}{3}w = \frac{10}{9}w$. The least number of people living in the town would be 9 women and 10 men because if we multiply $\frac{10}{9}$ by anything smaller than 9 we won't get a whole number.

b. The values of w that would produce whole numbers would be counting number multiples of 9. Some of the possible numbers of people living in the town are shown in the following table.

Possible Numbers of People in the Town

Women	Men
18	20
45	50
108	120

12. a. **Original price versus sale price**

Original price P	Sale Price S
10	7.30
20	14.60
30	21.90
40	29.20
50	36.50

b. $S(P) = P - 0.27P$ or $S(P) = 0.73P$

c.

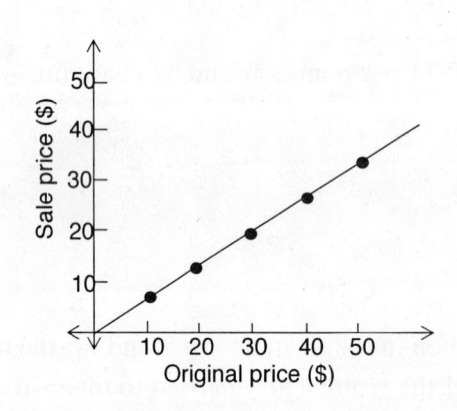

13. a. The Tasty & Juicy has 0.3125 ounces of concentrate per ounce of water while the Delicious Fruit Juice has 0.3 ounces of concentrate per ounce of water. The Tasty & Juicy is stronger.

b. It would take $11\frac{7}{13}$ ounces. Solve this algebraically by letting c represent the required amount of concentrate and w the necessary amount of water. We know that $\frac{c}{w} = \frac{3}{10}$

which can be rewritten to see that $w = \frac{10}{3}c$. And since the water and the concentrate together must add to 50 ounces w and c must satisfy $w + c = 50$. Replacing w with $\frac{10}{3}c$ gives us the equation $\frac{10}{3}c + c = 50$ to solve. This equation is equivalent to $\frac{13}{3}c = 50$ or $c = \frac{3}{13} \cdot 50 = \frac{150}{13} = 11\frac{7}{13}$.

    c. $11\frac{19}{21}$ ounces would be required.

14. a. **Price paid by store versus marked price**

Change in Price Paid by Store	Price Paid by Store ($)	Marked Price ($)	Change in Marked Price
5	10	14.5	7.25
5	15	21.75	7.25
	20	29	

    b. The rate of change of the marked price to the price the price the store paid is $\frac{7.25}{5} = 1.45$.

    c. If $M(p)$ is the marked price and p is the price the store paid then $M(p) = 1.45p$.

Section 5.3 Investigating Rational Number Operations

p. 400

2. a. $-\frac{3}{2}$ b. 0.25510204 approximately

    c. $\frac{11}{86}$ d. $\frac{1}{5x}$

    e. $-\frac{1}{3y}$ f. $\frac{1}{x+y}$

3. a. $5\left(\frac{1}{17}\right)$ b. $-3(11)$

c. $-2.3\left(\dfrac{6}{5}\right)$

d. $\dfrac{4}{7}\left(-\dfrac{1}{2}\right)$

e. $-9.2\left(-\dfrac{5}{11}\right)$

f. $x\left(\dfrac{1}{3y}\right)$

g. $5t(q)$

h. $\dfrac{3}{k}\left(\dfrac{c}{d}\right)$

i. $\dfrac{3x+4}{x-5} \cdot \dfrac{x+5}{9x}$

j. $\dfrac{y-2}{2x-6} \cdot \dfrac{4x+8}{2y+1}$

4. a. To add two rational numbers in fractional form, determine the common denominator and expand each fraction so all fractions have the same denominator. Then add the numerators and keep the denominator. Reduce if possible.

   b. To add two rational numbers in decimal form, align the decimal points and add digits with like place value.

5. a. The procedure is identical to that stated in the answer to 4a except that we find the difference, rather than the sum, of the numerators.

   b. The procedure is identical to that stated in the answer to 4b except that we subtract digits, rather than add, with like place value.

6. a. To multiply two rational numbers in fractional form, multiply the numerators to find the numerator of the product and multiply the denominators to find the denominator of the product. Reduce, if possible.

   b. To multiply two rational numbers in decimal form, multiply the numbers without the decimal points. The number of decimal places in the product equals the sum of the number of decimal places in the two factors.

7. a. $\left(\dfrac{7}{8}\right)\left(\dfrac{2}{3}\right) = \left(\dfrac{7}{4}\right)\left(\dfrac{1}{3}\right) = \dfrac{7}{12}$

   b. $\dfrac{7}{5} \div \left(-\dfrac{42}{17}\right) = \dfrac{7}{5}\left(-\dfrac{17}{42}\right) = \dfrac{1}{5}\left(-\dfrac{17}{6}\right) = -\dfrac{17}{30}$

   c. $-\dfrac{5}{8} + \dfrac{1}{12} = -\dfrac{15}{24} + \dfrac{2}{24} = -\dfrac{13}{24}$

   d. $\dfrac{9}{10} - \left(-\dfrac{3}{4}\right) = \dfrac{9}{10} + \dfrac{3}{4} = \dfrac{18}{20} + \dfrac{15}{20} = \dfrac{33}{20}$

   e. $\left(-\dfrac{2}{9}\right)\left(\dfrac{11}{5}\right) = -\dfrac{22}{45}$

f. $-7\frac{3}{4} - 4\frac{2}{3} = -\frac{31}{4} + \left(-\frac{14}{3}\right) = -\frac{93}{12} + \left(-\frac{56}{12}\right) = -\frac{149}{12} = -12\frac{5}{12}$

g. $\frac{5}{8} + \left(-\frac{7}{10}\right) = \frac{25}{40} + \left(-\frac{28}{40}\right) = -\frac{3}{40}$

h. $\left(-\frac{13}{12}\right) \div \left(-\frac{5}{9}\right) = -\frac{13}{12}\left(-\frac{9}{5}\right) = -\frac{13}{4}\left(-\frac{3}{5}\right) = \frac{39}{20} = 1\frac{19}{20}$

i. $\frac{2}{9} + \left(-\frac{7}{12}\right) = \frac{8}{36} + \left(-\frac{21}{36}\right) = -\frac{13}{36}$

j. $\frac{7}{9} - \left(\frac{5}{3}\right)\left(-\frac{3}{7}\right) = \frac{7}{9} - \left(\frac{5}{1}\right)\left(-\frac{1}{7}\right) = \frac{7}{9} - \left(-\frac{5}{7}\right) = \frac{7}{9} + \frac{5}{7} = \frac{49}{63} + \frac{45}{63} = \frac{94}{63} = 1\frac{31}{63}$

8. a. $\frac{2x+7}{x+5}$ b. $\frac{3y-11}{2y-1}$

c. $\frac{3x-2}{4x+7} + \frac{x+1}{4x+7} = \frac{4x-1}{4x+7}$ d. $\frac{2t+9}{3t-2}$

e. $\frac{8}{10x} + \frac{x^2}{10x} = \frac{8+x^2}{10x}$ f. $\frac{9x^2}{24x} - \frac{28}{24x} = \frac{9x^2-28}{24x}$

g. $\frac{3(x+3)}{(x+2)(x+3)} + \frac{4(x+2)}{(x+2)(x+3)} = \frac{3x+9+4x+8}{(x+2)(x+3)} = \frac{7x+17}{(x+2)(x+3)}$

h. $\frac{(3x-1)(x-4)}{(2x+5)(x-4)} - \frac{(2x+3)(2x+5)}{(2x+5)(x-4)} = \frac{(3x^2-13x+4)-(4x^2+16x+15)}{(2x+5)(x-4)} =$

$\frac{3x^2-13x+4-4x^2-16x-15}{(2x+5)(x-4)} = \frac{-x^2-29x-11}{(2x+5)(x-4)}$

9. a. $x = \frac{22}{4} = \frac{11}{2}$ b. $y = -\frac{13}{3}$

c. $x = -\frac{22}{6} = -\frac{11}{3}$ d. $y = \frac{4.8-7.63}{3.41} \approx -0.8299$

e. $5x = 3 - \frac{2}{7} = \frac{19}{7}$
$x = \frac{19}{7} \div 5 = \frac{19}{7} \cdot \frac{1}{5} = \frac{19}{35}$

f. $\frac{3}{2}x = 4 + \frac{2}{5} = \frac{22}{5}$
$x = \frac{2}{3} \cdot \frac{22}{5} = \frac{44}{15}$

g. $\frac{5}{6}x = \frac{2}{3} + \frac{7}{9} = \frac{13}{9}$

$x = \frac{6}{5} \cdot \frac{13}{9} = \frac{2}{5} \cdot \frac{13}{3} = \frac{26}{15}$

h. $x = -\frac{5}{6} - \frac{7}{4} = -\frac{31}{12}$

10. a. $\left(\frac{1}{4} + \frac{1}{3}\right) \div 2 = \left(\frac{3}{12} + \frac{4}{12}\right) \div 2 = \frac{7}{12} \div 2 = \frac{7}{12}\left(\frac{1}{2}\right) = \frac{7}{24}$

   b. $\left(\frac{1}{8} + \frac{1}{7}\right) \div 2 = \left(\frac{7}{56} + \frac{8}{56}\right) \div 2 = \frac{15}{56} \div 2 = \frac{15}{56}\left(\frac{1}{2}\right) = \frac{15}{112}$

   c. There is a rational number between any two rational numbers because the average of the two numbers is a rational number between the two given rational numbers.

   d. There is no integer between two consecutive integers.

11. Let $T(r) = 5r - 2$.

   a. $T\left(\frac{3}{5}\right) = 5\left(\frac{3}{5}\right) - 2 = 3 - 2 = 1$. Here we are evaluating the function.

   b. We are solving the equation $5r - 2 = 16$ to get $r = \frac{18}{5}$.

12. If $P(x) = 3x^2 - x - 4$ then

   $P\left(-\frac{2}{7}\right) = 3\left(-\frac{2}{7}\right)^2 - \left(-\frac{2}{7}\right) - 4 = 3\left(\frac{4}{49}\right) + \frac{2}{7} - 4 = \frac{12}{49} + \frac{2}{7} - 4 =$

   $\frac{12}{49} + \frac{14}{49} - \frac{196}{49} = \frac{26}{49} - \frac{196}{49} = -\frac{170}{49}$. We are evaluating the function.

13. a. **Table 4: More Reciprocal Exploration**

Number	Reciprocal
8	$\frac{1}{8} = 0.125$
7	$\frac{1}{7} \approx 0.143$

Number	Reciprocal
6	$\frac{1}{6} \approx 0.167$
5	$\frac{1}{5} = 0.2$
4	$\frac{1}{4} = 0.25$
3	$\frac{1}{3} \approx 0.333$
2	$\frac{1}{2} = 0.5$

b. The value of the reciprocal is increasing toward 1 as the number decreases toward 1.

14. The width must be $\frac{2}{3}$ of the length so if w w stands for the width and l stands for the length the equation $w = \frac{2}{3}l$ must be satisfied. For our width of $23\frac{7}{8}$ inches we must solve the equation $23\frac{7}{8} = \frac{2}{3}l$. Multiply both sides by $\frac{3}{2}$ to get $l = \frac{3}{2} \cdot 23\frac{7}{8} = \frac{3}{2} \cdot \frac{191}{8} = \frac{573}{16} = 35\frac{13}{16}$ inches.

15. $7\frac{3}{4}$ miles is $3\frac{1}{2}$ times as far as Bill can walk in the same amount of time. To find how far Bill walked we divide $7\frac{3}{4}$ miles by $3\frac{1}{2}$. $\left(7\frac{3}{4}\right) \div \left(3\frac{1}{2}\right) = \frac{31}{4} \div \frac{7}{2} = \frac{31}{4} \cdot \frac{2}{7} = \frac{31}{2} \cdot \frac{1}{7} = \frac{31}{14}$ So, Bill walked $2\frac{3}{14}$ miles.

16. a. The integers are closed under the operations of addition, subtraction, and multiplication.

    b. The rational numbers are closed under the operations of addition, subtraction, multiplication, and division.

Section 5.4 Reciprocal and Power Functions

p. 416

2. a. For example, the problems that dealt with the triangular numbers, car rentals, students and professors, concert tickets, and the Bulls T-shirts all contained functions which were discussed.

b. See previous sections in text dependent on the functions you chose.

3. $Power(x) = x^n$ where n is odd.

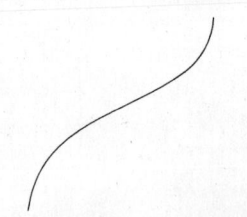

4. $Power(x) = x^n$ where n is even.

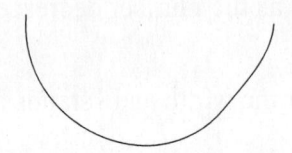

5. a. The amount you must study as a function of the number of credit hours you take. The input is the number of credit hours and the output is the amount of study time.

   b. The amount of free time that you have as a function of the number of credit hours you take. The input is the number of credit hours and the output is the amount of free time.

   c. The area of a cross section of a tree trunk is a function of the diameter of the tree trunk. The input is the diameter and the output is the area of the cross section.

   d. The average daily temperature in Chicago as a function of the date. The input is the date and the output is the average daily temperature.

   e. The cost of mailing a letter as a function of the letter's weight. The input is the weight of the letter and the output is the postal charge.

6. This problem is not possible since it will take 4 hours to go 120 miles if the average speed is 30 m.p.h. The total trip cannot then be made in 4 hours.

7. a. x^3 b. x^4

   c. x^5 d. x^5

8. x^{m+n}.

9. a. x^3 b. x^6

   c. x^8 d. x^6

10. x^{mn}

11. a. x b. x^2

 c. x^3 d. x

 e. x^2 f. x^3

12. x^{m-n}

 Property: If $m > n$ then $\dfrac{x^m}{x^n} = x^{m-n}$.

13. a. **Table 3: Width versus length of a rectangle with area 10 square feet.**

Width	Length	Area
1	10	10
2	5	10
4	2.5	10
5	2	10
10	1	10
w	$\dfrac{10}{w}$	$w \cdot \dfrac{10}{w} = 10$

 b. Find the length by dividing the area, 10, by the width.

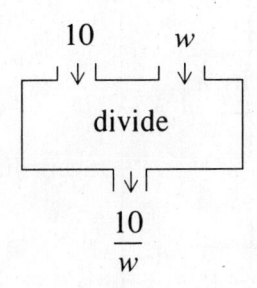

 c.

 $$10 \quad w \to \text{divide} \to \dfrac{10}{w}$$

 d. $l(w) = \dfrac{10}{w}$

e.

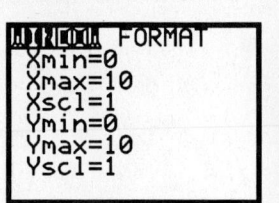

```
WINDOW FORMAT
Xmin=0
Xmax=10
Xscl=1
Ymin=0
Ymax=10
Yscl=1
```

f. Some examples appear in the table below.

Width	Length
5.106383	1.958333
1.3829787	7.2307692
7.3404255	1.3623188

g. If the length is 3.8 feet then the width is about 2.63 feet.

14. a. $x = \dfrac{18}{5}$ b. $y = \dfrac{28}{3}$

c. $\dfrac{4}{5}t = \dfrac{2}{3} + 3 = \dfrac{11}{3}$

$t = \dfrac{5}{4} \cdot \dfrac{11}{3} = \dfrac{55}{12}$

d. $y = -\dfrac{1}{9}$

15. a.

x

Multiply by a

ax

Add b

$ax+b$

$ax + b = \text{lit}(x)$

lit

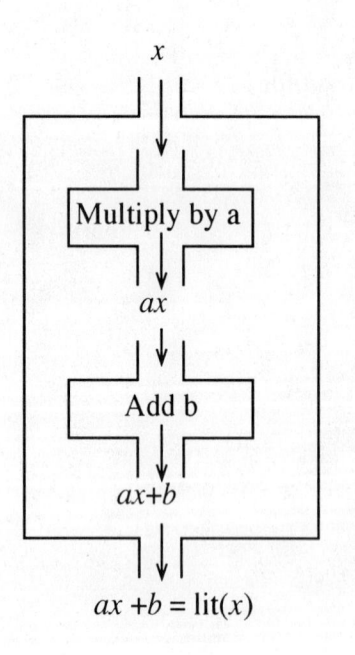

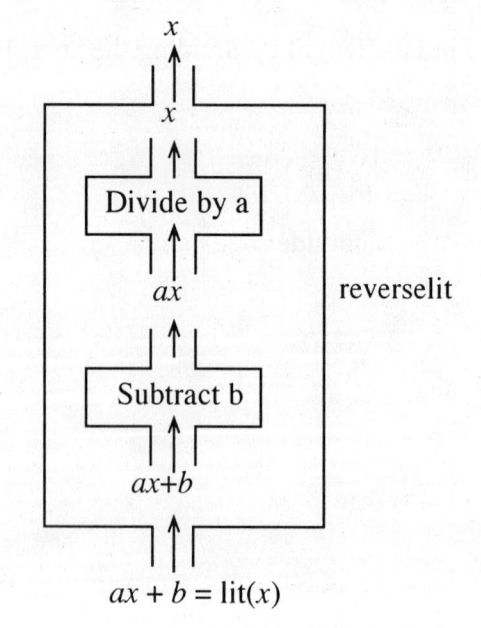

b.

x

x

Divide by a

ax

Subtract b

$ax+b$

$ax + b = \text{lit}(x)$

reverselit

16. a. $w = \dfrac{A}{l}$
    b. $2l = P - 2w$
       $l = \dfrac{P - 2w}{2}$
    c. $\dfrac{9}{5}C = F - 32$
       $C = \dfrac{5}{9}(F - 32)$

Section 5.5 Integer Exponents

p. 427

2. a. $8^{-2} = \dfrac{1}{8^2} = \dfrac{1}{64}$
   b. $7^{-1} = \dfrac{1}{7^1} = \dfrac{1}{7}$
   c. $k^{-5} = \dfrac{1}{k^5}$

3. a. $\dfrac{k^7}{k^2} = k^{7-2} = k^5$
   b. $\dfrac{t^4}{t^{11}} = t^{4-11} = t^{-7} = \dfrac{1}{t^7}$
   c. $\dfrac{z^5}{z} = z^{5-1} = z^4$
   d. $\dfrac{x}{x^3} = x^{1-3} = x^{-2} = \dfrac{1}{x^2}$
   e. $\dfrac{x^4}{y^2}$ has different bases
   f. $\dfrac{y^{17}}{y^{17}} = y^{17-17} = y^0 = 1$

4. a. $(a^4 b)^3 = (a^4)^3 (b)^3 = a^{12} b^3$
   b. $\left(\dfrac{x^3}{y^7}\right)^2 = \dfrac{(x^3)^2}{(y^7)^2} = \dfrac{x^6}{y^{14}}$

5. a. $(-3x^4 y^2)(2xy^5) = -6x^5 y^7$
   b. $(2a^5 b^4)^3 (ab^5)^2 = 2^3 a^{15} b^{12} a^2 b^{10} = 8a^{17} b^{22}$
   c. $(-x^2)^3 = (-1)^3 x^6 = -x^6$
   d. $(-x^3 y^5)(xy^7)(x^4 y) = -x^8 y^{13}$
   e. $(xy^{-5})^{-3} = x^{-3} y^{15} = \dfrac{1}{x^3}(y^{15}) = \dfrac{y^{15}}{x^3}$

f. $-5^2 = -(5)(5) = -25$ g. $(-5)^2 = (-5)(-5) = 25$

h. $-5^{-2} = -\dfrac{1}{5^2} = -\dfrac{1}{25}$ i. $(-5)^{-2} = \dfrac{1}{(-5)^2} = \dfrac{1}{25}$

j. $(7^{-1})^{-1} = 7^{(-1)(-1)} = 7^1 = 7$

k. $-(-11)^0 = -1$ l. $(3^{-4})^3 = 3^{-12} = \dfrac{1}{3^{12}}$

m. $(x^2)^{-3} = x^{-6} = \dfrac{1}{x^6}$

n. $\dfrac{a^{-4}b^3}{a^{-7}b^4} = a^{-4-(-7)}b^{3-4} = a^{-4+7}b^{-1} = a^3 b^{-1} = a^3\left(\dfrac{1}{b}\right) = \dfrac{a^3}{b}$

o. $(a^4 b^{-7})(a^{-3} b^{-2}) = a^{4+(-3)} b^{-7+(-2)} = a b^{-9} = a\left(\dfrac{1}{b^9}\right) = \dfrac{a}{b^9}$

6. -3^2: square 3 and take the opposite of the answer. Result: -9.

   $(-3)^2$: square -3. Result: 9.

7. Again, -3^2 means to first square 3 and then take the opposite of the answer. Result: -9. 3^{-2}, however, means to take the reciprocal of 3^2. So, first we convert to a positive exponent $3^{-2} = \dfrac{1}{3^2}$, then we simplify the 3^2 in the denominator. Result: $\dfrac{1}{9}$.

8. A negative exponent directs us to take the reciprocal of the base and raise the result to the opposite of the original exponent. For example, in 3^{-2} above the reciprocal of the base is $\dfrac{1}{3}$. The opposite of the original exponent would be 2 to get $\left(\dfrac{1}{3}\right)^2 = \dfrac{1}{3^2}$ or $\dfrac{1}{9}$.

9. a. The answers in the table are 32, 16, 8, 4, 2, 1, $\dfrac{1}{2}$, $\dfrac{1}{4}$.

   b. We divide the previous answer by 2 to get the next answer.

   c. The definitions of zero and negative exponents allow the pattern to hold for integer exponents.

Section 5.6 Review

p. 431

1. a. $\dfrac{1}{7}$ b. $-\dfrac{1}{18}$

    c. $-\dfrac{9}{5}$ d. $\dfrac{10}{87} \approx 0.11494\ldots$

    e. $\dfrac{3}{14}$ f. $-\dfrac{1}{8t}$

    g. $\dfrac{1}{2x-3}$ h. $2y+7$

    i. $\dfrac{4x-1}{3x+2}$ j. $\dfrac{-5(Y-2)}{-x+1}$

2. a. $14 \cdot \dfrac{1}{3}$ b. $-8 \cdot \dfrac{5}{2}$

    c. $-5.8 \cdot \dfrac{4}{3}$ d. $\dfrac{4}{5} \cdot -\dfrac{1}{8}$

    e. $-6.4 \cdot \dfrac{2}{7}$ f. $4a \cdot \dfrac{1}{5b}$

    g. $\dfrac{2x}{x-3} \cdot \dfrac{1}{6x}$ h. $\dfrac{3x+12}{4x-1} \cdot \dfrac{x-1}{x+4}$

3. a. $\left(-\dfrac{5}{1}\right)\left(-\dfrac{2}{13}\right) = \dfrac{10}{13}$ b. $\dfrac{29}{3} - \dfrac{19}{5} = \dfrac{145}{15} - \dfrac{57}{15} = \dfrac{88}{15} = 5\dfrac{13}{15}$

    c. $-\dfrac{35}{45} + \left(-\dfrac{12}{45}\right) = -\dfrac{47}{45} = -1\dfrac{2}{45}$ d. $-\dfrac{11}{15}\left(-\dfrac{9}{10}\right) = -\dfrac{11}{5}\left(-\dfrac{3}{10}\right) = \dfrac{33}{50}$

    e. $\dfrac{8}{12x} - \dfrac{15}{12x} = -\dfrac{7}{12x}$ f. $\dfrac{3(y+2)+5(y-3)}{(y-1)(y-3)(y+2)} = \dfrac{8y-9}{(y-1)(y-3)(y+2)}$

    g. $\dfrac{2x}{x-3} \cdot \dfrac{1}{6x} = \dfrac{1}{x-3} \cdot \dfrac{1}{3} = \dfrac{1}{3x-9}$ h. $\dfrac{3(x+4)}{4x-1} \cdot \dfrac{x-1}{x+4} = \dfrac{3}{4x-1} \cdot \dfrac{x-1}{1} = \dfrac{3x-3}{4x-1}$

4. a.

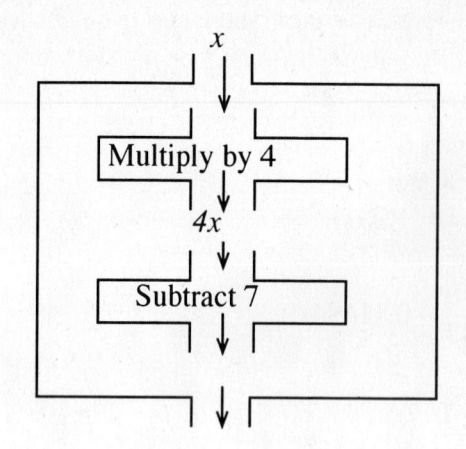

```
        x
        ↓
   ┌─────────────┐
   │ Multiply by 4│
   └─────────────┘
        ↓
        4x
        ↓
   ┌─────────────┐
   │  Subtract 7 │
   └─────────────┘
        ↓
```

b. $y = 4 \cdot \dfrac{3}{4} - 7 = 3 - 7 = -4$ We are evaluating the function.

$$4x - 7 = -\dfrac{2}{3}$$

c. $4x = -\dfrac{2}{3} + 7 = \dfrac{19}{3}$ We are solving the first equation shown here.

$x = \dfrac{1}{4} \cdot \dfrac{19}{3} = \dfrac{19}{12} = 1\dfrac{7}{12}$

d.
$4x - 7 = 5.47$
$4x = 12.47$ Solving an equation again.
$x = \dfrac{12.47}{4} = 3.1175$

e. $y = 4\left(-2\dfrac{6}{7}\right) - 7 = 4\left(-\dfrac{20}{7}\right) - 7 = -\dfrac{80}{7} - \dfrac{49}{7} = -\dfrac{129}{7} = -18\dfrac{3}{7}$ Evaluating.

5. a. **Original and Sale Price**

Original Price (P)	Sale Price (S)
10	6
15	9
20	12
25	15
30	18

b. $S(P) = 0.6P$

166

c. The graph as produced on the TI-83 is shown. The parametric mode has been chosen to display the graph because it allows us to view individual points easily using the trace feature. One of the points on the graph is displayed along with the equations entered into the Y= menu.

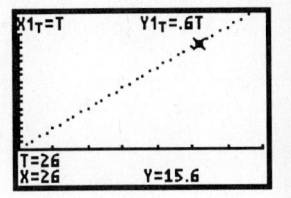

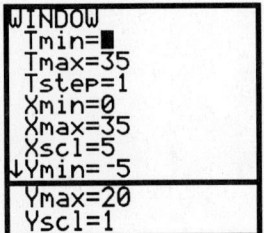

6. Let d represent distance in miles and t represent time in hours.

   a. 10–12: $\dfrac{\Delta d}{\Delta t} = \dfrac{78334 - 78239}{12 - 10} = \dfrac{95}{2} = 47.5$ m.p.h.

   12–1:15: $\dfrac{\Delta d}{\Delta t} = \dfrac{78334 - 78334}{1\frac{1}{4}} = 0$ m.p.h.

   1:15–2:30: $\dfrac{\Delta d}{\Delta t} = \dfrac{78383 - 78334}{1\frac{1}{4}} = \dfrac{49}{\frac{5}{4}} = 49\left(\dfrac{4}{5}\right) = 39.2$ m.p.h.

   2:35–5:10: $\dfrac{\Delta d}{\Delta t} = \dfrac{78489 - 78383}{2\frac{7}{12}} = \dfrac{106}{\frac{29}{12}} = 106\left(\dfrac{12}{29}\right) \approx 44$ m.p.h.

   b. $\dfrac{\Delta d}{\Delta t} = \dfrac{78489 - 78239}{7\frac{1}{6}} = \dfrac{250}{\frac{43}{6}} = 50\left(\dfrac{6}{43}\right) \approx 34.9$ m.p.h. The average speed over the entire trip was about 35 m.p.h.

7. a. slope $= \dfrac{4 - (-1)}{2 - 5} = \dfrac{5}{-3} = -\dfrac{5}{3}$

   b. slope $= \dfrac{-6 - (-9)}{11 - (-4)} = \dfrac{3}{15} = \dfrac{1}{5}$

   c. Since the change in the first coordinates is zero the line determined by these two points is not defined. The line is vertical. In general, vertical lines do not have a slope.

8. a. −1   b. 9

c. $\frac{2}{3}$ d. $-\frac{6}{5}$

9.   a. k^{99} b. 1

  c. -1 d. $\frac{1}{p^{23}}$

  e. a^7 f. $\frac{1}{b^{16}}$

  g. Can't be simplified because the bases are different.

  h. $\frac{1}{(x^4y^7)^2} = \frac{1}{x^8y^{14}}$ i. $\left(\frac{b^3}{a}\right)^4 = \frac{b^{12}}{a^4}$

10.   a. $(-a^{20}b^{35})(a^{14}b^6) = -a^{34}b^{41}$ b. $\frac{1}{x^7y^5}$

  c. x^2y^3 d. $(5^{-2}x^2y^{-6})(3^4x^{10}y^2) = \frac{81x^{12}}{25y^4}$

  e. $(x^2y^3)^3 = x^6y^9$ f. $\frac{1}{x+y}$

  g. $\frac{1}{x} + \frac{1}{y}$

11. You could compare the strength of each pot of coffee by doing a multiplicative comparison. Compare the ratio of the number of spoons of coffee to the amount of water in each of the pots. The one with the largest coffee to water ratio is the strongest.

12. The ratios to consider are $\frac{1.19}{12}$ or about $0.099 per ounce and $\frac{1.49}{16}$ which is about $0.093 per ounce. The soup in the 16 ounce can is cheaper per ounce. Other factors which might influence your decision are the amount of soup in the can, you may not want 16 ounces, or the kind of soup.

13. The gas mileage of the basic car is $\frac{275}{12}$ or $22\frac{11}{12}$ miles per gallon. The gas mileage for the deluxe version is $\frac{371}{18}$ or $20\frac{11}{18}$ miles per gallon. As expected the basic version gets better mileage.

CHAPTER 6

Real Numbers: Completing a Mathematical System

Section 6.1 Real Numbers and the Algebraic Extension

p. 446

2. a. The whole numbers are 3, 5, and 0.

　　b. The integers are $3, -7, 5, -\sqrt{9}, -\frac{15}{3}, 0$.

　　c. The rational numbers are $3, -7, 5, -\sqrt{9}, -\frac{15}{3}, 0, -\frac{2}{3}, \frac{3}{5}, \frac{13}{7}, -\frac{6}{11}, \sqrt{\frac{9}{4}}, \sqrt{0.25}, 2^{-3}, 1.3^2$, $|2.3|, |-3.1|, -|4.6|, -|-3.7|$.

　　d. The irrational numbers are $\sqrt{7}, -\sqrt{2}, \sqrt{\frac{2}{3}}, \sqrt{5.72}$.

　　e. All the numbers in the collection are real numbers.

3. a. A: $3 = 3$; B: $|2.3| = 2.3$; C: $-7 = -7$; D: $-\frac{2}{3} \approx -0.6667$; E: $\frac{3}{5} = 0.6$

　　　　F: $\sqrt{7} \approx 2.6458$; G: $|-3.1| = 3.1$; H: $5 = 5$; I: $-\sqrt{2} \approx -1.414$; J: $\frac{13}{7} \approx 1.8571$;

　　　　K: $-\frac{6}{11} \approx -0.5454$; L: $\sqrt{\frac{2}{3}} \approx 0.8165$; M: $-\sqrt{9} = -3$; N: $\sqrt{5.72} \approx 2.3917$;

　　　　O: $-|4.6| = -4.6$; P: $-\frac{15}{3} = -5$ Q: $0 = 0$; R: $\sqrt{\frac{9}{4}} = 1.5$; S: $\sqrt{0.25} = 0.5$;

　　　　T: $2^{-3} = 0.125$; U: $1.3^2 = 1.69$; V: $-|-3.7| = -3.7$.

4.

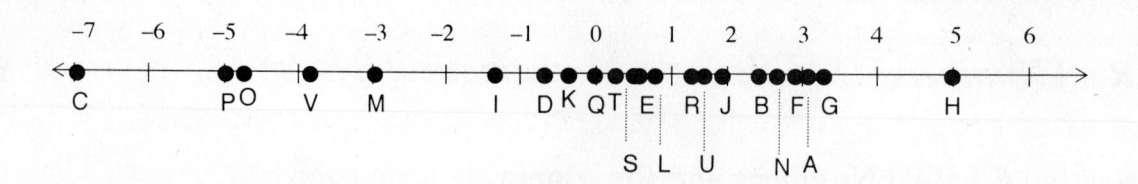

```
     -7  -6  -5  -4  -3  -2  -1   0   1   2   3   4   5   6
  <--•---+--••-•+---•--+---+--•+••••+•••+••••+---+--•+---+-->
     C     PO  V   M      I  DKQT E  R J B F G     H
                                  S L U    N A
```

5. C, P, O, Y, M, I, D, K, Q, T, S, E, L, R, U, J, B, N, F, A, G, H. Note: These letters are the labels used in numbers 3 and 4, above.

6. a. The whole numbers are closed with respect to addition and multiplication.

   b. The integers are closed with respect to addition, subtraction, multiplication, and opposing.

   c. The rational numbers are closed with respect to addition, subtraction, multiplication, division, opposing, and reciprocaling.

   d. The real numbers are closed with respect to addition, subtraction, multiplication, division, opposing and reciprocaling.

7. For all parts, all numbers displayed are rational numbers. There are an infinite number of real numbers between any two real numbers though the calculator can only display a finite number of them.

8. Some common numbers used to estimate π are 3.14 and $\frac{22}{7}$.

   a. Comparing the approximations to the first two hundred digits of π printed in the section, the rational number $\frac{355}{113}$ is the best approximation of the ones mentioned. Both the approximation 3.14 and $\frac{22}{7} \approx 3.142857$ are correct to two decimal places. But $\frac{355}{113} \approx 3.14159292$ is correct to six decimal places.

   b. The more decimal places in the approximation that match the actual decimal value, the better the approximation.

   c. The number π is an irrational number. Its decimal representation is non–terminating, non–repeating. Thus we can never write the complete decimal value. The only way to represent π exactly is to write π.

d. Approximations of π are usually rational as we usually use a fraction or a terminating decimal to approximate π. These are rational numbers.

9. a. $p = -\frac{11}{5}$ Rational numbers. b. $t = 0$ Whole numbers.

   c. $y = 1$ Whole numbers. d. $x = 1$ Whole numbers.

10. a. $(2x^{-5})(3^2 x^2) = 18x^{-3} = \frac{18}{x^3}$

    b. $(4x^{-1}y^4)^3(3xy^{-4})^{-2} = (4^3)x^{-3}y^{12}3^{-2}x^{-2}y^8 = \frac{64y^{20}}{9x^5}$

    c. $\frac{3y^2}{4x^5}$ d. $\frac{a^2}{2b}$

Section 6.2 The Square Root Function

p. 457

2. a. $\sqrt{7} = 2.6457513$ b. $-\sqrt{12} = -3.4641016$

   c. $\sqrt{81} = 9$ d. $-\sqrt{49} = -7$

   e. $\sqrt{-49}$ undefined f. $2\sqrt{11} = 6.6332496$

   g. $-3\sqrt{5} = -6.7082039$ h. $4\sqrt{9} = 12$

   i. $3\sqrt{-2}$ undefined j. $-7\sqrt{121} = -77$

3. Function machine for $3\sqrt{-2}$.

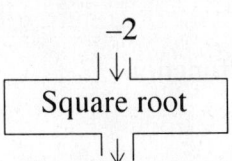

   −2
   ↓
   ┌──────────────┐
   │ Square root │
   └──────────────┘
   ↓
   No output since −2 is not in the domain of the function.

Function machine for $-7\sqrt{121}$.

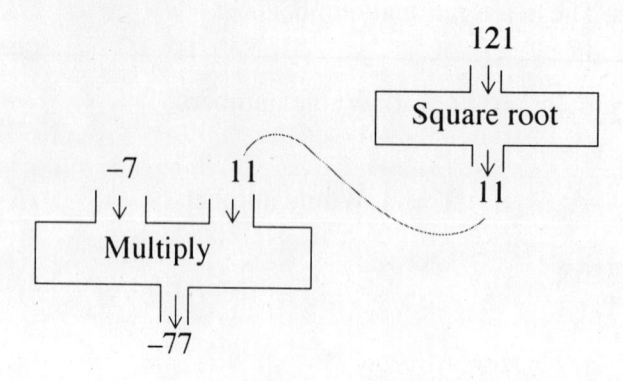

```
          121
           ↓
      ┌─────────┐
      │Square root│
      └─────────┘
           ↓
  -7   11   11
   ↓    ↓    ↓
  ┌──────────┐
  │ Multiply │
  └──────────┘
       ↓
      -77
```

4. a. **Table 2: Absolute Value Versus the Square Root of a Square**

x	$abs(x)$	$\sqrt{x^2}$
-7	7	7
-5	5	5
-3	3	3
-1	1	1
0	0	0
2	2	2
4	4	4
6	6	6

b. and c.

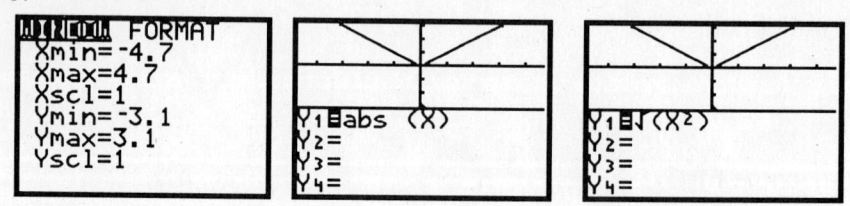

```
WINDOW FORMAT
Xmin=-4.7
Xmax=4.7
Xscl=1
Ymin=-3.1
Ymax=3.1
Yscl=1
```

Y1=abs (X)

Y1=√(X²)

d. The function $abs(x)$ and the function $\sqrt{x^2}$ are the same function.

5. The absolute value function.

6. a. **Table 3: Rate of Change of the Squaring Function**

x	x^2	$\Delta(x^2)$
1	1	3
2	4	5
3	9	7
4	16	9
5	25	

b. **Table 4: Rate of Change of the Square Root Function**

x	$\sqrt{x}$	$\Delta(\sqrt{x})$
1	1	.41421
2	1.4142	.31784
3	1.7321	.26795
4	2	.23607
5	2.23607	

c. The outputs are both increasing since the finite differences are positive.

d. The finite differences in part a are increasing at a constant rate meaning the outputs are increasing faster and faster. The finite differences in part b are decreasing meaning the outputs are increasing a slower and slower.

7. a. Comparing square roots with raising a number to the one-half power

x	$\sqrt{x}$	$x^{1/2}$
1	1	1
2	1.4142	1.4142
3	1.7321	1.7321
4	2	2
5	2.23607	2.23607
6	2.44949	2.44949
7	2.64575	2.64575
8	2.828427	2.828427

b. and c.

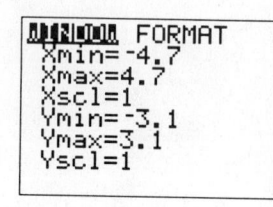

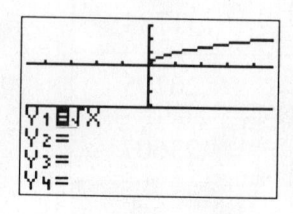

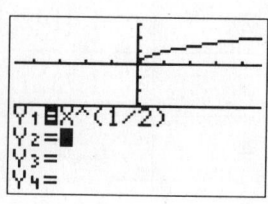

d. They are the same.

e. $(x^{1/2})^2 = (x^{1/2})(x^{1/2}) = x^{\frac{1}{2}+\frac{1}{2}} = x^1 = x$. So $x^{1/2}$ is the square root of x.

So $x^{1/2} = \sqrt{x}$.

8. a. The answers are getting closer to 1.

b. These also approach 1 as the number of repetitions is increased.

9. a. $3^{1/2} \approx 1.73205$ b. $16^{1/2} = 4$

c. $100^{1/2} = 10$ d. $17^{1/2} \approx 4.12311$

e. $-5^{1/2} \approx -2.23607$ f. $-49^{1/2} = -7$

g. $(-4)^{1/2}$ undefined

10. a. $(3, 8)$ and $(5, 2)$. $\Delta x = 5 - 3 = 2$, $\Delta y = 2 - 8 = -6$

$$distance = \sqrt{(2)^2 + (-6)^2} = \sqrt{4 + 36} = \sqrt{40} \approx 6.32455$$

b. $(-6, 3)$ and $(4, -9)$. $\Delta x = 4 - (-6) = 10$, $\Delta y = -9 - 3 = -12$

$$distance = \sqrt{(10)^2 + (-12)^2} = \sqrt{100 + 144} = \sqrt{244} \approx 15.6205$$

c. $(-3, -5)$ and $(-7, 2)$. $\Delta x = -7 - (-3) = -4$, $\Delta y = 2 - (-5) = 7$

$$distance = \sqrt{(-4)^2 + (7)^2} = \sqrt{16 + 49} = \sqrt{65} \approx 8.06226$$

d. $(8, -5)$ and $(2, -5)$. $\Delta x = 2 - 8 = -6$, $\Delta y = -5 - (-5) = 0$

$$distance = \sqrt{(-6)^2 + (0)^2} = \sqrt{36 + 0} = \sqrt{36} = 6$$

e. $(4, 3)$ and $(4, -6)$. $\Delta x = 4 - (4) = 0$, $\Delta y = -6 - 3 = -9$

$$distance = \sqrt{(0)^2 + (-9)^2} = \sqrt{0 + 81} = \sqrt{81} = 9$$

11. a. Let $d = 30$, then $t(30) = \sqrt{\dfrac{30}{16}} \approx 1.4$. It will take about 1.4 seconds to fall 30 feet.

b. One mile equals 5280 feet. So $d = 5280$. Then $t(5280) = \sqrt{\dfrac{5280}{16}} \approx 18$. It will take about 18 seconds to fall one mile.

c. distance fallen $= 9500 - 3000 = 6500$. So $d = 6500$. Then $t(6500) = \sqrt{\dfrac{6500}{16}} \approx 20$. The skydiver will free fall for about 20 seconds.

12. $S(d) = \sqrt{30fd}$. In this problem, $f = 0.83$. So $S(d) = \sqrt{30(0.83)d}$.

   a. Let $d = 100$. Then $S(100) = \sqrt{30(0.83)(100)} \approx 50$. The minimum speed of the car was about 50 m.p.h. if the skid marks are 100 feet long.

   b. Let $d = 60$. Then $S(60) = \sqrt{30(0.83)(60)} \approx 38.7$. The drive was going about 39 m.p.h. so a ticket is in order.

   c. Use trial and error to find d if $S(d) = 60 \Rightarrow \sqrt{30(0.83)d} = 60$. The length of the skid marks is about 145 feet.

13. a. $9y^2 - 49z^2$　　　　　　　　　b. $9y^2 + 42yz + 49z^2$

    c. $-14z$　　　　　　　　　　　　d. $x^3 + 1$

14. a. $\dfrac{6a}{3(a+3b)} = \dfrac{2a}{a+3b}$　　　b. $\dfrac{2(2x-4)}{2(3x-6)} = \dfrac{2x-4}{3x-6}$

    c. $\dfrac{x+3}{7(x-3)}$　　　　　　　　　d. This one is reduced.

Section 6.3 Classes of Basic Functions

p. 480

2. a. Linear, quadratic, absolute value, and square root.

   b. Oppositing and reciprocal.

   c. Quadratic.　　　　　　　　　d. Square root.

   e. Oppositing.　　　　　　　　　f. Reciprocal.

3. a. $x = 17$.　　　　　　　　　　b. $x = -4, x = 4$.

   c. $x = -13, x = 13$.　　　　　　d. $x = -13$

   e. $x = \dfrac{17}{2}$　　　　　　　　　　f. $x = 9$

4. Constant function: The tuition problem in which students were charged a flat rate for all the credit hours over 12 taken in a semester.

Linear function: Okimbe's Bulls T–shirts.

Quadratic: The travels of the falling corned beef sandwich.

Opposite: Calculator golf.

Absolute value: The stock market problem.

Reciprocal: The 240–mile trip.

Square root: The area of the dog pen.

5. The algebraic representation for the Okimbie profit function is $y = 10x - 1000$. See exploration number 4 on page 307 in Section 4.4.

   a. The vertical intercept is $(0, -1000)$. This means that if Okimbie sells no shirts he is in the hole $1000, the cost of the shirts.

   b. The horizontal intercept is found by solving the equation $10x - 1000 = 0$ to get $x = 100$. This is the number of shirts Okimbie must sell to just break even.

6. a. **Table 12: Input-Output Table**

x	–5	–4	–3	–2	–1	0	1	2	3	4	5
$y(x)$	–6	–4	–2	0	2	4	6	8	10	12	14

   b. Vertical intercept: 4   Horizontal intercept: –2.

   c. The graph of $y(x) = 2x + 4$ appears in the standard viewing window.

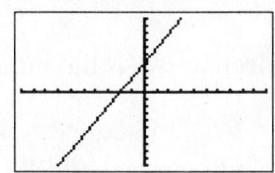

   d. Increasing since, as the input increases, the output also increases. This can be determined by reading the table of values from left to right.

7. a. **Table 13 Input-Output Table**

x	–5	–4	–3	–2	–1	0	1	2	3	4	5
$y(x)$	16	14	12	10	8	6	4	2	0	–2	–4

   b. Vertical intercept: 6   Horizontal intercept: 3.

c. The graph of $y(x) = -2x + 6$ appears in the standard viewing window.

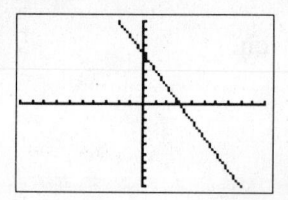

d. Decreasing since, as the input increases, the output decreases. This can be determined by reading the table of values from left to right.

e. The graph is a line that intersects the horizontal axis at three and the vertical axis at six.

8. a. **Table 14: Input-Output Table**

x	−5	−4	−3	−2	−1	0	1	2	3	4	5
$y(x)$	14	6	0	−4	−6	−6	−4	0	6	14	24

b. Vertical intercept: −6   Horizontal intercepts: −3 and 2.

c. The graph of $y(x) = x^2 + x - 6$ appears in the standard viewing window.

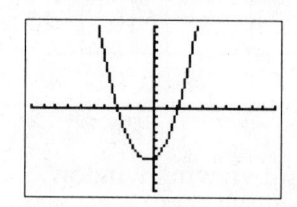

d. Neither. As the input increases up to −1, the output decreases. As the input increases from zero, the output also increases.

e. The inputs are −5, −4, −3, −2, −1, 0, 1, 2, 3, 4, 5. The outputs are −6, −4, 0, 6, 14, 24.

f. The domain is all real numbers.

g. The range is all real numbers greater than or equal to approximately −6.

9. The functions in Explorations 5 and 6 have one horizontal intercept while the function in Exploration 7 has two. This occurs since the functions given in Explorations 5 and 6 are linear functions while the function given in Exploration 7 is a quadratic function.

10. As x increases, $\frac{1}{x}$ decreases to 0. This is true since we are increasing the size of the denominator making the value of the fraction smaller and smaller.

11. a. In y stages the spaceship will travel $11y$.

   b. The function $Distance(y) = 11y$ is a linear function.

   c. The new distance function is $Earth(y) = 11y + 4$.

Section 6.4 Linear Functions

p. 504

2. a. slope = 3   b. vertical intercept: $(0, -12)$

   c. horizontal intercept: $(4, 0)$ d. The graph is increasing.

   e. $3x - 12 = 7$
   $X = \dfrac{19}{3}$

3. a. slope = -5 b. vertical intercept: $(0, 10)$

   c. horizontal intercept: $(2, 0)$ d. The graph is decreasing.

   e. $-5x + 10 = 7$
   $X = \dfrac{3}{5}$

4. a. slope = 0   b. vertical intercept: $(0, -3)$

   c. There is no horizontal intercept.   d. The graph is neither increasing nor decreasing.

   e. The output cannot be 7 for this function.

5. a. slope = -2 b. vertical intercept: $(0, -5)$

   c. horizontal intercept: $\left(-\dfrac{5}{2}, 0\right)$ d. The graph is decreasing.

   e. $-2x - 5 = 7$
   $x = -6$

6. a. slope = $\dfrac{2}{3}$ b. vertical intercept: $(0, -5)$

   c. horizontal intercept: $\left(\dfrac{15}{2}, 0\right)$ d. The graph is increasing.

e. $\frac{2}{3}x - 5 = 7$

$x = 18$

7. a. slope = 7 b. vertical intercept: $\left(0, \frac{8}{5}\right)$

c. horizontal intercept: $\left(-\frac{8}{35}, 0\right)$ d. The graph is increasing.

$$7p + \frac{8}{5} = 7$$

e. $7p = 7 - \frac{8}{5} = \frac{35}{5} - \frac{8}{5} = \frac{27}{5}$

$$p = \frac{27}{5} \cdot \frac{1}{7} = \frac{27}{35}$$

8. a. The graph of $y(x) = 3x + 4$ appears in the standard viewing window.

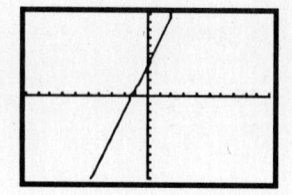

b.

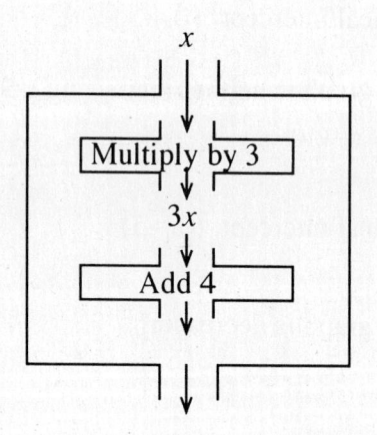

```
        x
        ↓
   ┌─────────┐
   │Multiply by 3│
   └─────────┘
        ↓
        3x
        ↓
   ┌─────────┐
   │  Add 4  │
   └─────────┘
        ↓
```

c. $y(x) = 3x + 4$ d. horizontal intercept: $\left(-\frac{4}{3}, 0\right)$

e. $y(x) = 3x + 6$ f. $y(x) = -\frac{1}{3}x + 4$

9. a. The graph of $y(x) = -2x + 3$ appears in the standard viewing window.

   b.

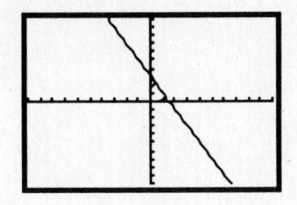

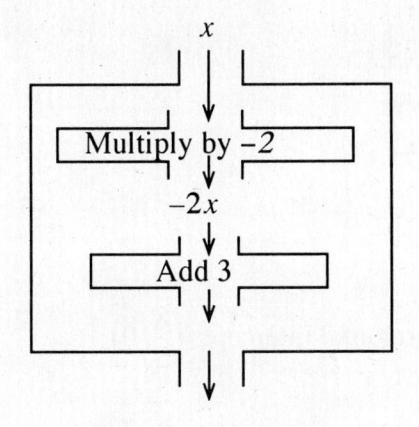

   c. $y(x) = -2x + 3$

   d. horizontal intercept: $\left(\frac{3}{2}, 0\right)$

   e. $y(x) = -2x + 5$

   f. $y(x) = \frac{1}{2}x + 3$

10. a. The graph of $y(x) = \frac{3}{2}x - 4$ appears in the standard viewing window.

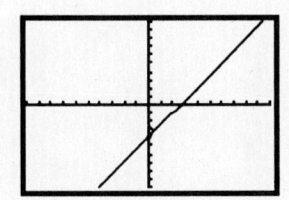

b.

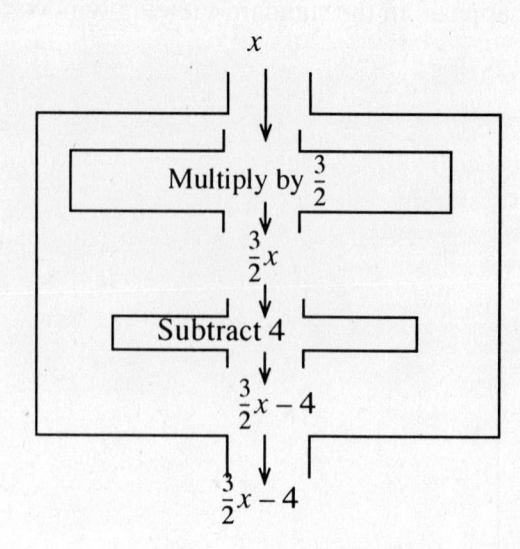

```
        x
        ↓
  ┌──────────────┐
  │ Multiply by 3/2 │
  └──────────────┘
        ↓
       (3/2)x
        ↓
  ┌──────────────┐
  │   Subtract 4   │
  └──────────────┘
        ↓
     (3/2)x − 4
        ↓
     (3/2)x − 4
```

c. $y(x) = \frac{3}{2}x - 4$

d. horizontal intercept: $\left(\frac{8}{3}, 0\right)$

e. $y(x) = \frac{3}{2}x - 2$

f. $y(x) = -\frac{2}{3}x - 4$

11. a. The graph of $y(x) = -\frac{3}{5}x - 1$ appears in the standard viewing window.

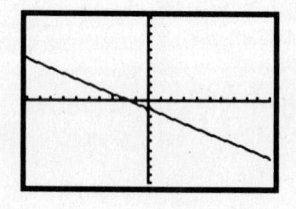

b.

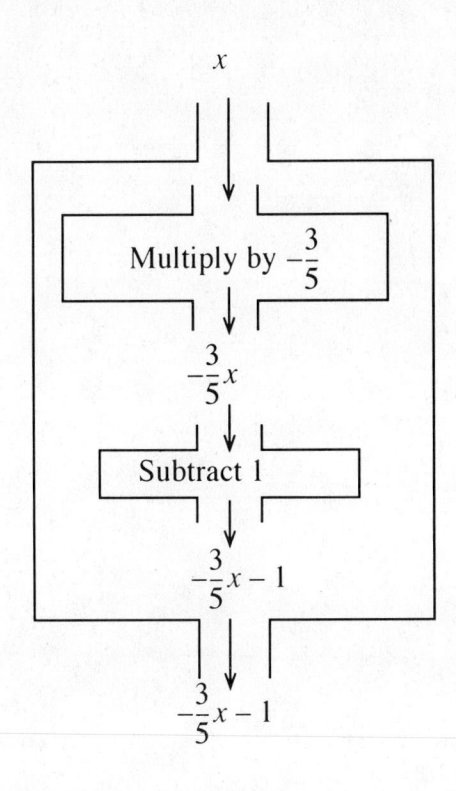

```
        x
        ↓
   ┌─────────────┐
   │ Multiply by -3/5 │
   └─────────────┘
        ↓
      -3/5 x
        ↓
   ┌─────────────┐
   │  Subtract 1 │
   └─────────────┘
        ↓
     -3/5 x - 1
        ↓
     -3/5 x - 1
```

c. $y(x) = -\dfrac{3}{5}x - 1$ d. horizontal intercept: $\left(-\dfrac{5}{3}, 0\right)$

e. $y(x) = -\dfrac{3}{5}x + 1$ f. $y(x) = \dfrac{5}{3}x - 1$

12. a. The graph of $y(x) = \dfrac{7}{4}x - \dfrac{3}{2}$ appears in the standard viewing window.

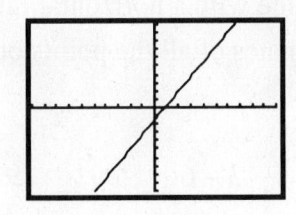

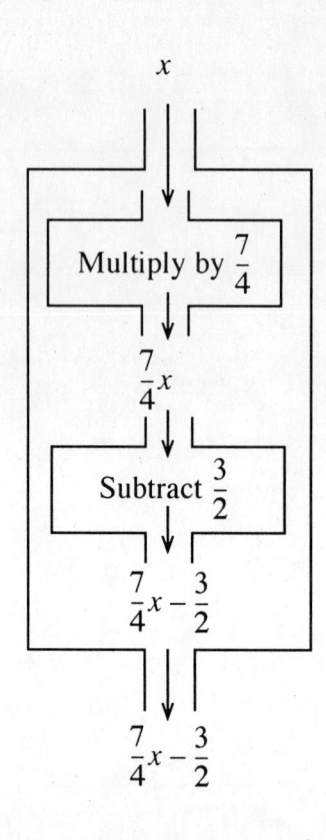

b.

```
           x
           ↓
   ┌───────────────┐
   │ Multiply by 7/4 │
   └───────────────┘
           ↓
         (7/4)x
           ↓
   ┌───────────────┐
   │  Subtract 3/2  │
   └───────────────┘
           ↓
      (7/4)x − 3/2
           ↓
      (7/4)x − 3/2
```

c. $y(x) = \dfrac{7}{4}x - \dfrac{3}{2}$ d. horizontal intercept: $\left(\dfrac{6}{7}, 0\right)$

e. $y(x) = \dfrac{7}{4}x + \dfrac{1}{2}$ f. $y(x) = -\dfrac{4}{7}x - \dfrac{3}{2}$

13. a. $y(x) = 0$ b. $x = 0$

   c. The horizontal axis is a horizontal line. It has a vertical intercept of $(0, 0)$. In addition all of the y-coordinates of the points on the horizontal axis are 0. Thus, the equation is $y = 0$. The vertical axis is a vertical line with a horizontal intercept of $(0, 0)$. One should also observe that the x-coordinates of all the points on the vertical axis are 0. The equation is $x = 0$.

14. a. $m = \dfrac{\Delta y}{\Delta x} = \dfrac{4-7}{2-3} = \dfrac{-3}{-1} = 3$ b. $m = \dfrac{\Delta y}{\Delta x} = \dfrac{-3-6}{5-(-2)} = \dfrac{-9}{7} = -\dfrac{9}{7}$

   c. $m = \dfrac{\Delta y}{\Delta x} = \dfrac{6-2}{9-4} = \dfrac{4}{5}$

15. The output at the horizontal intercept for the cookie problem is 120 cookies. This represents the breakeven point.

16. a. Vertical intercept is (0, –10). b. Slope is approximately 5.

c.

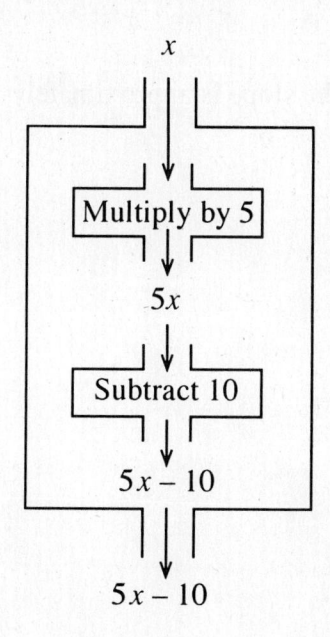

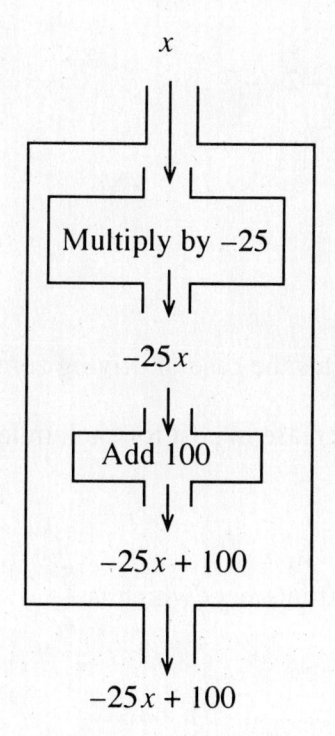

```
         x
         ↓
    ┌─────────────┐
    │ Multiply by 5│
    └─────────────┘
         ↓
         5x
         ↓
    ┌─────────────┐
    │  Subtract 10 │
    └─────────────┘
         ↓
       5x – 10
         ↓
       5x – 10
```

d. Approximate equation is $y(x) = 5x - 10$.

e. The horizontal intercept is (2, 0).

17. a. Vertical intercept is (0, 100). b. Slope is approximately –25.

c.

```
          x
          ↓
    ┌──────────────┐
    │Multiply by –25│
    └──────────────┘
          ↓
        –25x
          ↓
    ┌──────────────┐
    │   Add 100    │
    └──────────────┘
          ↓
      –25x + 100
          ↓
      –25x + 100
```

d. The approximate equation is $y(x) = -25x + 100$.

e. The horizontal intercept is $(4, 0)$.

18. a. The vertical intercept is –3.   b. The slope is approximately –2.

c.

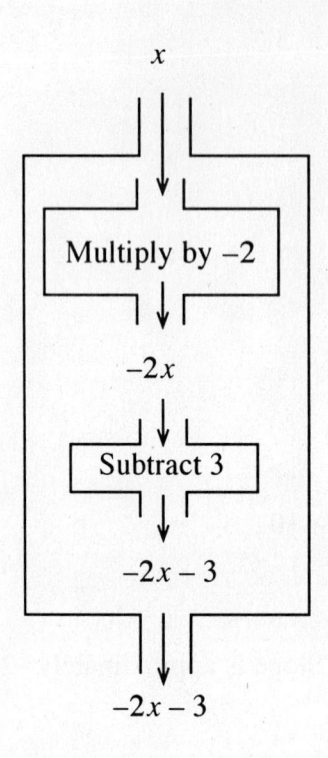

```
         x
         ↓
  ┌─────────────┐
  │ Multiply by –2 │
  └─────────────┘
         ↓
        –2x
         ↓
  ┌─────────────┐
  │  Subtract 3   │
  └─────────────┘
         ↓
       –2x – 3
         ↓
       –2x – 3
```

d. The approximate equation is $y(x) = -2x - 3$.

e. The horizontal intercept is $(-1.5, 0)$.

19. a. $T(m) = 0.20m + 15$

b. The vertical intercept is 15. It represents the case of driving zero miles.

c. The slope is 0.20. It represents the increase in cost for each mile driven.

20. a. $y(x) = 0$ means we want the vertical intercept which is $\frac{5}{2}$.

b. Solving $2x - 5 = 3$ we get $x = 4$.

c. Solving $2x - 5 = 13$ gives us $x = 9$.

21. a. $P(c) = 10 \Rightarrow 0.25c - 30 = 10 \Rightarrow c = 160$. So we must sell 160 cookies to realize a profit of $10.

    b. $P(c) = 100 \Rightarrow 0.25c - 30 = 100 \Rightarrow c = 520$. So we must sell 520 cookies to realize a profit of $100.

22. a–d. Yes to both.

    e. The domain is all real numbers.

    f. The range is all real numbers.

Section 6.5 Quadratic Functions

p. 521

2. a. Narrower.
   b. Shifted up 4 units.
   c. Reflected about the x–axis and narrower.
   d. Shifted down 7 units.
   e. Shifted up one unit and wider.
   f. Shifted down 4 units and narrower.
   g. Reflected about the x–axis and shifted down 3 units.

3. $x - (-6)$ can be written as an addition by changing the operation to addition and finding the opposite of -6. Doing both results in $x + 6$. Therefore, $x - (-6)$ and $x + 6$ are equivalent.

4. a. Zeros: 6, –6. x–value of vertex: 0.
   b. Zeros: 9, –7. x–value of vertex: 1, found by computing $\frac{9 + (-7)}{2} = \frac{2}{2} = 1$.
   c. Zeros: –3, –5. x–value of vertex: –4.
   d. Zeros: 11, 7. x–value of vertex: 9.
   e. Zeros: –5, 8. x–value of vertex: 1.5.
   f. Zeros: 2. x–value of vertex: 2.

5. a. $y(x) = (x - 1)(x - 4)$ So $y(x) = x^2 - 5x + 4$

b. $y(x) = (x+3)(x+2)$ So $y(x) = x^2 + 5x + 6$

c. $y(x) = (x-4)(x+6)$ So $y(x) = x^2 + 2x - 24$

d. $y(x) = (x+5)(x-7)$ So $y(x) = x^2 - 2x - 35$

e. $y(x) = (x+5)(x-5)$ So $y(x) = x^2 - 25$

f. $y(x) = (x-7)(x-7)$ So $y(x) = x^2 - 4x + 49$

6. a. Zeros: −4, 4

   $y(x) = (x+4)(x-4)$. So $y(x) = x^2 - 16$.

   b. Zeros: 1, 6.

   $y(x) = (x-1)(x-6)$. So $y(x) = x^2 - 7x + 6$.

   c. Zeros: −5, −2.

   $y(x) = (x+5)(x+2)$. So $y(x) = x^2 + 7x + 10$.

   d. The graph is a parabola that opens upward. The vertical intercept is at ten. The horizontal intercepts are at −2 and −5. The vertex is at approximately (−3.5, −5).

7. The perimeter would be given by the value of $2n$.

8. a. $x^2 + 3x + 3x + 9 = x^2 + 6x + 9$

   b. $x + 3$ See part f, below for the diagram.

   c. $x^2 + 3x + 3x + 9 = x^2 + 6x + 9$

   d. They are the same.

   e. $x^2 + 3x + 2x + 6$ It is not a square since the lengths of the sides are not all the same.

	x	3
x	x^2	$3x$
2	$2x$	6

f.

	x	3
x	x²	3x

3	3x

9

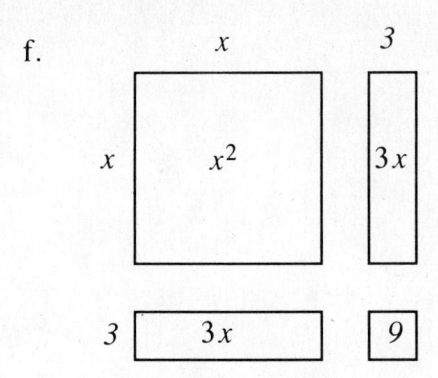

They do form a square when put together. The length of the sides for the square would be $x + 3$.

9. a. $4x - 7 = -9$

$x = -\dfrac{1}{2}$

b. $5 - 3x = \dfrac{2}{3}$

$-3x = \dfrac{2}{3} - 5 = -\dfrac{13}{3}$

$x = \left(-\dfrac{13}{3}\right)\left(-\dfrac{1}{3}\right) = \dfrac{13}{9}$

c. $\dfrac{2}{5}t + 3 = \dfrac{5}{7}$

$\dfrac{2}{5}t = \dfrac{5}{7} - 3 = -\dfrac{16}{7}$

$t = -\dfrac{16}{7} \cdot \dfrac{5}{2} = -\dfrac{8}{7} \cdot \dfrac{5}{1} = -\dfrac{40}{7}$

d. $\dfrac{7}{6} - 2y = -5$

$-2y = -5 - \dfrac{7}{6} = -\dfrac{37}{6}$

$y = -\dfrac{37}{6} \cdot \left(-\dfrac{1}{2}\right) = \dfrac{37}{12}$

10. a. Slope = 2   Vertical Intercept: (0, 7)   Horizontal Intercept: $\left(-\dfrac{7}{2}, 0\right)$

b. Slope = –9   Vertical Intercept: (0, 5)   Horizontal Intercept: $\left(\dfrac{5}{9}, 0\right)$

c. Slope = $\dfrac{3}{4}$ Vertical Intercept: $\left(0, -\dfrac{2}{5}\right)$ Horizontal Intercept: $\left(\dfrac{8}{15}, 0\right)$

d. Slope = 5   Vertical Intercept: (0, –11)   Horizontal Intercept: $\left(\dfrac{11}{5}, 0\right)$

Section 6.6 Review

p. 526

1. a. The whole numbers are 0 and 18.

   b. Integers: 0, 18, -9, $-\sqrt{4}$, $-\dfrac{24}{6}$

   c. Rational Numbers: 0, 18, -9, $-\sqrt{4}$, $-\dfrac{24}{6}$, $-\dfrac{4}{5}$, $\dfrac{24}{9}$, $\dfrac{3}{5}$, $|-5.4|$, 5^{-2}, $\sqrt{\dfrac{4}{9}}$, 0.8^2, $\sqrt{0.49}$

   d. Irrational Numbers: $\sqrt{8}$, $-\sqrt{5}$, $\sqrt{\dfrac{5}{6}}$, $\sqrt{1.04}$

   e. All are real numbers.

2 & 3. The labels are for showing the order requested in exploration 3 and for reference in exploration 4, appearing below.

A: -9 B: $-\dfrac{24}{6} = -4$ C: $-\sqrt{5} \approx -2.236068$ D: $-\sqrt{4} = -2$

E: $-\dfrac{4}{5} = -0.8$ F: $-\dfrac{1}{12} = -0.08333...$ G: $0 = 0$ H: $5^{-2} = \dfrac{1}{25} = 0.04$

I: $\dfrac{3}{5} = 0.6$ J: $0.8^2 = 0.64$ K: $\sqrt{\dfrac{4}{9}} = \dfrac{2}{3} = 0.666...$

L: $\sqrt{0.49} = 0.7$ M: $\sqrt{\dfrac{5}{6}} \approx 0.912871$ N: $\sqrt{1.04} \approx 1.01983903$

O: $\dfrac{24}{9} = 2\dfrac{2}{3} = 2.666...$ P: $\sqrt{8} \approx 2.8284271$ Q: $|-5.4| = 5.4$ R: $18 = 18$

4. The letters used to label the numbers above are used to indicate their position in the graph below.

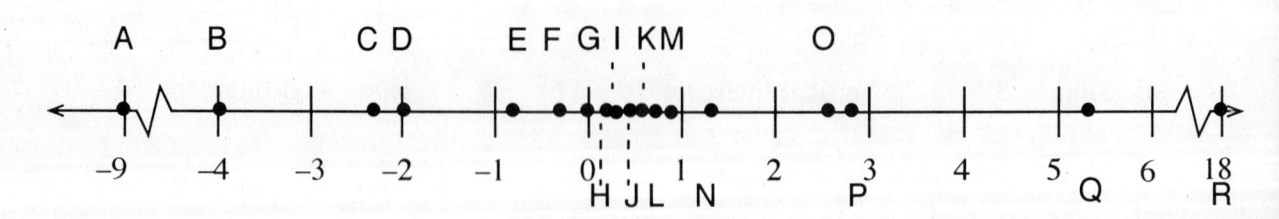

```
   A      B        C D    E F G I K M       O
───/\/────┼────────┼─┼────┼─┼─┼─┼─┼─┼───────┼────────┼────┼──/\/──
  -9     -4       -3 -2  -1  0   1    2    3        4    5   6  18
                               H J L N          P             Q   R
```

The position of the values on the number line is approximate.

5. The only one that is not a real number is g because the square of any real number must be greater than or equal to zero. So, there is no real number we could square and get −64.

   a. $\sqrt{3} \approx 1.732$
   b. $5\sqrt{17} \approx 20.616$
   c. $\sqrt{36} = 6$
   d. $-\sqrt{64} = -8$
   e. $-2\sqrt{6} \approx -4.899$
   f. $-\sqrt{22} \approx -4.690$
   g. not a real number
   h. $7\sqrt{4} = 14$

   6 represents the only one which is not a real number because there is no real number we could square to get −64.

6. a. $3^{1/2} \approx 1.732$
   b. $-64^{1/2} = -8$
   c. $19^{1/2} \approx 4.359$
   d. $144^{1/2} = 12$
   e. not a real number
   f. $64^{1/2} = 8$
   g. $-6^{1/2} \approx 2.449$

7. We use the formula $\sqrt{(\Delta x)^2 + (\Delta y)^2}$ to find the length of each line segment.

   a. $\sqrt{(-6-9)^2 + (7-5)^2} = \sqrt{(-15)^2 + 2^2} = \sqrt{225 + 4} = \sqrt{229} \approx 15.133$

   b. $\sqrt{(4-(-7))^2 + (-8-(-15))^2} = \sqrt{11^2 + 7^2} = \sqrt{121 + 49} = \sqrt{170} \approx 13.038$

8. a. $x = 52$
   b. $x^2 = 5$
      $x = \pm\sqrt{5}$
   c. $|x| = 4$
      $x = \pm 4$
   d. $-x = -(9)$
      $x = 9$
   e. $\dfrac{1}{x} = \dfrac{13}{8}$
      $x = \dfrac{8}{13}$
   f. $\sqrt{x} = 4$
      $x = 16$

9. a. **Table 1: Input-Output Table**

x	−5	−4	−3	−2	−1	0	1	2	3	4	5
$y(x)$	−34	−29	−24	−19	−14	−9	−4	1	6	11	16

b. Vertical Intercept: (0, −9)   Horizontal Intercept: $\left(\frac{9}{5}, 0\right)$

c. The graph as produced on a TI-83 with is shown.

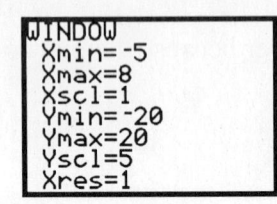

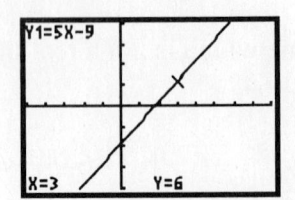

```
WINDOW
Xmin=-5
Xmax=8
Xscl=1
Ymin=-20
Ymax=20
Yscl=5
Xres=1
```

Y1=5X−9, X=3, Y=6

d. The function is increasing. As the x-coordinates increase the y-coordinates also increase.

10. a. **Table 2: Input-Output Table**

x	−5	−4	−3	−2	−1	0	1	2	3	4	5
$y(x)$	20	9	0	−7	−12	−15	−16	-15	-12	-7	0

b. Vertical Intercept: (0, −15)   Horizontal Intercepts: (−3, 0) and (5, 0)

c. The graph is shown as it appears on a TI-83. The vertex has been highlighted using the Trace item in the calculate menu and setting the value of x to be 1.

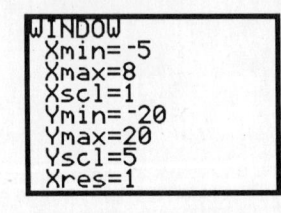

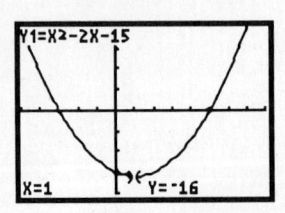

```
WINDOW
Xmin=-5
Xmax=8
Xscl=1
Ymin=-20
Ymax=20
Yscl=5
Xres=1
```

Y1=X²−2X−15, X=1, Y=-16

d. The function is neither increasing nor decreasing. As the x-coordinates increase the y-coordinates decrease until x is 1 and then begin to increase.

11. a. Slope = 4   b. output at the vertical intercept: −11

c. input at the horizontal intercept: $\frac{11}{4}$

d. The function is increasing. The graph is shown with the horizontal intercept point shown.

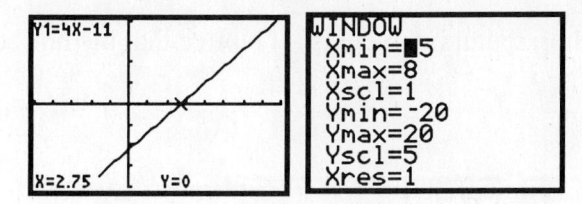

12. a. Slope = $\dfrac{2}{7}$ b. output at the vertical intercept: 5

c. input at the horizontal intercept: $-\dfrac{35}{2}$

d. The function is increasing. The graph is shown with the horizontal intercept highlighted.

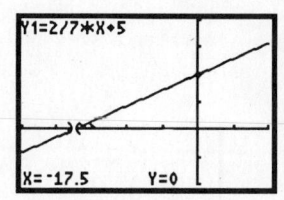

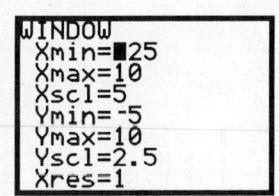

13. a. Slope = $-\dfrac{3}{5}$ b. output at the vertical intercept: $\dfrac{5}{4}$

c. input at the horizontal intercept: $\dfrac{25}{12}$

d. The function is decreasing. The graph is shown with the horizontal intercept highlighted.

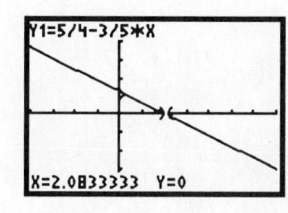

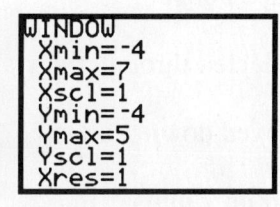

14. a. Slope = 0   b. output at the vertical intercept: 0

c. All real numbers.   d. The graph is the horizontal axis.

15. a.

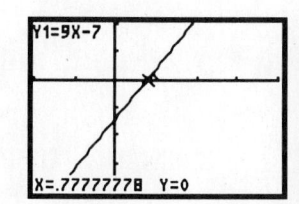

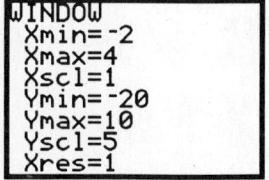

b. $y(x) = 9x - 7$

c. The input at the horizontal intercept is $\frac{7}{9}$. Notice that the horizontal intercept is shown on the graph.

16. a.

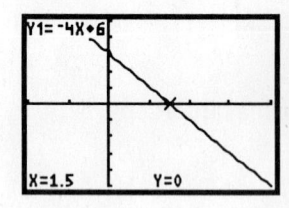

    Y1=-4X+6, X=1.5, Y=0
    WINDOW Xmin=-2, Xmax=4, Xscl=1, Ymin=-10, Ymax=10, Yscl=2, Xres=1

    b. $y(x) = -4x + 6$

    c. The input at the horizontal intercept is $\frac{3}{2}$. Notice that the horizontal intercept is shown on the above graph.

17. a.

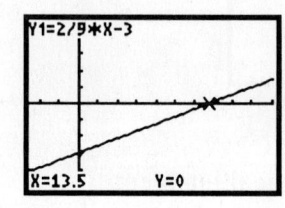

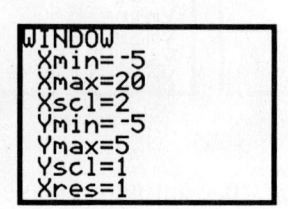

    Y1=2/9*X-3, X=13.5, Y=0
    WINDOW Xmin=-5, Xmax=20, Xscl=2, Ymin=-5, Ymax=5, Yscl=1, Xres=1

    b. $y(x) = \frac{2}{9}x - 3$

    c. The input at the horizontal intercept is $\frac{27}{2}$. Notice that the horizontal intercept is shown on the graph above.

18. a. Narrower and reflected through the x-axis.

    b. Narrower and moved down 3 units.

    c. Wider and moved up 7 units.

    d. Narrower, reflected through the x-axis and moved down 4 units.

19. a. Zeros: 5, −5   x-value of vertex: 0

    b. Zeros: 2, −7   x-value of vertex: $-\frac{5}{2}$

    c. Zeros: −1, −9   x-value of vertex: −5

d. Zeros: 6, 3   x-value of vertex: $\dfrac{9}{2}$

20. a. $y(x) = (x-4)(x+7) = x^2 + 3x - 28$

   A display of a table of values of both expressions is shown.

   b. $y(s) = (x+8) \cdot (x+6) = x^2 + 14x + 48$

   The table values shown computed from each expression are the same.

   c. $y(x) = (x-5) \cdot (x+5) = x^2 - 25$

   d. $y(x) = (x+6)^2 = x^2 + 12x + 36$

21. a. $-6x^3 - 7x^2 + 7x + 3$ b. $35x^3y^5 - 21xy^6$

   c. $16x^3 - 36x^2y - 29xy^2 + 35y^3$ d. $4a^2 - 9b^2$

   e. $4x^2 - 28x + 49$ f. $25w^2 + 80wz + 64z^2$

22. a. The terms are $4x$ and -11.   b. The numerical coefficients are 4 and -11.

23. a. The terms are $5x^2$, $-7x$ and -8. b. The numerical coefficients are 5, -7 and -8.

24. a. $-2x^5 + 8x^4 - x^3 + x^2 + 11x - 3$

25. a. $\dfrac{x}{4}$ would be a linear function which would have outputs which are the length of the square.

   b. The area of the square would be given by the quadratic function $\left(\dfrac{x}{4}\right)^2 = \dfrac{x^2}{16}$.

195

CHAPTER 7

Answering Questions with Linear and Quadratic Functions

Section 7.1 Linear Equations and Inequalities in One Variable

p. 551

2.   a.   2   b.   –5

       c. $\dfrac{8}{7}$ d. $-\dfrac{4}{3}$

       e. $-\dfrac{2.7}{4.3} = -\dfrac{27}{43}$ f. $-\dfrac{-4}{\frac{3}{5}} = 4\left(\dfrac{5}{3}\right) = \dfrac{20}{3}$

3.   a. $x - 2 = 0$ b. $x + 5 = 0$
            $x - 2 + 2 = 0 + 2$ $x + 5 - 5 = 0 - 5$
                   $x = 2$ $x = -5$

       c. $7t - 8 = 0$ d. $3r + 4 = 0$
            $7t - 8 + 8 = 0 + 8$ $3r + 4 - 4 = 0 - 4$
                  $7t = 8$ $3r = -4$
                  $\dfrac{7t}{7} = \dfrac{8}{7}$ $\dfrac{3r}{3} = \dfrac{-4}{3}$
                     $t = \dfrac{8}{7}$ $r = -\dfrac{4}{3}$

e.
$$4.3z + 2.7 = 0$$
$$4.3z + 2.7 - 2.7 = 0 - 2.7$$
$$4.3z = -2.7$$
$$\frac{4.3z}{4.3} = \frac{-2.7}{4.3}$$
$$z = -\frac{27}{43}$$

f.
$$\frac{3}{5}x - 4 = 0$$
$$\frac{3}{5}x - 4 + 4 = 0 + 4$$
$$\frac{3}{5}x = 4$$
$$\frac{\frac{3}{5}x}{\frac{3}{5}} = \frac{4}{\frac{3}{5}}$$
$$x = 4\left(\frac{5}{3}\right)$$
$$x = \frac{20}{3}$$

4. a.
$$5x - 3 = 2$$
$$5x - 3 + 3 = 2 + 3$$
$$5x = 5$$
$$\frac{5x}{5} = \frac{5}{5}$$
$$x = 1$$

b.
$$x + 7 = 4$$
$$x + 7 - 7 = 4 - 7$$
$$x = -3$$

c.
$$3x - 13 = -27$$
$$3x - 13 + 13 = -27 + 13$$
$$3x = -14$$
$$\frac{3x}{3} = \frac{-14}{3}$$
$$x = -\frac{14}{3}$$

d.
$$5.2x + 11 = 4$$
$$5.2x + 11 - 11 = 4 - 11$$
$$5.2x = -7$$
$$x = \frac{-7}{5.2} = -\frac{70}{52} = -\frac{35}{26}$$

e.
$$-\frac{2}{3}x - 5 = -\frac{1}{4}$$
$$-\frac{2}{3}x - 5 + 5 = -\frac{1}{4} + 5$$
$$-\frac{2}{3}x = -\frac{1}{4} + \frac{20}{4}$$
$$-\frac{2}{3}x = \frac{19}{4}$$
$$\frac{-\frac{2}{3}x}{-\frac{2}{3}} = \frac{\frac{19}{4}}{-\frac{2}{3}}$$
$$x = \left(\frac{19}{4}\right)\left(-\frac{3}{2}\right)$$
$$x = -\frac{57}{8}$$

5. a.
$$5x - 3 < 2$$
$$5x - 3 + 3 < 2 + 3$$
$$5x < 5$$
$$\frac{5x}{5} < \frac{5}{5}$$
$$x < 1$$

b.
$$x + 7 > 4$$
$$x + 7 - 7 > 4 - 7$$
$$x > -3$$

c.
$$3x - 13 \le -27$$
$$3x - 13 + 13 \le -27 + 13$$
$$3x \le -14$$
$$\frac{3x}{3} \le \frac{-14}{3}$$
$$x \le -\frac{14}{3}$$

d.
$$5.2x + 11 \ge 4$$
$$5.2x + 11 - 1 \ge 4 - 11$$
$$5.2x \ge -7$$
$$\frac{5.2x}{5.2} \ge \frac{-7}{5.2}$$
$$x \ge -\frac{35}{26}$$

e. $\quad -\dfrac{2}{3}x - 5 < -\dfrac{1}{4}$

$\quad\quad -\dfrac{2}{3}x - 5 + 5 < -\dfrac{1}{4} + 5$

$\quad\quad\quad\quad -\dfrac{2}{3}x < -\dfrac{1}{4} + \dfrac{20}{4}$

$\quad\quad\quad\quad -\dfrac{2}{3}x < \dfrac{19}{4}$

$\quad\quad\quad\quad \dfrac{-\dfrac{2}{3}x}{-\dfrac{2}{3}} > \dfrac{\dfrac{19}{4}}{-\dfrac{2}{3}}$

$\quad\quad\quad\quad\quad x > \left(\dfrac{19}{4}\right)\left(-\dfrac{3}{2}\right)$

$\quad\quad\quad\quad\quad x > -\dfrac{57}{8}$

---

6. a. $\quad 4x + y = -13$ $\quad\quad$ Slope = -4 $\quad\quad$ Vertical Intercept: $(0, -13)$

$\quad\quad 4x + y - 4x = -13 - 4x$

$\quad\quad\quad y = -4x - 13$

b. $\quad 2x - 9y = 17$ $\quad\quad$ Slope = $\dfrac{2}{9}$ $\quad\quad$ Vertical Intercept: $\left(0, -\dfrac{17}{9}\right)$

$\quad\quad 2x - 9y - 2x = 17 - 2x$

$\quad\quad\quad -9y = -2x + 17$

$\quad\quad\quad \dfrac{-9y}{-9} = \dfrac{-2x + 17}{-9}$

$\quad\quad\quad y = \dfrac{2}{9}x - \dfrac{17}{9}$

c. $\quad x + 5y = 0$ $\quad\quad$ Slope = $-\dfrac{1}{5}$ $\quad\quad$ Vertical Intercept: $(0, 0)$

$\quad\quad x + 5y - x = 0 - x$

$\quad\quad\quad 5y = -x$

$\quad\quad\quad y = -\dfrac{1}{5}x$

d.
$$4x - 3y = -1$$
$$4x - 3y - 4x = -1 - 4x$$
$$-3y = -4x - 1$$
$$\frac{-3y}{-3} = \frac{-4x - 1}{-3}$$
$$y = \frac{4}{3}x + \frac{1}{3}$$

Slope = $\frac{4}{3}$ Vertical Intercept: $\left(0, \frac{4}{3}\right)$

7. The equation must be of the form $y = -5x + b$. So, if we substitute -3 for x and 1 for y we can solve the resulting equation for b. Solving $1 = -5(-3) + b$ gives us $b = -14$. The requested function is defined by the equation $y(x) = -5x - 14$.

8. a. The slope of the line determined by the given points is $\frac{7-3}{4-(-6)} = \frac{4}{10} = \frac{2}{5}$. The equation must be of the form $y = \frac{2}{5}x + b$. So, we solve the equation we get when we substitute 4 for x and 7 for y. $7 = \frac{2}{5} \cdot 4 + b$ gives us $b = \frac{27}{5}$. The linear function is given by the equation $y(x) = \frac{2}{5}x + \frac{27}{5}$.

   b. Solving $-3 = \frac{2}{5} \cdot 1 + b$ gives us $b = -\frac{17}{5}$. The requested equation is
   $y(x) = \frac{2}{5}x - \frac{17}{5}$.

   c. The equation would be of the form $y = -\frac{5}{2}x + b$. So, we must solve $-3 = -\frac{5}{2} \cdot 1 + b$ to find b. We get $b = -\frac{1}{2}$. The requested equation is $y(s) = -\frac{5}{2}x - \frac{1}{2}$

9. a. First solve $2x + 3y = 4$ for y. The slope is the coefficient of x in that equation.

   $3y = -2x + 4$ $y = -\frac{2}{3}x + \frac{4}{3}$ The slope is $-\frac{2}{3}$. The requested equation must contain the point $(-3, 5)$ so, we solve $5 = -\frac{2}{3}(-3) + b$ to get $b = 3$. The requested equation is $y = -\frac{2}{3}x + 3$.

b. In this equation the slope should be $\frac{3}{2}$, so we solve $5 = \frac{3}{2}(-3) + b$ to get $b = \frac{19}{2}$.

The requested equation is $y(x) = \frac{3}{2}x + \frac{19}{2}$.

10. a. $T(m) = 0.2m + 15$

b.
$$\begin{aligned} T(m) &= 40 \\ 0.2m + 15 &= 40 \\ 0.2m + 15 - 15 &= 40 - 15 \\ 0.2m &= 25 \\ \frac{0.2m}{0.2} &= \frac{25}{0.2} \\ m &= 125 \end{aligned}$$

125 miles were driven for a total charge of $40.

c.
$$\begin{aligned} T(m) &= 100 \\ 0.2m + 15 &= 100 \\ 0.2m + 15 - 15 &= 100 - 15 \\ 0.2m &= 85 \\ \frac{0.2m}{0.2} &= \frac{85}{0.2} \\ m &= 425 \end{aligned}$$

425 miles were driven for a total charge of $100.

11. a.
$$\begin{aligned} 0.25c - 30 &= 75 \\ 0.25c - 30 + 30 &= 75 + 30 \\ 0.25c &= 105 \\ \frac{0.25c}{0.25} &= \frac{105}{0.25} \\ c &= 420 \end{aligned}$$

420 cookies must be sold to realize a profit of $75.

b.
$$\begin{aligned} 0.25c - 30 &= 100 \\ 0.25c - 30 + 30 &= 100 + 30 \\ 0.25c &= 130 \\ \frac{0.25c}{0.25} &= \frac{130}{0.25} \\ c &= 520 \end{aligned}$$

520 cookies must be sold to realize a profit of $100.

12. $y(x) = 5x + 2$

    a. $5x + 2 = 14$
        $5x = 12$
        $x = \dfrac{12}{5}$

    b. $5x + 2 = -4$
        $5x = -6$
        $x = \dfrac{-6}{5}$

    c. $5x + 2 \leq 2$
        $5x \leq 0$
        $x \leq 0$

    d. $5x + 2 > 4$
        $5x > 2$
        $x > \dfrac{2}{5}$

13. $y(x) = 3 - 2x$

    a. $3 - 2x = 0$
        $-2x = -3$
        $x = \dfrac{3}{2}$

    b. $3 - 2x = 7$
        $-2x = 4$
        $x = -2$

    c. $3 - 2x < 3$
        $-2x < 0$
        $x > 0$

When multiplying or dividing an inequality by a negative number, the sense of the inequality reverses (Compare step 2 to step 3).

    d. $3 - 2x \geq 4$
        $-2x \geq 1$
        $x \leq -\dfrac{1}{2}$

When multiplying or dividing an inequality by a negative number, the sense of the inequality reverses (Compare step 2 to step 3).

14. $y(x) = 7x + 2$

    a. $y(3) = 7(3) + 2 = 23$

    b. $7x + 2 = 3$
        $7x = 1$
        $x = \dfrac{1}{7}$

15. a. $y(1) \approx -3$ b. $y(-3) \approx -14$

c. $y(x) = 0 \Rightarrow x \approx 2$ d. $y(x) = 9 \Rightarrow x \approx 5$

e. It appears that $y(x) = 5$ when $x \approx 3.5$. Since the outputs are increasing as the input increases, $y(x) > 5$ when $x > 3.5$.

f. It appears that $y(x) = -1$ when $x \approx 1.5$. Since the outputs are increasing as the input increases, $y(x) \leq -1$ when $x \leq 1.5$.

Section 7.2 Systems of Equations

p. 569

2. a. $\begin{matrix} x + y = 8 \\ x - y = 4 \end{matrix}$ Solve $x + y = 8$ for x by subtracting y from both sides. The equation becomes $x = 8 - y$. Substitute this for x in $x - y = 4$.

We get	$(8 - y) - y = 4$
Simplifying the left side:	$8 - 2y = 4$
Solve for y	
Subtract 8 from both sides	$-2y = -4$
Divide both sides by -2	$y = 2$
Substitute into $x = 8 - y$	$x = 8 - 2 = 6$

The solution to the system is $(6, 2)$.

b. $\begin{matrix} x + 2y = 5 \\ x + y = 3 \end{matrix}$ Solve $x + y = 3$ for x by subtracting y from both sides. The equation becomes $x = 3 - y$. Substitute this for x in $x + 2y = 5$.

We get	$(3 - y) + 2y = 5$
Simplifying the left side	$3 + y = 5$
Solve for y	
Subtract 3 from both sides	$y = 2$
Substitute into $x = 3 - y$	$x = 3 - 2 = 1$ The solution to the system is $(1, 2)$.

c. $\begin{array}{l}5x - y = 13\\2x + 3y = 12\end{array}$. Solve $5x - y = 13$ for y by adding y to both sides to get $5x = 13 + y$. Then subtract 13 from both sides to get $5x - 13 = y$. Substitute for y in $2x + 3y = 12$.

We get $\qquad 2x + 3(5x - 13) = 12$

Simplify the left side $\qquad 2x + 15x - 39 = 12$

$\qquad 17x - 39 = 12$

Solve for x

Add 39 to both sides $\qquad 17x = 51$

Divide both sides by 17 $\qquad x = 3$

Substitute into $5x - 13 = y$ $\qquad y = 5(3) - 13 = 15 - 13 = 2$

The solution to the system is $(3, 2)$.

d. $\begin{array}{l}x + 3y = 5\\2x - 3y = -8\end{array}$. Solve $x + 3y = 5$ for x by subtracting $3y$ from both sides to obtain $x = 5 - 3y$. Substitute for x in $2x - 3y = -8$

We get $\qquad 2(5 - 3y) - 3y = -8$

Simplify the left side $\qquad 10 - 6y - 3y = -8$

$\qquad 10 - 9y = -8$

Subtract 10 from both sides $\qquad -9y = -18$

Divide both sides by -9 $\qquad y = 2$

Substitute into $x = 5 - 3y$ $\qquad x = 5 - 3(2) = 5 - 6 = -1$

The solution to the system is $(-1, 2)$.

e. $\begin{array}{l}3x - 2y = 0\\5x + 10y = 4\end{array}$. Solve $3x - 2y = 0$ for y. Add $2y$ to both sides obtaining $3x = 2y$.

Divide both sides by 2 obtaining $y = \dfrac{3}{2}x$. Substitute for y into $5x + 10y = 4$.

204

We get $5x + 10\left(\frac{3}{2}x\right) = 4$

Simplify the left side $5x + 15x = 4$

$20x = 4$

Divide both sides by 20 $x = \frac{1}{5}$

Substitute into $y = \frac{3}{2}x$ $y = \left(\frac{3}{2}\right)\left(\frac{1}{5}\right) = \frac{3}{10}$

The solution to the system is $\left(\frac{1}{5}, \frac{3}{10}\right)$.

f. The second equation is obtained from the first by multiplying both sides of the first by two. These represent the same relationship and the solution to the system is all ordered pairs that satisfy the relationship.

g. There are no ordered pairs in the solution to the system. The graphs of the two equations do not intersect since they are parallel lines.

h. $\begin{array}{l} 5x + y = -2 \\ 2x + 7y = 3 \end{array}$ Solve the first equation for y. We get $y = -5x - 2$. Substitute this expression for y in the second equation to get $2x + 7(-5x - 2) = 3$. Solving this equation gives us $2x - 35x - 14 = 3$ which is equivalent to $-33x - 14 = 3$. Adding 14 to both sides and then dividing by -33 gives $x = -\frac{17}{33}$. Substitute this value for x in the first equation where we solved for y to get

$y = -5\left(-\frac{17}{33}\right) - 2 = \frac{85}{33} - 2 = \frac{85}{33} - \frac{66}{33} = -\frac{19}{33}$. The solution is $\left(-\frac{17}{33}, -\frac{19}{33}\right)$.

3. The equations in 1f are dependent since they represent the same relationship.

4. The equations in 1g are inconsistent since they have no common solutions. The system in 1f has an infinite number of solutions while the system in 1g has no solutions.

5. Answer is based on each person's individual experience.

6. a. The graphs do not intersect since the slopes are the same and the horizontal intercepts are different.

   b. There are no solutions to this system of equations because the two graphs do not intersect. We could say that the solution set is the empty set.

c. The graphs of your two equations should have the same slope and different horizontal intercepts.

7. a.
$$2x - 3 = x + 4$$
$$2x - x - 3 + 3 = x - x + 4 + 3$$
$$x = 7$$

b.
$$7x + 2 = 4x - 5$$
$$7x - 4x + 2 - 2 = 4x - 4x - 5 - 2$$
$$3x = -7$$
$$x = -\frac{7}{3}$$

c.
Original inequality	$x - 8 > 2 - 3x$
Add 3x to both sides	$x - 8 + 3x > 2 - 3x + 3x$
Simplify	$4x - 8 > 2$
Add 8 to both sides	$4x - 8 + 8 > 2 + 8$
Simplify	$4x > 10$
Divide both sides by 4	$\dfrac{4x}{4} > \dfrac{10}{4}$
Simplify	$x > \dfrac{5}{2}$

d.
Original inequality	$5x + 4 \leq 3x$
Subtract 5x from both sides	$5x + 4 - 5x \leq 3x - 5x$
Simplify	$4 \leq -2x$
Divide both sides by –2 and reverse the direction of the inequality.	$\dfrac{4}{-2} \geq \dfrac{-2x}{-2}$
Simplify	$-2 \geq x$

e. $6t + 11 = 6 - 5t$
   $11t = -5$
   $t = -\dfrac{5}{11}$

f. $2 - z = 11z + 7$
   $-5 = 12z$
   $z = -\dfrac{5}{12}$

g. $4k - 9 < 7k + 2$
   $-11 < 3k$
   $-\dfrac{11}{3} < k$

h. $9 - 5p < 2p - 3$
   $12 < 7p$
   $\dfrac{12}{7} < p$

i. $3z + 1 < 2 - 7z$
   $10z < 1$
   $z < \dfrac{1}{10}$

j. $a - 1 < 4a + 1$
   $-2 < 3a$
   $-\dfrac{2}{3} < a$

8. In the equation $8B = 40$, B represents the number of books, not the price of each book.

9. The system for the basket ball problem is $\begin{array}{l}2x + 3y = 58\\ x + y = 27\end{array}$ where x is the number of two–point shots made and y is the number of three point shots made. The numerical and graphical solutions appear below as displayed on a TI-82.

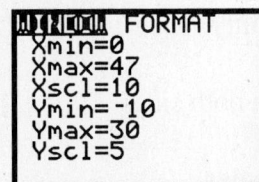

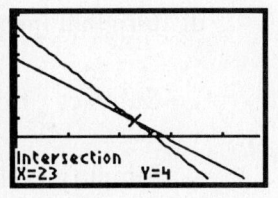

Solving algebraically, solve the second equation for y.

$$y = 27 - x$$

Substitute $(27 - x)$ for y in the first equation $2x + 3(27 - x) = 58$

Simplify left side $2x + 81 - 3x = 58$

$-x + 81 = 58$

Subtract 81 from both sides $-x + 81 - 81 = 58 - 81$

Simplify $-x = -23$

Divide both sides by -1. $\dfrac{-x}{-1} = \dfrac{-23}{-1}$

207

Simplify $\qquad x = 23$

If $x = 23$ then $y = 27 - 23 = 4$

Thus, the Bulls made 23 two–point shots and 4 three–point shots.

10. a. $2x - 3 = x + 4$

   b.
Original equation	$2x - 3 = x + 4$
Subtract x from both sides	$2x - 3 - x = x + 4 - x$
Simplify both sides	$x - 3 = 4$
Add 3 to both sides	$x - 3 + 3 = 4 + 3$
Simplify both sides	$x = 7$

   The outputs of the two functions are equal when the input is 7.

   c. $2x - 3 > x + 4$

   d.
Original inequality	$2x - 3 > x + 4$
Subtract x from both sides	$2x - 3 - x > x + 4 - x$
Simplify both sides	$x - 3 > 4$
Add 3 to both sides	$x - 3 + 3 > 4 + 3$
Simplify both sides	$x > 7$

   The output of $y(x) = 2x - 3$ exceeds the output of $y(x) = x + 4$ when the input exceeds 7.

11. a. Equation $\qquad 3x - 7 = 4 - 5x$

    b.
Add $5x$ to both sides	$3x - 7 + 5x = 4 - 5x + 5x$
Simplify both sides	$8x - 7 = 4$
Add 7 to both sides	$8x - 7 + 7 = 4 + 7$
Simplify both sides	$8x = 11$

Divide both sides by 8 $\quad\dfrac{8x}{8} = \dfrac{11}{8}$

Simplify both sides $\quad x = \dfrac{11}{8}$

So the outputs are equal when the input is $\dfrac{11}{8}$.

c. Inequality $\quad 3x - 7 \leq 4 - 5x$

d. We use the same steps as above to solve the inequality. Since we never multiply or divide both sides by a negative number, the direction of the inequality remains the same throughout the solving process. The output of $y = 3x - 7$ will be no larger than the output of $y = 4 - 5x$ when the input is less than or equal to $\dfrac{11}{8}$.

Section 7.3 Finding Zeros of Quadratic Functions by Factoring

p. 589

2.   a. The zeros are -2 and 6.

     b. The factors are $(x - (-2)) = (x + 2)$ and $(x - 6)$.

     c. The function is $y(x) = (x + 2)(x - 6) = x^2 - 4x - 12$.

3.   a. We must solve $x^2 + 11x + 12 = -16$. Add -16 to both sides to place zero on the right hand side and get the equation $x^2 + 11x + 28 = 0$. Factor the left side: $(x + 7)(x + 4) = 0$. Now we can see that the solutions are -7 and -4. Using a table of ouputs for the function $y(x) = x^2 + 11x + 12$ we see that the inputs of -7 and -4 do produce an output of -16. The graph shows this too.

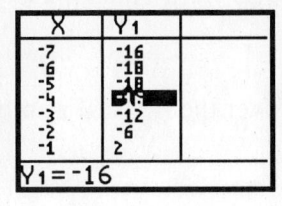

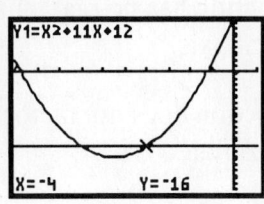

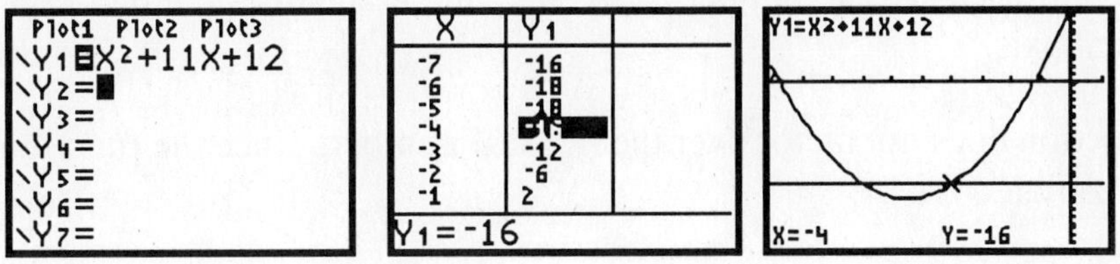

b. We must solve the equation $x^2 - 2x + 1 = 16$. First subtract 16 form both sides $x^2 - 2x - 15 = 0$. Factor the left side: $(x-5)(x+3) = 0$. The factors and hence the product will be zero if $x = 5$ or $x = -3$.

If we enter the original function into the Y= menu and produce a table we see that the inputs of 5 and −3 produce an output of 16. The graph shows the solutions too.

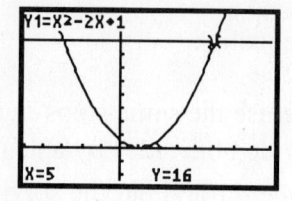

c. We must solve $x^2 - 10x + 10 = -6$. First add 6 to both sides: $x^2 - 10x + 16 = 0$. Now factor the left side: $(x-8)(x-2) = 0$. We can now see that $x = 8$ and $x = 2$ make the two factors and hence the product equal to zero. The table and graph appear below.

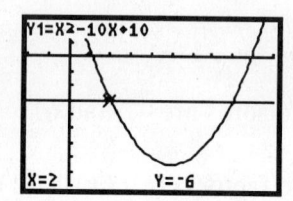

d. Solve the equation $x^2 - 5x + 9 = 9$. Subtract 9 form both sides: $x^2 - 5x = 0$. Factor the left hand side: $x(x-5) = 0$. We see that $x = 0$ and $x = 5$ make the factors and hence the product equal to zero. The table and graph are displayed below as a check.

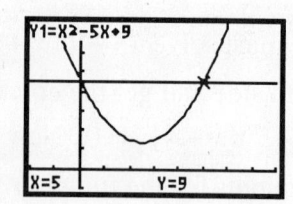

4.   a. This function has no real number zeros since the graph does not intersect the horizontal axis.

    b. This function does not factor over the rational numbers since the function has no real number zeros.

5.   a. Zeros are 0 and 2. $9x^2 - 18x = 9x(x-2)$

    b. Zeros are 0 and $\frac{2}{3}$. $21y^4 - 14y^3 = 7y^3(3y-2)$

c. Zeros are 0, –6, and 1. $5x^3 + 25x^2 - 30x = 5x(x^2 + 5x - 6) = 5x(x + 6)(x - 1)$

d. Zeros are –1 and –8. $\qquad x^2 + 9x + 8 = (x + 1)(x + 8)$

e. Zeros are 7 and –1. $\qquad p^2 - 6p - 7 = (p - 7)(p + 1)$

f. No rational number zeros. $\qquad x^2 + 12x + 13$ is prime.

g. Zeros are 6 and –5. $\qquad x^2 - x - 30 = (x - 6)(x + 5)$

h. No rational number zeros. $\qquad y^2 - 2y - 45$ is prime.

i. Zeros are –5 and –3. $\qquad x^2 + 8x + 15 = (x + 5)(x + 3)$

j. Zeros are 4 and 2. $\qquad x^2 - 6x + 8 = (x - 4)(x - 2)$

k. Zeros are –11 and 2. $\qquad x^2 + 9x - 22 = (x + 11)(x - 2)$

l. Zeros are –5 and 3. $\qquad p^2 + 2p - 15 = (p + 5)(p - 3)$

m. Zeros are 8 and –7. $\qquad x^2 - x - 56 = (x - 8)(x + 7)$

n. Zeros are 4 and 11. $\qquad x^2 - 15x + 44 = (x - 4)(x - 11)$

o. Zero is –7. $\qquad x^2 + 14x + 49 = (x + 7)(x + 7)$

p. Zero is 6. $\qquad y^2 - 12y + 36 = (y - 6)(y - 6)$

6. a. $x^2 + 3x = 70$.

Make right side zero by subtracting 70 from both sides: $x^2 + 3x - 70 = 0$.

Factor: $(x + 10)(x - 7) = 0$.

The zeros are –10 and 7. So the width is 7 yards and the length is 10 yards.

b. $x^2 + 3x = 108$.

Make right side zero by subtracting 108 from both sides: $x^2 + 3x - 108 = 0$.

Factor: $(x + 12)(x - 9) = 0$.

The zeros are −12 and 9. So the width is 9 yards and the length is 12 yards.

7. a. $x+7$
    b. $A(x) = x(x+7) = x^2 + 7x$

   c. $A(9) = 9^2 + 7(9) = 81 + 63 = 144$. So the area is 144 square feet when the width is 9 feet.

   d. If the length is 12 feet then the width is 5 feet.

   So $A(5) = 5^2 + 7(5) = 25 + 35 = 60$. The area is 60 square feet when the length is 12 feet.

   e. $A(x) = 78$. So $x^2 + 7x = 78$. Make the right side zero by subtracting 78 from both sides to get $x^2 + 7x - 78 = 0$. Factor: $(x+13)(x-6) = 0$. The zeros are −13 and 6. So the width is 6 feet and the length is 13 feet when the area is 78 square feet.

8. a. $\dfrac{x^2 - 8x + 15}{x^2 - 2x - 15} = \dfrac{(x-5)(x-3)}{(x-5)(x+3)} = \dfrac{x-3}{x+3}$

   b. $\dfrac{x^2 + 9x + 14}{x^2 + 2x - 35} = \dfrac{(x+2)(x+7)}{(x+7)(x-5)} = \dfrac{x+2}{x-5}$

   c. $\dfrac{3x - 6}{x^2 - x - 2} = \dfrac{3(x-2)}{(x-2)(x+1)} = \dfrac{3}{x+1}$

   d. Since the denominator cannot be factored the fraction is in reduced form, as is.

9. a. $\dfrac{5}{2x-10} - \dfrac{7}{3x-15} = \dfrac{5}{2(x-5)} - \dfrac{7}{3(x-5)} = \dfrac{15}{6(x-5)} - \dfrac{14}{6(x-5)} = \dfrac{1}{6(x-5)}$

   b. $\dfrac{2x}{(x-2)(x+1)} - \dfrac{5}{(x+3)(x+1)} = \dfrac{2x(x+3)}{(x-2)(x+1)(x+3)} - \dfrac{5(x-2)}{(x+3)(x+1)(x-2)} =$

   $\dfrac{2x(x+3) - 5(x-2)}{(x-2)(x+1)(x+3)} = \dfrac{2x^2 + 6x - 5x + 10}{(x-2)(x+1)(x+3)} = \dfrac{2x^2 + x + 10}{(x-2)(x+1)(x+3)}$

10. a. $5x + 7 = 11x + 5$
    $2 = 6x$
    $x = \dfrac{2}{6} = \dfrac{1}{3}$

   b. $5x + 7 > 11x + 5$
   $2 > 6x$
   $\dfrac{1}{3} > x$

c. $\begin{aligned} 5x+7 &< 11x+5 \\ 2 &< 6x \\ \frac{1}{3} &< x \end{aligned}$

d. $\begin{aligned} 7-3t &= 8t+6 \\ 1 &= 11t \\ \frac{1}{11} &= t \end{aligned}$

e. $\begin{aligned} 3x-11 &< 5x-1 \\ -2x &< 10 \\ x &> -5 \end{aligned}$

f. $\begin{aligned} 9x &> 3-4x \\ 13x &> 3 \\ x &> \frac{3}{13} \end{aligned}$

g. $7-t = 4-t$ has no solution.

h. $\begin{aligned} 5+3x &= 4-7x \\ 10x &= -1 \\ x &= -\frac{1}{10} \end{aligned}$

i. $\begin{aligned} \frac{2}{3}z - \frac{5}{6} &= 4z + \frac{3}{4} \\ 12\left(\frac{2}{3}z - \frac{5}{6}\right) &= 12\left(4z + \frac{3}{4}\right) \\ 8z - 10 &= 48z + 9 \\ -19 &= 40z \\ -\frac{19}{40} &= z \end{aligned}$

j. $\begin{aligned} 1 + \frac{1}{2}x &= 5x - \frac{2}{3} \\ 6\left(1 + \frac{1}{2}x\right) &= 6\left(5x - \frac{2}{3}\right) \\ 6 + 3x &= 30x - 4 \\ 10 &= 27x \\ \frac{10}{27} &= x \end{aligned}$

11. a. $\begin{aligned} 2x+y &= 7 \\ 5x-2y &= -1 \end{aligned}$ First solve the first equation for y to get $y = -2x+7$. Substitute in the second equation and solve for x.

$\begin{aligned} 5x - 2(-2x+7) &= -1 \\ 5x + 4x - 14 &= -1 \\ 9x - 14 &= -1 \\ 9x &= 13 \\ x &= \frac{13}{9} \end{aligned}$

Substitute this value for x in the first equation and solve for y.

$y = -2\left(\frac{13}{9}\right) + 7 = -\frac{26}{9} + 7 = -\frac{26}{9} + \frac{63}{9} = \frac{37}{9}$ The solution to the system is $\left(\frac{13}{9}, \frac{37}{9}\right)$.

b. $\begin{aligned} y &= 5x - 2 \\ 5x - y &= 17 \end{aligned}$ This system has no solution. Each of the equations has a graph which is

213

c. $\begin{array}{l}3x+4y = 9\\ x-5y = 0\end{array}$ First solve the second equation for x to get $x = 5y$. Substitute into the first equation and solve for y.

$$3(5y)+4y = 9$$
$$19y = 9$$
$$y = \frac{9}{19}$$

and $x = 5\left(\frac{9}{19}\right) = \frac{45}{19}$. The solution to the system is $\left(\frac{45}{19}, \frac{9}{19}\right)$.

d. $\begin{array}{l}x = 2-4y\\ 3x+y = 7\end{array}$ Replace x in the second equation with $2-4y$ and solve for y.

$$3(2-4y)+y = 7$$
$$6-12y+y = 7$$
$$-11y = 1$$
$$y = -\frac{1}{11}$$

And $x = 2-4\left(-\frac{1}{11}\right) = 2+\frac{4}{11} = \frac{26}{11} = 2\frac{4}{11}$. The solution to the system is $\left(\frac{26}{11}, -\frac{1}{11}\right)$.

12. Your system should be of the form $\begin{array}{l}ax+by = c\\ ax+by = d\end{array}$ where c and d are different numbers.

13. We must solve $3x-2 = 7x+9$ first. The solution is $x = -\frac{11}{4}$. To find the y-coordinate find the output for either function using this value as the input.

$y = 3\left(-\frac{11}{4}\right)-2 = -\frac{33}{4}-\frac{8}{4} = -\frac{41}{4}$ The intersection point is $\left(-\frac{11}{4}, -\frac{41}{4}\right)$.

14. a. $x+2y = 7$ is equivalent to $y = -\frac{1}{2}x+\frac{7}{2}$. So, the slope of the given line is $-\frac{1}{2}$. The equation of the requested line must be of the form $y = -\frac{1}{2}x+b$. To find b we substitute the coordinates $(-1, 2)$ into the equation and solve for b. $2 = -\frac{1}{2}(-1)+b$ has the solution $b = \frac{3}{2}$. The requested equation is $y = -\frac{1}{2}x+\frac{3}{2}$.

b. $4x - 3y = -1$ is equivalent to $y = \frac{4}{3}x + \frac{1}{3}$ so, the slope of the given line is $\frac{4}{3}$. The slope of a line perpendicular to this line must be $-\frac{3}{4}$. Solve $0 = -\frac{3}{4}(-4) + b$ to get $b = -3$. The equation of the requested line is $y = -\frac{3}{4}x - 3$.

Section 7.4 Additional Factoring Experiences

p. 607

2. a. prime
    b. $(2x + 1)(3x - 1)$
    c. $10(4y^2 + 3)$
    d. $(x + 8)(x + 5)$
    e. prime
    f. $(y - 2)(2y - 1)$
    g. $(y - 3)(8y + 7)$
    h. $(2x + 3)(3x - 2)$
    i. $(x + 4)(3x - 1)$
    j. $(5y - 11)(5y + 11)$
    k. prime
    l. prime
    m. $6(3x + 2)^2$
    n. $(x + 2)(x + 7)$
    o. $(3a + 5)(2a - 1)$
    p. $(x + 3)(x - 9)$
    q. $2(y - 4)(2y - 1)$
    r. $(x + 4)(4x - 1)$
    s. $t(t - 11)$
    t. $(2x + 5)(3x - 7)$
    u. $(2x - 5)(4x + 1)$
    v. $(x - 5)(4x + 9)$
    w. $(x + 8)^2$
    x. $(3x - 1)(3x + 1)$

3. a. $x + 5$
    b. $A(x) = x(x + 5) = x^2 + 5x$

    c. $A(11) = 11(11 + 5) = 11(16) = 176$ The area is 176 square feet when the width is 11 feet.

    d. We must solve $x(x + 5) = 14$. Simplifying the left side and subtracting 14 from both sides we get $x^2 + 5x - 14 = 0$. Factoring the left side we get $(x + 7)(x - 2) = 0$. The two solutions to this equation are -7 and 2. 2 is the only one that makes sense in this problem. So, the requested width is 2 feet and the length is 7 feet.

e. Solve this in a way similar to part d. 4 feet would be the width and 9 feet would be the length. Notice that the length is 5 feet more than the width and the area would be 36 square feet.

4. a. $2x - 1$ b. $A(x) = x(2x - 1)$

   c. $A(5) = 5(2 \cdot 5 - 1) = 5 \cdot 9 = 45$ The area would be 45 square feet.

   d. Solve $x(2x - 1) = 1$. The equation is equivalent to $2x^2 - x - 1 = 0$. Factoring the left hand side gives us $(2x + 1)(x - 1) = 0$. The only solution which makes sense in the problems $x = 1$. So, the width is 1 foot and the length is also 1 foot.

   e. Set up the equation $x(2x - 1) = 6$. The only solution to this equation which makes sense is $x = 2$. The width is 2 feet and the length is 3 feet.

5. **Table 3: Factoring**

Factored Form	Expanded Form
$(x-2)(x^2+2x+4)$	$x^3 - 8$
$(x+2)(x^2-2x+4)$	$x^3 + 8$
$(x-3)(x^2+3x+9)$	$x^3 - 27$
$(x+3)(x^2-3x+9)$	$x^3 + 27$
$(a-b)(a^2+ab+b^2)$	$a^3 - b^3$
$(a+b)(a^2-ab+b^2)$	$a^3 + b^3$

6. a. $(x-4)(x^2+4x+8)$ b. $(y+5)(y^2-5y+25)$

   c. $(3a+b)(9a^2-3ab+b^2)$ d. $2(2x-3y)(4x^2+6xy+9y^2)$

7. a. $2x - \dfrac{7}{3} = \dfrac{9}{5}$

   $2x = \dfrac{62}{15}$

   $x = \dfrac{31}{15}$

   b. $2 - 7z = 5z + 9$

   $-7 = 12z$

   $-\dfrac{7}{12} = z$

c.
$$x^2 + 3x - 4 = 0$$
$$(x+4)(x-1) = 0$$
$$x = -4 \quad \text{or} \quad x = 1$$

d.
$$4x > 7x - 1$$
$$-3x > -1$$
$$x < \frac{1}{3}$$

e.
$$9y + 5 \le 2y - 7$$
$$7y \le -12$$
$$y \le -\frac{12}{7}$$

f.
$$6x^2 - 11x - 10 = 0$$
$$(2x+5)(3x+2) = 0$$
$$x = \frac{5}{2} \quad \text{or} \quad x = -\frac{2}{3}$$

g.
$$\frac{4}{7}x - 1 = \frac{2}{5} + 3x$$
$$35\left(\frac{4}{7}x - 1\right) = 35\left(\frac{2}{5} + 3x\right)$$
$$20x - 35 = 14 + 105x$$
$$-49 = 85x$$
$$-\frac{49}{85} = x$$

h.
$$28y^2 + 43y = -10$$
$$28y^2 + 43y + 10 = 0$$
$$(4y+5)(7y+2) = 0$$
$$y = -\frac{5}{4} \quad \text{or} \quad y = -\frac{2}{7}$$

i.
$$3 + 10y \ge 6y - 2$$
$$4y \ge -5$$
$$y \ge -\frac{5}{4}$$

j.
$$4x^2 - 7x + 2 = 2$$
$$4x^2 - 7x = 0$$
$$x(4x - 7) = 0$$
$$x = 0 \text{ or } x = \frac{7}{4}$$

k.
$$4t^2 + 2t - 9 = 2t$$
$$4t^2 - 9 = 0$$
$$(2t-3)(2t+3) = 0$$
$$t = \frac{3}{2} \text{ or } t = -\frac{3}{2}$$

l.
$$16x^2 + 24x + 9 = 0$$
$$(4x+3)^2 = 0$$
$$x = -\frac{3}{4}$$

m.
$$\frac{4}{3}z - \frac{1}{6} \ge \frac{3}{4} - z$$
$$12\left(\frac{4}{3}z - \frac{1}{6}\right) \ge 12\left(\frac{3}{4} - z\right)$$
$$8z - 2 \ge 9 - 12z$$
$$20z \ge 11$$
$$z \ge \frac{11}{20}$$

n.
$$2p^2 + 15 = 13p$$
$$2p^2 - 13p + 15 = 0$$
$$(2p-3)(p-5) = 0$$
$$p = \frac{3}{2} \text{ or } p = 5$$

8. a.
$$4x - 7(2x - 3) = 1$$
$$4x - 14x + 21 = 1$$
$$-10x + 21 = 1$$
$$-10x = -20$$
$$x = 2$$

$y = 2(2) - 3 = 1$ The intersection point is $(2, 1)$.

b.
$$y = 5(3y + 2) - 1$$
$$y = 15y + 10 - 1$$
$$-9 = 14y$$
$$-\frac{9}{14} = y$$

$x = 3\left(-\frac{9}{14}\right) + 2 = -\frac{27}{14} + \frac{28}{14} = \frac{1}{14}$

The intersection point is $\left(\frac{1}{14}, -\frac{9}{14}\right)$.

c. Solve $2x - y = 7$ for y to get $y = 2x - 7$.

Substitute into the second equation and solve
$$3x + 5(2x - 7) = -2$$
$$3x + 10x - 35 = -2$$
$$13x = 33$$
$$x = \frac{33}{13}$$

Then find the y-coordinate by evaluating $y = 2\left(\frac{33}{13}\right) - 7 = \frac{66}{13} - \frac{91}{13} = -\frac{25}{13}$.

The intersection is at $\left(\frac{33}{13}, -\frac{25}{13}\right)$.

d. $x + 3y = 4$ is equivalent to $x = -3y + 4$. Substituting we solve

$$5(-3y + 4) - 8y = -1$$
$$-15y + 20 - 8y = -1$$
$$-23y = -21$$
$$y = \frac{21}{23}$$

$x = -3\left(\frac{21}{23}\right) + 4 = -\frac{63}{23} + \frac{92}{23} = \frac{29}{23}$

The intersection point is $\left(\frac{29}{23}, \frac{21}{23}\right)$.

9. a. $\dfrac{2x+8}{x^2+3x-4} = \dfrac{2(x+4)}{(x+4)(x-1)} = \dfrac{2}{x-1}$; $x \neq -4, x \neq 1$

  b. $\dfrac{2x^2+x}{4x^2-1} = \dfrac{x(2x+1)}{(2x+1)(2x-1)} = \dfrac{x}{2x-1}$; $x \neq -\dfrac{1}{2}, x \neq \dfrac{1}{2}$

  c. $\dfrac{x^2-10x+25}{x^2+4x+5}$ is reduced since the denominator cannot be factored.

  d. $\dfrac{4x^2-4x-8}{6x^2-8x-8} = \dfrac{4(x-2)(x+1)}{2(3x+2)(x-2)} = \dfrac{2(x+1)}{3x+2}$; $x \neq 2, x \neq -\dfrac{2}{3}$

10. a. $2x - 7y = 9$ is equivalent to $y = \dfrac{2}{7}x - \dfrac{9}{7}$. So the slope is $\dfrac{2}{7}$ and the vertical intercept is $\left(0, -\dfrac{9}{7}\right)$.

  b. $x + 5y - 8 = 0$ is equivalent to $y = -\dfrac{1}{5}x + \dfrac{8}{5}$. The slope is $-\dfrac{1}{5}$ and the vertical intercept is $\left(0, \dfrac{8}{5}\right)$.

11. a. $y = \dfrac{2}{3}x + 1$ is the requested equation.

  b. The slope of the line is $\dfrac{7-2}{3-(-1)} = \dfrac{5}{4}$. The equation of the line is of the form $y = \dfrac{5}{4}x + b$. Find b by substituting -1 for x and 2 for y. $2 = \dfrac{5}{4}(-1) + b$ is equivalent to $b = \dfrac{13}{4}$. The requested equation is $y = \dfrac{5}{4}x + \dfrac{13}{4}$.

  c. $4x - y = 2$ is equivalent to $y = 4x - 2$, so the slope of the given line and the requested line is 4. To find b we solve $4 = 4(-3) + b$ which gives us $b = 16$. The equation of the requested line is $y = 4x + 16$.

  d. $x + 6y = -5$ is equivalent to $y = -\dfrac{1}{6}x - \dfrac{5}{6}$. The slope of the given line is $-\dfrac{1}{6}$ which makes the slope of any line perpendicular to this line 6. To find b we solve $-2 = 6(-1) + b$ which gives us $b = 4$. The equation of the requested line is $y = 6x + 4$.

12. a. $\dfrac{4}{x} - \dfrac{3}{7x} = \dfrac{28}{7x} - \dfrac{3}{7x} = \dfrac{25}{7x}$

b. $\dfrac{1}{x-2} + \dfrac{2}{x+3} = \dfrac{x+3}{(x-2)(x+3)} + \dfrac{2(x-2)}{(x-2)(x+3)} = \dfrac{x+3+2(x-2)}{(x-2)(x+3)} =$

$\dfrac{3x-1}{(x-2)(x+3)}$

c. $\dfrac{9}{(2x-5)(3x+1)} - \dfrac{2x}{(2x-5)(x-4)} =$

$\dfrac{9(x-4)}{(2x-5)(3x+1)(x-4)} - \dfrac{2x(3x+1)}{(2x-5)(x-4)(3x+1)} =$

$\dfrac{9(x-4) - 2x(3x+1)}{(2x-5)(3x+1)(x-4)} = \dfrac{9x - 36 - 6x^2 - 2x}{(2x-5)(3x+1)(x-4)} = \dfrac{-6x^2 + 7x - 36}{(2x-5)(3x+1)(x-4)}$

d. $\dfrac{3x}{2(x-1)} + \dfrac{7}{4(x+1)} = \dfrac{2(3)x(x+1)}{4(x-1)(x+1)} + \dfrac{7(x-1)}{4(x+1)(x-1)} = \dfrac{6x^2 + 6x + 7x - 7}{4(x-1)(x+1)} =$

$\dfrac{6x^2 + 13x - 7}{4(x-1)(x+1)}$

13. It appears from the graph that (2, 0) and (0, 6) are points on the line. This would make the slope of the line $\dfrac{6-0}{0-2} = -3$. The equation of the line would be $y = -3x + 6$.

Section 7.5 Review

p. 613

1. a. Slope = 7   b. Vertical Intercept: (0, 21)

   c. Horizontal Intercept: (−3, 0)

   d. The function is increasing.

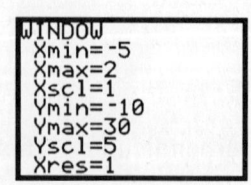

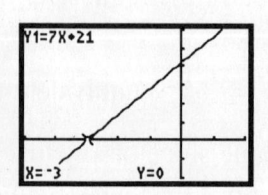

```
WINDOW
Xmin=-5
Xmax=2
Xscl=1
Ymin=-10
Ymax=30
Yscl=5
Xres=1
```

```
Y1=7X+21



X=-3     Y=0
```

2. a. Slope = $\dfrac{3}{7}$ b. Vertical Intercept: (0, 2)

220

c. Horizontal Intercept: $\left(-\frac{14}{3}, 0\right)$

d. The function is increasing.

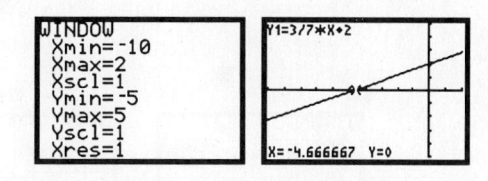

3. a. Slope = 0
   b. Vertical Intercept: (0,0)
   c. Horizontal Intercept: (0,0)
   d. The graph is the x-axis and the function is neither increasing nor decreasing.

4. a. Slope = $-\frac{3}{4}$
   b. Vertical Intercept: $\left(0, \frac{2}{3}\right)$
   c. Horizontal Intercept: $\left(\frac{8}{9}, 0\right)$
   d. The function is decreasing.

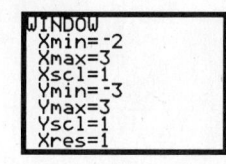

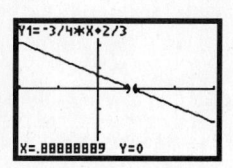

5. a. Slope = $\frac{5}{7}$
   b. Vertical Intercept: $\left(0, -2\frac{3}{4}\right)$
   c. Horizontal Intercept: $\left(\frac{77}{20}, 0\right)$
   d. The function is increasing.

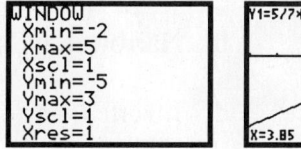

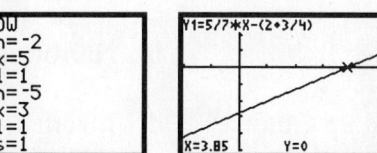

6. a.

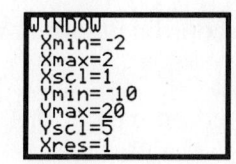

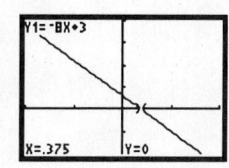

b. $y = -8x + 3$　　　　　c. Input at the horizontal intercept: $\dfrac{3}{8}$

d. $y = -8x$　　　　　e. $y = \dfrac{1}{8}x + 3$

7. a.

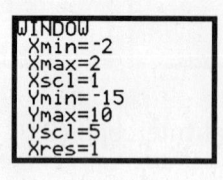

    ```
    WINDOW
    Xmin=-2
    Xmax=2
    Xscl=1
    Ymin=-15
    Ymax=10
    Yscl=5
    Xres=1
    ```

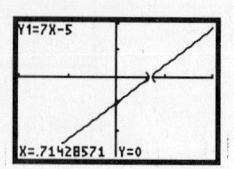

    Y1=7X-5, X=.71428571, Y=0

    b. $y = 7x - 5$　　　　　c. Input at the horizontal intercept: $\dfrac{5}{7}$

    d. $y = 7x - 8$　　　　　e. $y = -\dfrac{1}{7}x - 5$

8. a.

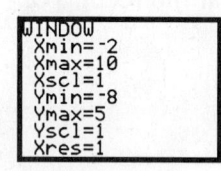

    ```
    WINDOW
    Xmin=-2
    Xmax=10
    Xscl=1
    Ymin=-8
    Ymax=5
    Yscl=1
    Xres=1
    ```

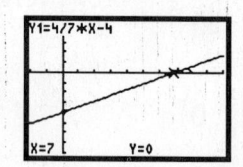

    Y1=4/7*X-4, X=7, Y=0

    b. $y = \dfrac{4}{7}x - 4$　　　　　c. Input at the horizontal intercept: 7

    d. $y = \dfrac{4}{7}x - 7$　　　　　e. $y = -\dfrac{7}{4}x - 4$

9. a. $y = x - 3$ assuming that the points $(0, -3)$ and $(4, 1)$ are on the graph.

    b. $y = 5x + 7$

    c. $y = -2x + 3$

    d. $y = -4x + 6$ assuming that the points $(0, 6)$ and $(2, -2)$ are on the graph.

10. a. Narrower　　　　　b. Narrower and moved down 5 units.

    c. Wider and moved up 4 units.　　　　　d. Inverted, narrower and moved up 11 units.

11. a. Zeros: 3 and −3　　　　　x-coordinate of the vertex: 0

    b. Zeros: 5 and −4　　　　　x-coordinate of the vertex: $\dfrac{1}{2}$

    c. Zeros: −7 and −6　　　　　x-coordinate of the vertex: $-\dfrac{13}{2}$

    d. Zeros: 8 and 2　　　　　x-coordinate of the vertex: 5

12. a. $y = (x+8)(x-7) = x^2 + x - 56$

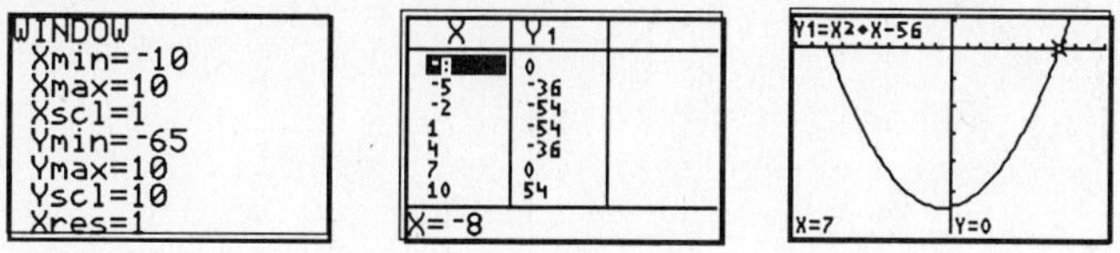

   b. $y = (x+9)(x+4) = x^2 + 13x + 36$

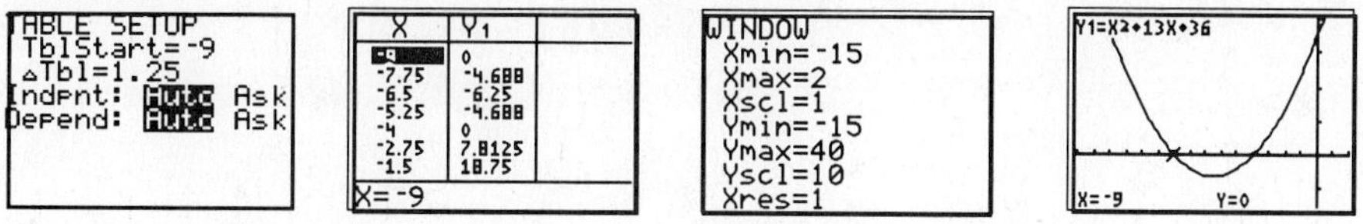

   c. $y = (x-4)(x+4) = x^2 - 16$

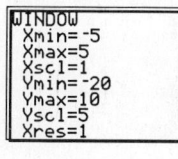

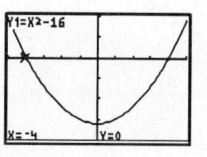

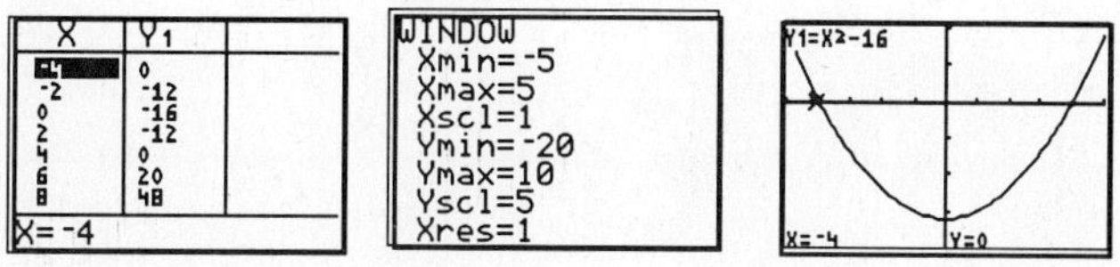

   d. $y = (x+7)^2 = x^2 + 14x + 49$

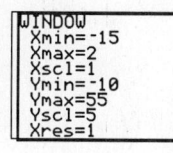

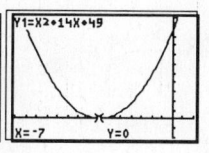

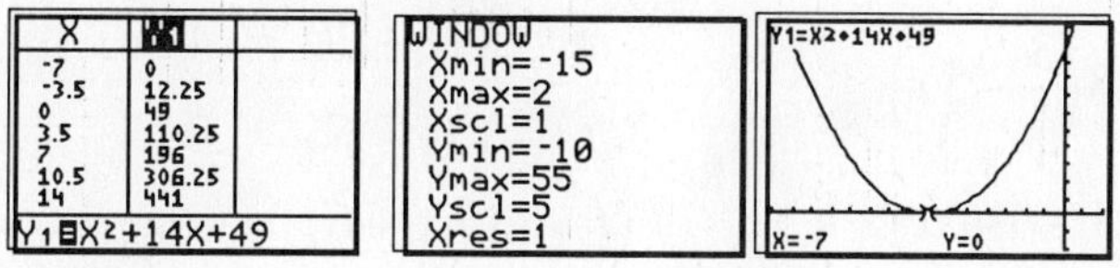

13. a. $(x-10)(x+10)$ b. $(y-9)^2$

   c. prime   d. $2(y-3)(y+8)$

   e. $4(6x+5)$ f. $(t-11)(t-6)$

   g. $5(x+3)^2$ h. $7(x^2+4)$

   i. $(x+8)(x+9)$ j. $4(2x+3)^2$

14. a. $y(-2) = 3(-2) - 5 = -6 - 5 = -11$

   b. $3x - 5 = -2 \quad\quad 3x = 3 \quad\quad x = 1$

15. $0 = -2x - 7 \quad\quad 7 = -2x \quad\quad -\dfrac{7}{2} = x$

16.

a.
$$5x - 7 = 15$$
$$5x = 22$$
$$x = \frac{22}{5}$$

b.
$$3 - 11t = 24$$
$$-11t = 21$$
$$t = -\frac{21}{11}$$

c.
$$4x - 3 > 9$$
$$4x > 6$$
$$x > \frac{3}{2}$$

d.
$$y(y + 18) = -81$$
$$y^2 + 18y = -81$$
$$y^2 + 18y + 81 = 0$$
$$(y + 9)^2 = 0$$
$$y = -9$$

e.
$$3 + 2y = 12y - 6$$
$$9 = 10y$$
$$\frac{9}{10} = y$$

f.
$$1 - 2y < 17$$
$$-2y < 16$$
$$y > -8$$

g.
$$\frac{2}{7}x + 1 = \frac{3}{4} - 5x$$
$$28\left(\frac{2}{7}x + 1\right) = 28\left(\frac{3}{4} - 5x\right)$$
$$8x + 28 = 21 - 140x$$
$$148x = -7$$
$$x = -\frac{7}{148}$$

h.
$$t^2 - 17t + 70 = 4$$
$$t^2 - 17t + 66 = 0$$
$$(t - 11)(t - 6) = 0$$
$$t = 11 \text{ or } t = 6$$

i.
$$3p - 8 = 4 - 5p$$
$$8p = 12$$
$$p = \frac{12}{8} = \frac{3}{2}$$

j.
$$2y^2 = 10y + 48$$
$$2y^2 - 10y - 48 = 0$$
$$2(x - 8)(x + 3) = 0$$
$$x = 8 \text{ or } x = -3$$

k.
$$\frac{5}{2}t - \frac{4}{3} < 9 + \frac{8}{6}t$$
$$6\left(\frac{5}{2}t - \frac{4}{3}\right) < 6\left(9 + \frac{8}{6}t\right)$$
$$15t - 8 < 54 + 8t$$
$$7t < 62$$
$$t < \frac{62}{7}$$

l. No value of x will satisfy this inequality.

17. a. The system is equivalent to the system $\begin{array}{l} y = \frac{3}{5}x - \frac{7}{5} \\ y = -4x - 2 \end{array}$. These functions may be graphed in the standard viewing window and the intersect item on the **Calculate** menu can be used to find an approximation to the intersection point.

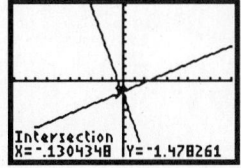

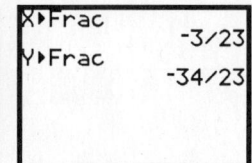

By returning to the home screen and entering X and Y along with the Frac item from the **Math** menu these values can be converted to fraction form. It is difficult to use a table to find the intersection point.

To find the solution by substitution one may solve the equation $\frac{3}{5}x - \frac{7}{5} = -4x - 2$ to get $x = -\frac{3}{23}$ and evaluate one of the functions using this input value to get $y = -\frac{34}{23}$.

b. Solving each equation for y gives us two functions which we can graph. We get $y = -3x + \frac{9}{2}$ and $y = -3x - 1$. We can see that these two functions have graphs which are straight lines with the same slope and different vertical intercepts. The lines are therefore parallel and there is no intersection point. The graphs appear in the standard viewing window.

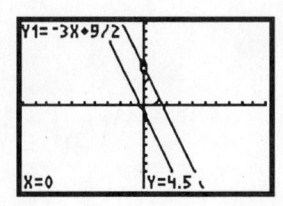

c. We solve each equation for y so that we may enter them into the Y= menu on our TI-83 or 82's. $y = 4x + 8$ and $y = -\frac{3}{2}x + \frac{7}{2}$. The graphs and the intersection point found by using the intersect item on the **Calculate** menu are shown. The standard viewing window has been chosen. The fraction form of the coordinates are also shown.

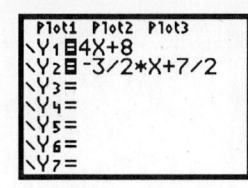

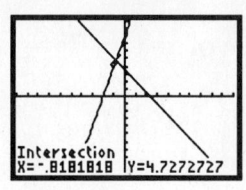

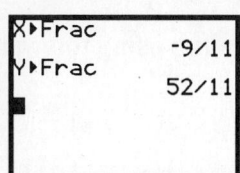

To solve the system by substitution we solve the equation $4x + 8 = -\frac{3}{2}x + \frac{7}{2}$ for x to get

$$2(4x+8) = 2\left(-\frac{3}{2}x+\frac{7}{2}\right) \qquad 8x+16 = -3x+7 \qquad 11x = -9 \qquad x = -\frac{9}{11}.$$

The y value is found by evaluating one of the functions:

$$y = 4\left(-\frac{9}{11}\right)+8 = -\frac{36}{11}+\frac{88}{11} = \frac{52}{11}$$

d. The two equations are equivalent to $\begin{array}{c} y = -\frac{1}{5}x+\frac{7}{5} \\ y = 5x-2 \end{array}$. Enter these two functions into the Y= menu and use the intersect item on the **Calculate** menu to find the intersection point graphically.

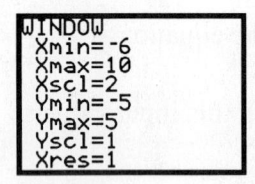

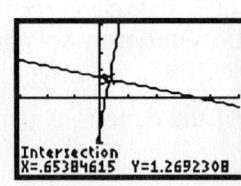

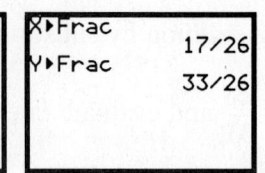

```
WINDOW
Xmin=-6
Xmax=10
Xscl=2
Ymin=-5
Ymax=5
Yscl=1
Xres=1
```

```
Intersection
X=.65384615  Y=1.2692308
```

```
X▶Frac
          17/26
Y▶Frac
          33/26
```

To find the intersection point by substitution first solve $-\frac{1}{5}x+\frac{7}{5} = 5x-2$ for x.

$$5\left(-\frac{1}{5}x+\frac{7}{5}\right) = 5(5x-2)$$

$$-x+7 = 25x-10$$

$$17 = 26x$$

$$\frac{17}{26} = x$$

$$y = 5\left(\frac{17}{26}\right)-2 = \frac{85}{26}-\frac{52}{26} = \frac{33}{26}$$

18. Find the x-coordinate by solving $7x+2 = -5x+3$ to get $x = \frac{1}{12}$. The y-coordinate is found by evaluating using this value of x: $y = 7\left(\frac{1}{12}\right)+2 = \frac{7}{12}+\frac{24}{12} = \frac{31}{12} = 2\frac{7}{12}$.

19.
$$x^2 + 3x = 4$$
$$x^2 + 3x - 4 = 0$$
$$(x+4)(x-1) = 0$$
$$x = -4 \text{ or } x = 1$$

The only solution that makes sense in the problem is the positive one. The width is 1 foot and the length is 4 feet.

20. a. $y = \dfrac{1}{15}x + \dfrac{19}{15}$ b. $y = -\dfrac{13}{4}x + \dfrac{49}{2}$

 c. $y = 3$ d. $x = -5$

21. a. $y = -\dfrac{4}{7}x + \dfrac{9}{7}$ Slope $= -\dfrac{4}{7}$

 Vertical Intercept: $\left(0, \dfrac{9}{7}\right)$ Horizontal Intercept: $\left(\dfrac{9}{4}, 0\right)$

 b. $y = \dfrac{4}{3}x + 5$ Slope $= \dfrac{4}{3}$

 Vertical Intercept: $(0, 5)$ Horizontal Intercept: $\left(-\dfrac{15}{4}, 0\right)$

22. a.

Number of Sweatshirts Sold	Net Profit ($)
0	-500
10	-300
20	-100
30	100
40	300
50	500
60	700

b. $y(x) = 20x - 500$ where x represents the number of sweatshirts sold and $y(x)$ represents the net profit.

c. The ratio of the change in the net profit to the number of sweatshirts sold is $20 per sweatshirt.

d. The graph is shown with the vertical intercept displayed using the Trace feature. The vertical intercept indicates that when they have sold no sweatshirts they will be in the whole $500.

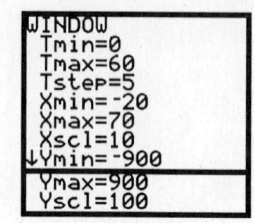

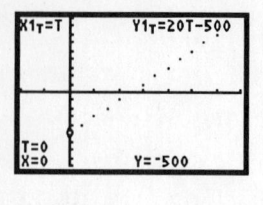

23. a. $\dfrac{12x-8}{3x^2+10x-8} = \dfrac{4(3x-2)}{(3x-2)(x+4)} = \dfrac{4}{x+4}; \; x \neq -4, \; x \neq \dfrac{2}{3}$

b. $\dfrac{9x^2+24x+16}{9x^2-16} = \dfrac{(3x+4)^2}{(3x+4)(3x-4)} = \dfrac{3x+4}{3x-4}; \; x \neq \dfrac{4}{3}, \; x \neq -\dfrac{4}{3}$

24. a. $\dfrac{2}{x-7} - \dfrac{5}{x+4} = \dfrac{2(x+4)}{(x-7)(x+4)} - \dfrac{5(x-7)}{(x-7)(x+4)} = \dfrac{2(x+4)-5(x-7)}{(x-7)(x+4)} =$

$\dfrac{2x+8-5x+35}{(x-7)(x+4)} = \dfrac{-3x+43}{(x-7)(x+4)}; \; x \neq 7, \; x \neq -4$

b. $\dfrac{3x}{2x^2+5x+2} + \dfrac{4}{2x^2+7x+3} = \dfrac{3x(x+3)}{(2x+1)(x+2)(x+3)} + \dfrac{4(x+2)}{(2x+1)(x+2)(x+3)} =$

$\dfrac{3x(x+3)+4(x+2)}{(2x+1)(x+2)(x+3)} = \dfrac{3x^2-9x+4x+8}{(2x+1)(x+2)(x+3)} = \dfrac{3x^2-5x+8}{(2x+1)(x+2)(x+3)}$

$x \neq -\dfrac{1}{2}, \; x \neq -2, \; x \neq -3$

Assessment:
Group Exams
and
Individual Skills Exams

PROBLEM SET 1A (GROUP)

Names of people in team:

_____

_____

_____

_____

Clearly record all answers in the space provided. Write in complete sentences and be sure all parts of the question are answered.

1. For each part, indicate if the variable has a domain that is a subset of the whole numbers. Justify your answers.

    a. Let $s=$ how many students are in a class.

    b. Let $t =$ how tall each student is.

    c. Let $b =$ how many buses are needed to transport 457 students.

2. Prove that a valid path in the S P C system can have a P-count of 25.

3. Use the S P C Rules to make the valid path S C C P P P P C as short as possible. Show each step using appropriate notation.

4. **Sandwich Purchase:** Luanne bought a sandwich from the vending machine for $1.65. The vending machine takes nickels, dimes, and quarters. She dropped 19 coins into the vending machine to pay for her sandwich. How many of each kind of coin did Luanne use to pay for the sandwich? Find all possible answers to this problem. Prove that there are no other answers.

PROBLEM SET IB (GROUP)

Names of people in team:

_____
_____
_____
_____
_____

Clearly record all answers in the space provided. Write in complete sentences and be sure all parts of the question are answered. Have fun!

1. One of your friends claims to be able to read minds. He asks you to pick a number between 1 and 10, add 2 to it, and square the result, subtract the square of your original number, divide the result by 4 and finally, subtract your original number.

    a. Demonstrate your understanding of the problem by selecting three values between 1 and 10 and calculating the answer.

    b. Write the calculation, using correct grouping symbols and operations.

    c. Generalize your result for any number between 1 and 10 by writing an algebraic statement of the computation, with x representing any number selected (between 1 and 10).

2. Explain how the distributive property is used "behind the scenes" in the FOIL method. Some mathematicians content that such short-cuts are the essence of mathematics. Others warn that short-cuts cause important concepts to be overlooked. What do you think about using short-cuts such as the FOIL method? Has it been a successful shortcut for you in the past? Why or why not?

3. Use the function machine to answer the following.

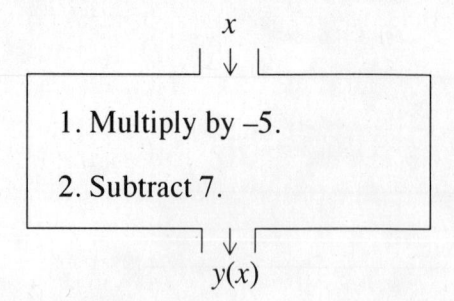

$$x$$
$$\downarrow$$

| 1. Multiply by –5. |
| 2. Subtract 7. |

$$\downarrow$$
$$y(x)$$

   a. Find the output if the input is –7.

   b. Write an algebraic representation (equation) for $y(x)$.

   c. Find the input if the output is 5. Describe how you found the answer.

4. a) Describe the difference between $2n$ and n^2.

   b) Describe the difference between n^2 and 2^n.

5. You need to purchase 15 prizes and you have $36 to spend. The items available for prizes cost either $2.00 or $3.00. How many different items at each price will you purchase?

   Solve this problem using two different strategies. For each strategy, show all work . Compare the two strategies and state which you prefer. Justify your selection.

   Strategy 1:

   Strategy 2:

6. Simplify each expression by performing the indicated operations. Indicate the order of simplification by recording your steps.

   a) $83 - 14(10) + 172$

   b) $192 - \sqrt{49} + 126 \div 3$

   c) $7(3) - \dfrac{5 - 2(2^3 - 7)^2}{3 + \sqrt{7 + 2}}$

Use these numbers to answer questions 8- 10.

$-\sqrt{7}$ $-\dfrac{5}{8}$ 1 $\sqrt{4}$ 2 51 $\dfrac{11}{13}$ $0.\bar{1}$ 0 $\sqrt{-7}$ -2

7. List all even numbers._____

8. List all prime numbers._____

9. List all whole numbers._____

10. Given cubes (or blocks) of two different colors:

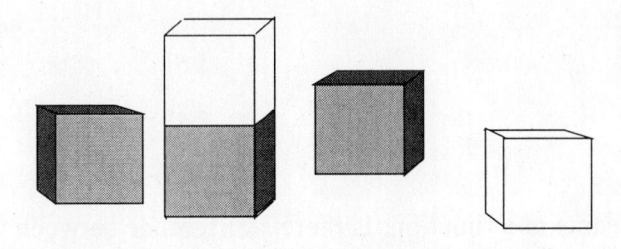

   a) build as many different towers as possible four cubes high.

   b) build as many different towers as possible five cubes high.

   c) determine whether you have built all possible towers without omitting or duplicating any.

   d) provide a convincing argument that all possible arrangements have been found.

   Describe your strategies and record all efforts, clearly documenting both your thinking process and your results.

PROBLEM SET IIA (GROUP)

Names of people in team **who worked on this problem set:**

_____

_____

_____

_____

_____

Clearly record all answers in the space provided. Write in complete sentences and be sure all parts of the question are answered.

1. Which of the following graphs (a., b., or c.) best depicts the distance of a runner from the starting line of a race over time if she must climb a large hill in the middle of the race? Explain.

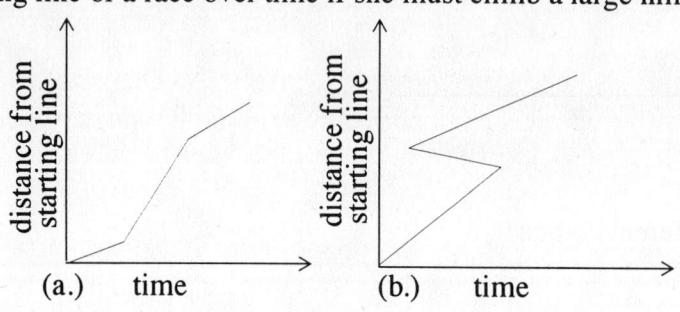

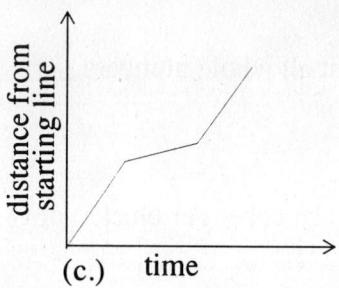

2. Assume that f is the name of a function. Is there a difference between $3f(2)$ and $2f(3)$? If yes, what is the difference? If no, explain why there is no difference. Provide an example to back up your answer.

3. A friend, Sue, comes to you with her definition of function. Is her definition acceptable? Why or why not? How does it "fit" with your definition?

   Sue's definition: A **function** is a **correspondence** that assigns to each element of one set one and only one element of a second set. A diagram appears below. h2

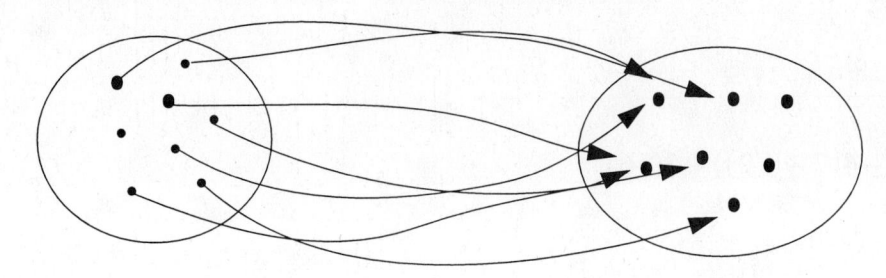

4. A room contains a group of people. Each shakes hands with all the other people in the room. The number of handshakes is a function of the number of people in the room. Let the number of people in the room be the input and the number of handshakes be the output. Complete the table and describe the pattern in the finite differences in the number of handshakes.

Number of people	Number of handshakes
2	
3	
4	
5	
6	
7	
8	

Problem Set IIB (Group)

1. Calculate. Show all work:

    a. $-19 - 7 =$

    b. $-|-4| + (-3) =$

    c. $3 - 4(7 - (-2)) + 5(-6) =$

2. Find the number of dots on an equilateral triangle that has 74 dots on a side. Write down the algebraic notation for your answer and describe what you did.

3. If $Z(p) = 2p^2 - 5p + 7$, find $Z(-3)$. Show all work and briefly state what you did.

4. You will rent a car for one day. It costs $23 plus 14¢ per mile. Let $C(m)$ represent the total cost of renting the car if you drive m miles.

    a. What is the input?

    b. What is the output?

    c. Write an algebraic representation (equation) for $C(m)$.

5. This question refers to the S P C system from Chapter 1.

    a. If you apply Rule 3 to the valid path S P P P P C, you get the valid path S C P C. What was the input to Rule 3? What was the output from Rule 3? Is Rule 3 acting like a function? Why or why not?

    b. Create your own example similar to part a, but using a different rule. Indicate the input, the rule used, and the output. Is your example a function? Why or why not?

6. In a given town, it has been observed that there are three U.S.-made cars for every foreign-made car.

   a. Complete Table 1.

Foreign-made cars	U.S.-made cars
1000	
2000	
3000	
4000	

   Table 1: Foreign versus U.S.-made cars

   b. Write an algebraic formula (equation) expressing the number of U.S.-made cars A as a function of the number of foreign-made cars F.

   c. Graph the relationship. Label axes. Label each point from Table 1 with an ordered pair.

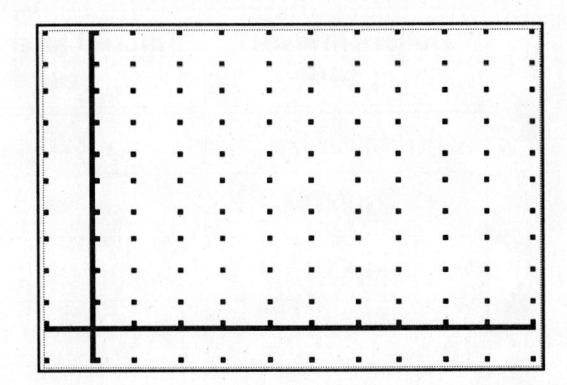

7. Which of the following graphs (a, b, or c) best represents the distance of a runner from the starting line of a race over time if she must climb a large hill in the middle of the race?

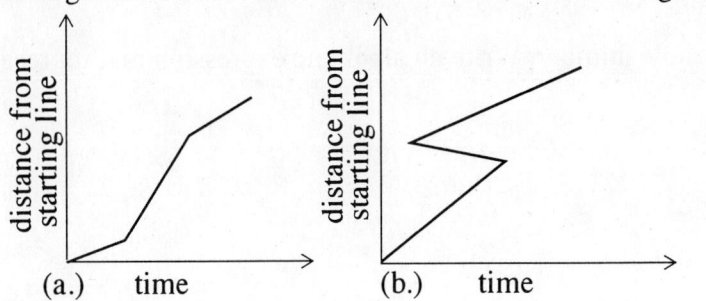

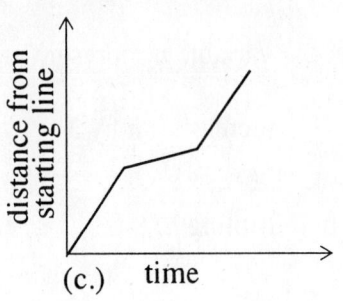

   (a.) (b.) (c.)

   Explain.

8. You just inherited $10,000 and you wish to invest it. You decide to invest part of the money in a somewhat risky account that pays 10% interest annually and the rest in a safer account that pays 8% interest annually. The graph below shows how the interest received at the end of one year depends on the amount invested at 10%.

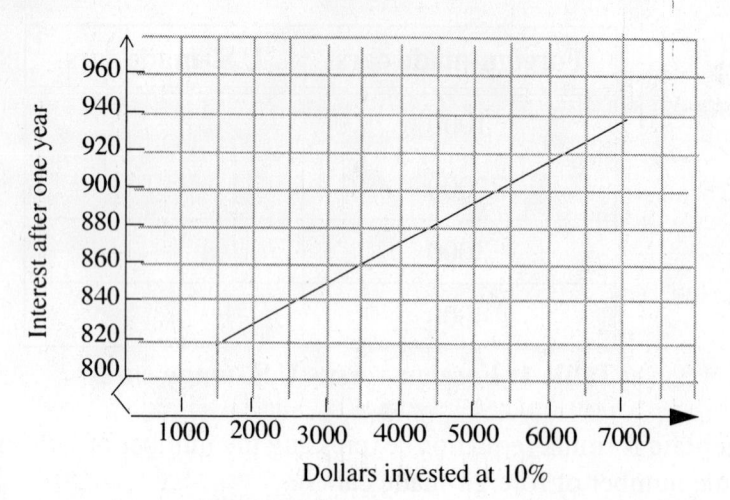

a. Use the graph to complete Table 2.

Dollars invested at 10%	Interest after one year
	820
3000	
4000	
	920
7000	

Table 2: Interest earnings

9. Let the variable n represent any whole number. Write an algebraic expression that represents

a. increasing n by 3.

b. tripling n.

c. cubing n.

10. A room contains a group of people. Each shakes hands with all the other people in the room. The number of handshakes is a function of the number of people in the room.

   a. Let the number of people in the room be the input and the number of handshakes be the output. Complete Table 3.

Number of people	Number of hand-shakes
2	
3	
4	
5	
6	
7	
8	

   Table 3: Handshakes

   b. Analyze the sequence of numbers displayed as outputs in Table 3. Record your observations, being as informative as possible.

11. Given the ordered pairs (-3, 7), (2, -18), (7, 27), (-1, -42), and (4, 38), state the six boundary values for a viewing window that would contain all the points corresponding to the ordered pairs.

   xmin = _____ ymin = _____

   xmax = _____ ymax = _____

   xscl = _____ yscl = _____

12. What is the opposite of $3x^2 - 7x + 2$? _____

13. Write the opposite and the absolute value for each of the following:

   a. -18 b. $7p$

PROBLEM SET IIC (GROUP)

In your group you should work together to answer the following questions. Write out your answers, but be prepared to explain what you did without undue reliance on your notes. You will also need to be able to answer related questions which are not explicitly asked here. Finally the instructor will determine which member of the group will begin each question, so each of you should be able to answer every question.

1. A major idea in Chapter 3 is function. Discuss everything which pertains to the function concept. Use examples of functions to illustrate different representations of the same function. Draw a concept map to illustrate your discussion.

2. Do a finite difference table for each of the functions: $f(x) = x^4$ and $g(x) = 3x^4 - 2x^2 - 1$. Use -4 to 6 as your input values. How are they alike? How are they different? Discuss the patterns you found in the tables.

3. There are eight colored pencils in a package. How many different ways can the pencils be lined up in the package? Suppose you only have four colors--red, blue, green and yellow. How many ways can they be lined up? List all of the ways. What function allows you to answer how many ways we can do this? Describe how you know that you have all of the ways in the list.

4. Represent the following expression with two different models of operations on integers:

    a. 5 + (-2)   b. (-3) + (-7)   c. 4 - 7   d. (-3) - 6   e. (-2) - (-7)

    f. 3(-4)   g. (-5)(-3)

5. Discuss the *absolute value* function. Show its definition, a table, function machine, algebraic expression, graph and a story problem which makes use of the idea.

6. Discuss the *opposite* function. Show its definition, a table, function machine, algebraic expression, graph and a story problem which makes use of the idea.

Clearly record all answers in the space provided. Write in complete sentences and be sure all parts of the question are answered. All work should be done by you, not someone else.

PROBLEM SET IIIA (GROUP)

1. The following statements are false. Correct each by changing the right side of "=".
   a. If x is larger than two then $|1 - x| = 1 + x$. Correct right side:_____

   b. $(3^2)(3^3) = 9^5$ Correct right side:_____

   c. $x + y - 3(z + w) = x + y - 3z + w$ Correct right side:_____

   d. $a^2 b^5 = (ab)^7$ Correct right side:_____

   e. $3x^{-1} = \dfrac{1}{3x}$ Correct right side:_____

   f. $(3a)^4 = 3a^4$ Correct right side:_____

   g. $-4^2 = 16$ Correct right side:_____

   h. $(x + 4)^2 = x^2 + 16$ Correct right side:_____

2. Find the value of the input that makes the outputs of the functions $f(x) = 3x - 5$ and $f(x) = 3 - 7x$ the same. Briefly describe what you did.

   a. Draw a function machine for the expression $-x^2$.

   b. Draw a function machine for the expression $(-x)^2$.

3. Consider the linear function $y(x) = 7x - 5$.

   a. What is the slope? _____

   b. What is the vertical intercept? _____

   c. What is the horizontal intercept? _____

   d. Find x so that $y(x) = 16$. Describe how you found the answer.

4. Consider the quadratic function $y(x) = x^2 + x - 20$.

   a. What are the zeros of the function? _____

   b. What is the x–coordinate of the vertex? _____

   c. What is the factored form of the function? _____

5. Consider the graph below.

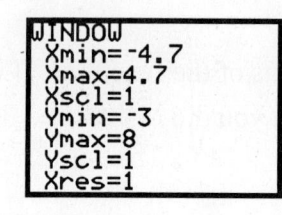

   ```
   WINDOW
   Xmin=-4.7
   Xmax=4.7
   Xscl=1
   Ymin=-3
   Ymax=8
   Yscl=1
   Xres=1
   ```

   a. What is the slope? _____

   b. What is the vertical intercept? _____

   c. What is the horizontal intercept? _____

   d. What is the equation of the line? _____

6. Consider the graph below.

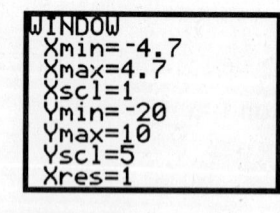

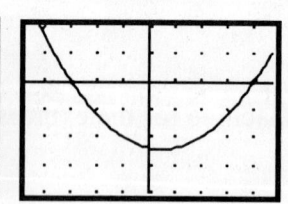

   ```
   WINDOW
   Xmin=-4.7
   Xmax=4.7
   Xscl=1
   Ymin=-20
   Ymax=10
   Yscl=5
   Xres=1
   ```

   a. Write the horizontal intercepts. _____

   b. Write the factors of the function. _____

7. A line has slope $\frac{2}{5}$ and a vertical intercept $(0, -3)$. Draw the graph of the line below.

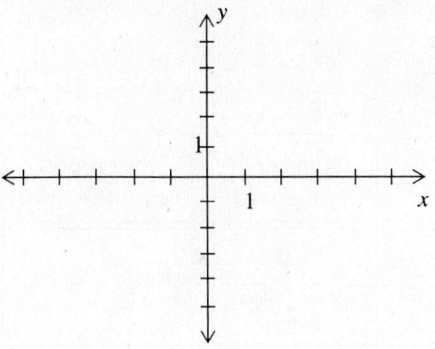

8. Describe how the slope effects the graph of a line. Address the cases when the slope is positive, when the slope is zero, and when the slope is negative. Also address the cases when the absolute value of the slope is large versus when the absolute value of the slope is small.

9. Use the graph of $y(x)$ below to answer parts a and b. The scale on both axes is one.

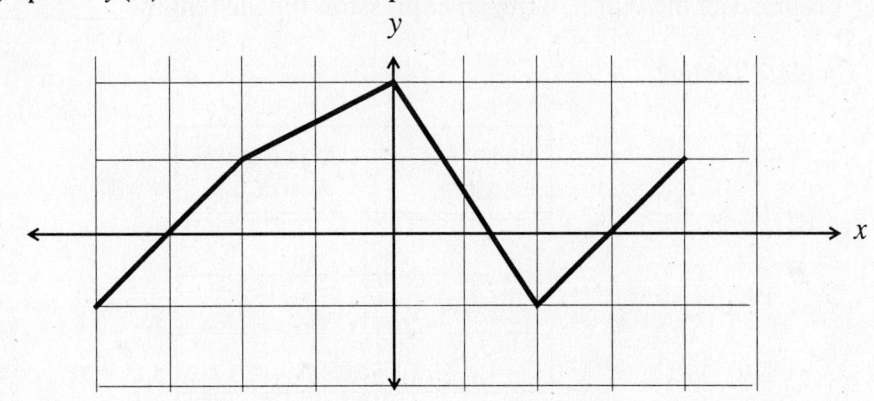

   a. What is $y(4)$? What did you do?

   b. If $y(x) = -1$. what is x? what did you do?

10. Jerry manufactures rings that she sells on weekends. For a month her supplies cost $200. She sells each ring for four dollars.

   a. Complete Table 1.

Number of rings sold	Net profit ($)
0	
25	
60	
100	

   Table 4: Jerry's profit for a month

   b. Write an equation that defines the net profit P as a function of the number r of rings sold.

   c. Find the breakeven point; that is the number of rings that most be sold so that Jerry's net profit for the month is zero. Describe how you found the answer.

   d. How many rings must be sold for Jerry to realize a net profit of $60? Describe how you found the answer.

11. My dog, Ocheslio, requires a rectangular dog pen that is two feet longer than it is wide.

   a. If x represents the width, write an expression for the length. _____

   b. Complete Table 2.

Width (feet) of pen	Area (square feet) of pen
5	
10	
15	
20	

   Table 5: Dimensions of Ocheslio's dog pen

   c. Estimate the dimensions of the pen if the area is 100 square feet. Describe what you did.

UNIT IA (INDIVIDUAL SKILLS)

Name_____

Clearly record all answers in the space provided. Write in complete sentences and be sure all parts of the question are answered.

1. Consider the expression $3 + 4(2)^3 - 8 \div 4$.

   a. Write the answer. _____
   b. List the order of operations.

   c. Insert parentheses in the original expression so that the addition is done last.

2. Is the path S P P P P P P P C P C P a valid path in the S P C system? Justify your answer.

3. Let the variable q represent any whole number. Write an algebraic expression for eight more than the product of q and 11.

4. Find the output of $46 - 3t + 4t^2$ if $t = 11$. Show your work.

5. Write a generalization for the numerical examples $6 + 2 = 2 + 6$, $18 + 23 = 23 + 18$, $1 + 0 = 0 + 1$, etc.

6. Write a numerical example of the distributive property of multiplication over addition.

   Write an example showing that the set of whole numbers is *not* closed under division.

7. If you apply Rule 3 to the valid path S C P P P C P C P, identify the sequence of letters that is represented by the variable x and the sequence of letters that is represented by the variable y.

8. Describe a situation that can be represented mathematically by the expression $75 - 2c$.

9. Are the whole numbers 113 and 1633 prime numbers? Provide proof for your answer.

10. Draw a function machine representation for the expression $12 - 3x$.

11. Explain why each number is or is not logically a member of each statement's domain.

   a. −20 cm. for the length of the side of a rectangle.

   b. −30 degrees for the temperature in Minnesota in January.

   c. 350 hours for the time spent studying algebra this week.

12. Given the function machine

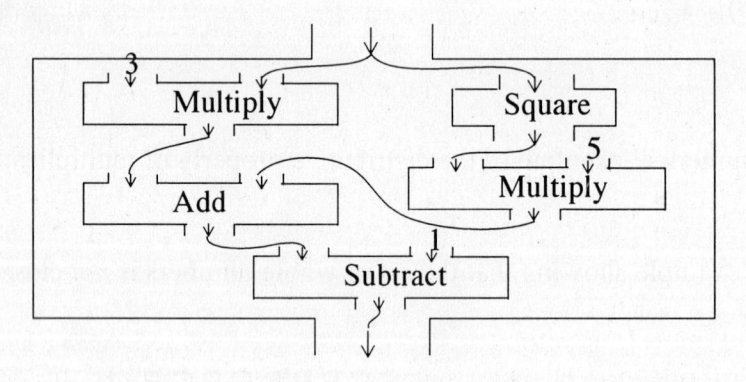

   a. Find the output if the input is 6.   Answer:

   b. Write the algebraic expression. Assume that the input variable is x.

UNIT IB (INDIVIDUAL SKILLS)

1. Given the expression $3 + 4(2)^3 - 8 \div 4$

   a. Write the answer. _____

   b. List the order of operations.

   c. Insert parentheses so that the addition is done last.

2. Is the path S P P P P P P P C P C P a *valid* path in the S P C system? Justify your answer.

3. Let the variable q represent any whole number. Write an algebraic expression for eight more than the product of q and 11.

4. Find the output of $46 - 3t + 4t^2$ if $t = 11$. Show your work.

5. Write a generalization for the numerical examples $6 + 2 = 2 + 6$, $18 + 23 = 23 + 18$, $1 + 0 = 0 + 1$, etc.

6. Write a numerical example of the distributive property of multiplication over addition.

7. Write an example showing that the set of whole numbers is *not* closed under division.

8. Refer to Figure 1 on page 31 of the text. If you apply Rule 3 to the path S P P P C P C P, identify the sequence of letters that is represented by the variable x. and the sequence of letters that is represented by the variable y.

9. Describe a situation that can be represented mathematically by the expression $75 - 2c$.

10. Beginning with the path S P, show that the path S C C P P C is a valid path in the SPC system. Show each step in the process and state the rule used at each step of the process.

11. Draw a function machine representation for the expression $12 - 3x$.

Do problem 12 on a separate sheet of paper.

12. The SPC system is an example of a logical system. The set of whole numbers is an example of a mathematical system. List the components of each system. Write a paragraph describing the similarities and differences between the systems.

Extra Credit: Is 191 a prime number or a composite number. Clearly justify your answer.

UNIT IIA (INDIVIDUAL SKILLS)

Clearly record all answers in the space provided. Write in complete sentences and be sure all parts of the question are answered.

1. Write the value:
    a. $15 + 5(4) \div 10 =$

    b. $-19 - 7 =$

    c. $3 - 4(7 - (-2)) + 5(-6) =$

2. Consider the diagram.

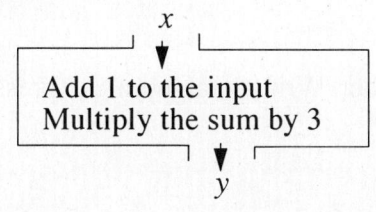

    x → [Add 1 to the input / Multiply the sum by 3] → y

    a. What are the output(s) if the input is 7? What did you do?

    b. What are the input(s) if the output is 18? What did you do?

3. Consider the equation $y = 3x - 7$.

    a. What are the output(s) if the input is 5? What did you do?

    b. What are the input(s) if the output is 0? What did you do?

4. Consider the following table copied from a TI–82 graphics calculator.

X	Y1
-3	12
-2	5
-1	0
0	-3
1	-4
2	-3
3	0

    X = -3

    a. What are the output(s) if the input is –2? What did you do?

    b. What are the input(s) if the output is –3? What did you do?

5. Given the function $Z(p) = 2p^2 - 5p + 7$, find:

    a. $Z(4)$

    b. $Z(-3)$

    c. $Z(t)$

6. Clearly describe how to find the column that 75967 will be in for the table on page 88 of your text.

7. Let the variable n represent any whole number. Write an algebraic expression that represents

    a. Increasing n by 3.
    b. Tripling n.
    c. Cubing n

8. In a given town, it has been observed that there are three U.S.–made cars for every foreign–made car.

    a. Complete Table 1.

Foreign–made cars	U.S.–made cars
1000	
2000	
3000	
4000	

    Table 6: Foreign versus U.S.–made cars

    b. Write an algebraic formula (equation) expressing the number of U.S.–made cars A as a function of the number of foreign–made cars F.

c. Graph the relationship. Label axes. Label each point from Table 1 with an ordered pair.

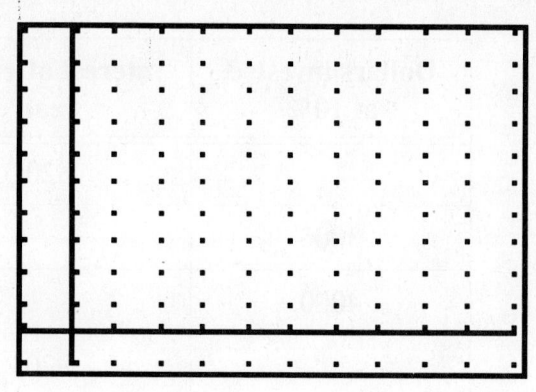

9. You will rent a car for one day. It costs $23 plus 14¢ per mile. Let $C(m)$ represent the total cost of renting the car if you drive m miles.

   a. What is the input and what is the output?

   b. Draw a function machine.

   c. Write an algebraic representation (equation) for $C(m)$.

10. You just inherited $10,000 and you wish to invest it. You decide to invest part of the money in a somewhat risky account that pays 10% interest annually and the rest in a safer account that pays 8% interest annually. The graph below show how the interest received at the end of one year depends on the amount invested at 10%.(5 pts.)

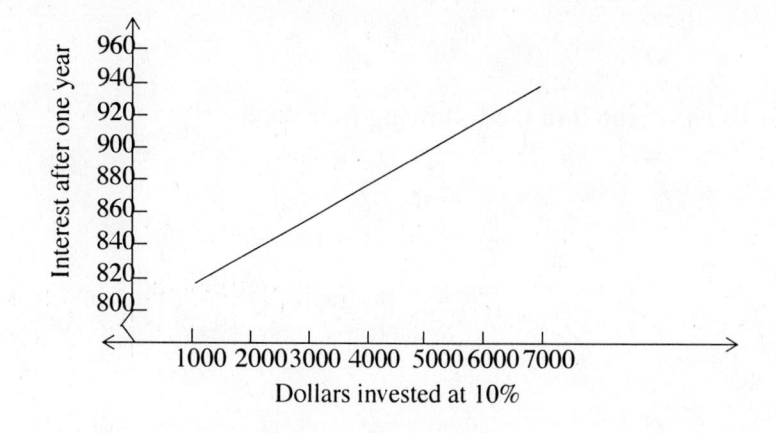

a. Use the graph to complete Table 2.

Dollars invested at 10%	Interest after one year
	820
3000	
4000	
	920
7000	

Table 7: Interest earnings

b. What quantity is the input and what quantity is the output?

11. Complete the following sentences:

a. A function is

b. The domain of a function is

c. The range of a function is

d. Functions can be represented in the following four ways:

UNIT IIB (INDIVIDUAL SKILLS)

Indicate those problems on which you used a calculator and state how you used a calculator for those problems on which the calculator was used (to check work, to find answer, etc).

1. Compute the value of $-9-(-8)(2)+6-(3+5)$.

2. Given $y(x) = x^2 - 5x - 6$, find

    a. $y(-2)$

    b. $y\left(\dfrac{2}{3}\right)$

3. Perform the indicated operations.

    a. $(3x^2 - 2x - 7) - (x^2 + 4x - 2)$

    b. $(3x - 4)(x^2 + 5x - 2)$

    c. $(3x - 5)^2$

4. If $C(n)$ represents the nth cube, find $C(5)$.

5. If $N(t) = 3t^2 - 4t + 2$, find $N(6)$.

6. Let the variable n represent any whole number. Write an algebraic expression that represents:

    a. Increasing x by 3.

    b. Doubling x.

    c. Squaring x.

d. Explain the difference between parts a, b, and c.

7. Write an expression for the number of ways to arrange 100 books vertically on a shelf.

8. A rectangle has a perimeter of 20 feet. Table 3 expresses the relationship between the length of the rectangle and the area of the rectangle.

Length of rectangle in feet	Area of rectangle in square feet
1	9
2	16
3	21
4	24
5	25
6	24
7	21
8	16
9	9

Table 8: Length and area of rectangles

a. Express the values in Table 1 as ordered pairs.

b. Complete the graph below. Label each axis, specify the units for each tick mark, and label the points corresponding to the data in Table 3.

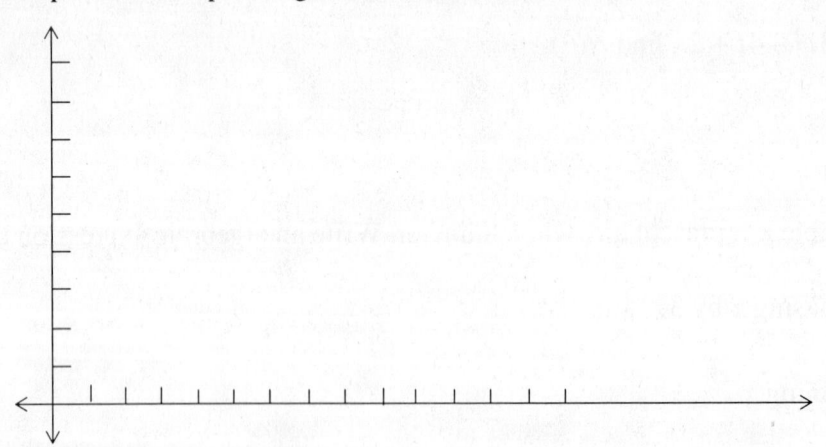

9. Is the set of prime numbers closed under multiplication? Explain.

FINAL EXAM REVIEW

1. Compute the value of $-9-(-8)(2)+6-(3+5)$.

2. Given $y(x) = x^2 - 5x - 6$, find

    a. $y(-2)$

    b. $y\left(\dfrac{2}{3}\right)$

3. Solve the following equations.

    a. $5x - 7 = 0$

    b. $3x + 11 = 23$

    c. $4x - 7 = 7x + 3$

4. Solve the following inequalities.

    a. $6x + 1 > 17$

    b. $5x - 3 < 2x + 7$

5. Simplify completely. Leave no negative exponents in your answers.

    a. $(2x^3y^{-4})(3x^7y)$

    b. $\dfrac{a^4b^2c^9}{ab^7c^5}$

    c. $2x^{-3}$

    d. $(2x^0y^3)^4$

6. Perform the indicated operations.

   a. $(3x^2 - 2x - 7) - (x^2 + 4x - 2)$

   b. $(3x - 4)(x^2 + 5x - 2)$

   c. $(3x - 5)^2$

7. Factor completely.

   a. $9x - x^2$

   b. $x^2 - 6x - 16$

   c. $2a^2 - 5a - 3$

   d. $25m^2 - 20m + 4$

8. For each of the following, identify the slope, the vertical intercept, and the horizontal intercept. Graph each on a separate sheet of paper.

   a. $y(x) = 2x - 6$

   b. $y(x) = \left(-\frac{3}{5}\right)x + 4$

9. Find the slope of the line through the points $(2, -3)$ and $(7, 4)$.

10. Find the equation of the line with slope 5 and a vertical intercept at –2.

11. For the parabola $y(x) = x^2 - 25$, identify the zeros and the x–value of the vertex.

FINAL EXAM (SKILLS)

1. Compute the value of $-1-3(-5)-2^3-(1+(-3))$. _____

2. Given $y(x) = x^2 - 2x + 3$, find

    a. $y(-4)$ _____

    b. $y\left(\dfrac{2}{5}\right)$ _____

3. Solve the following equations.

    a. $7x + 9 = 0$ _____

    b. $4x - 5 = -13$ _____

    c. $3x + 9 = 7x - 2$ _____

4. Solve the following inequalities.

    a. $4x - 3 < 5$ _____

    b. $9x + 2 > x - 4$ _____

5. Simplify completely. Leave no negative exponents in your answers.

    a. $(5a^5b^2)(2ab^{-4})$ _____

    b. $\dfrac{x^3yz^5}{xy^3z^4}$ _____

    c. $3b^{-2}$ _____

    d. $(5a^7b^0)^3$ _____

6. Perform the indicated operations.

   a. $(x^2 + 7x - 9) - (3x^2 - 2x + 5)$ _____

   b. $(2x + 3)(x^2 - x - 4)$ _____

   c. $(4x - 3)^2$ _____

7. Factor completely.

   a. $8x^2 - 18$ _____

   b. $x^2 - 10x - 24$ _____

   c. $3y^2 - y - 4$ _____

   d. $a^2 + 18a + 81$ _____

8. For each of the following, identify the slope, the vertical intercept, and the horizontal intercept. Graph each on a separate sheet of paper.

   a. $y(x) = -5x + 2$ _____

   b. $y(x) = \frac{2}{3}x - 6$ _____

9. Find the slope of the line through the points $(-5, 4)$ and $(2, 7)$. _____

10. Find the equation of the line with slope -7 and a vertical intercept at 3. _____

11. For the parabola $y(x) = x^2 - 6x + 5$, identify the zeros and the x-value of the vertex.

12. Are $1 + 2x$ and $3x$ the same? Justify your answer.

13. Given the graph

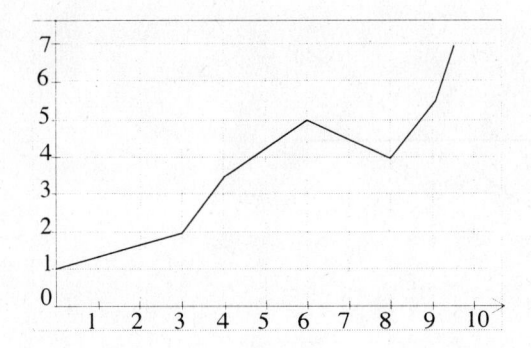

a) y(3) = _____

b) If y(x) = 5, what is x? _____

14. Compare the graph of $C(x) = -(2x-3)$ with the graph of $D(x) = -2x+3$.

a) Sketch both graphs on the axes below.

b) Describe the relationship between the two graphs below.

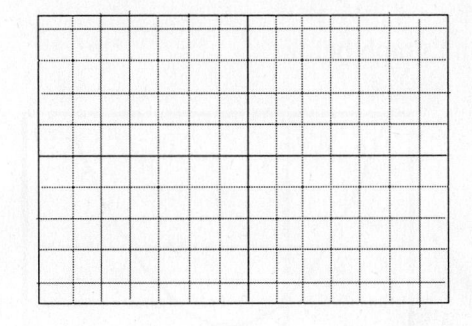

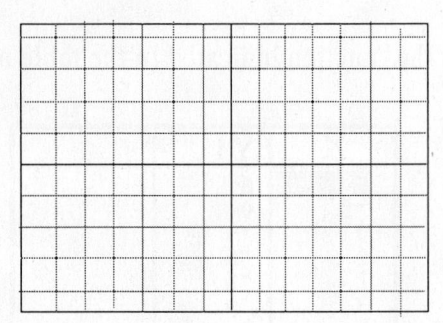

15. $8 + \left(\frac{1}{4}\right)^{-1} - 2^0 \times 5 - 2^3 =$ _____

16. Which is the larger number? $\sqrt{4}$ or 4^2 _____

17. What are the next three numbers in the sequence 0, 1, 3, 6, 10,... _____

Name the sequence _____

18. Evaluate -8^2 _____

259

19. Which of the following equations has the given graph? Justify your answer by showing all work to determine how you selected your answer.

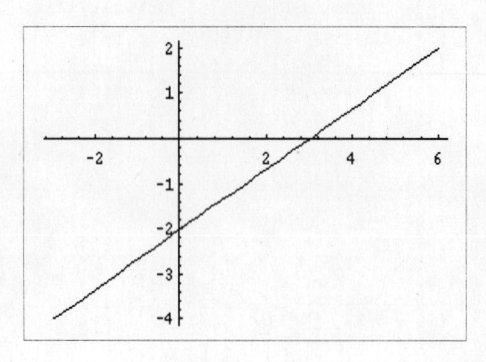

    a. $3x + 2y = 6$ b. $2x + 3y = 6$ c. $2x - 3y = 6$ d. $3x - 2y = 6$ e. none

20. For $f(x) = 3 - x$ and $g(x) = x^2 - 5$,

    a) How do you interpret $f(-1) + g(0)$?   _____

    b) Evaluate $f(-1) + g(0)$.   _____

21. Consider the function indicated in the table and graph below.

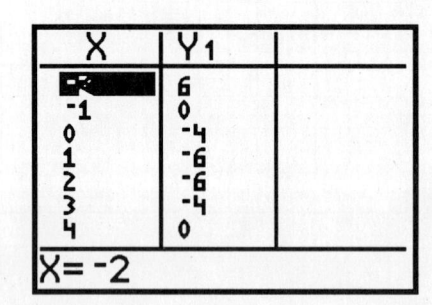

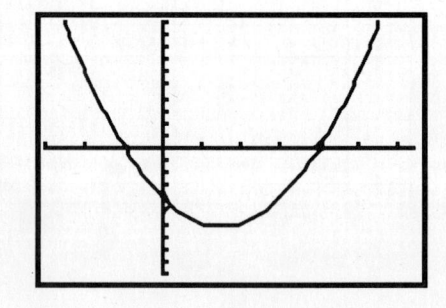

    a) Find the equation. Explain how you used the representations and your process.

    b) What is the same across all three representations?

    c) What information is best provided by each representation?

    d) Which representation do you feel most comfortable with? Why?

Alternative Forms of Assessment

CONCEPT MAPS

Purpose

Concepts maps are used to

- organize and reflect on the content learned.
- provide a tool for self-assessment and review.
- visualize and make explicit the connections between various concepts.
- record the development of richer understandings built on previously learned content.

Definition: Concept maps are a visual language for integrating thinking, learning, teaching, and assessment.

A concept map provides a visual picture of a whole topic or concept and shows how different ideas and/or processes are related to the main topic.

Students create concept maps by following a trail of thoughts from an initial idea and mapping these thoughts out on paper.

The concept map requires that students think about specific connections in their knowledge of a concept.

Students use concept maps

- to visualize connections between newly-acquired knowledge and previously-learned content.
- to organize and reflect on the content learned.
- for review.

The process of creating a concept map helps the student

- recall details.
- identify main points of topics discussed.

CREATING A CONCEPT MAP

- On a piece of paper, write down the main topic of the concept map, then list all of the words that you associate with that topic. Concepts, procedures, your feelings about the topic, previous knowledge and other representations such as function machines, graphs and/or tables can also be included on your list.

- Think in terms of making an outline, with a Main Idea, supporting ideas, and details. Making your list of concepts and procedures is kind of like making an outline of a book you've been assigned in English class.

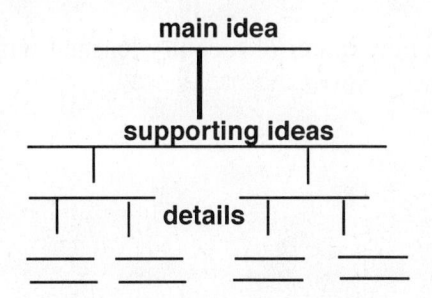

```
              main idea
                  |
         supporting ideas
         |                |
      |  |  details  |  |
      __ __          __ __
```

- Working from your list, use a highlighter and identify what you think are the most important supporting ideas for your topic. Under each of your main sub-groups, identify those words, representations, and procedures you associate with each of your main supporting categories.

- After you have finished analyzing your list, write each word on a separate post-it.

- Post the main topic in the middle of a piece of paper. Arrange the post-it for each key word or idea you've listed around the main topic.

- Near each key word, arrange post-it notes with the words, representations, and procedures you associate with that key word. Writing the words on post-it notes allows you to rearrange the words so that you can indicate the connections among words in that group and between words in other groups that you see as related.

- Build from the main topic in a way that makes sense to you.

- When you've posted and arranged all of your words, draw in the linkages and connections between words and between groups of words. Use arrows to indicate the direction of each link. Wherever possible, write in the relationship along the connecting link.

- Sometimes, you have a word that you connect with more than one key word. Arrange your key words so that the shared connecting word is located between them so you can draw connections to both key words.

- Be creative and personalize your map.

Assessing a concept map

Assessment and evaluation may be based on

- the variety of details included in the map.

- the number and complexity of links between pieces of information.

- the amount of integration of the current map with previously-learned material.

- any obvious key ideas or words missing from the concept map.

- connections between new material recently learned with content covered previously, both in and prior to this course.

Concept map diagrams

Sample diagrams for a concept map appear on the next three pages.

There is no one way to construct a concept map. The following examples show how one group developed their concept map on algebra and two students organized their thinking about functions.

An Example of a Student Concept Map

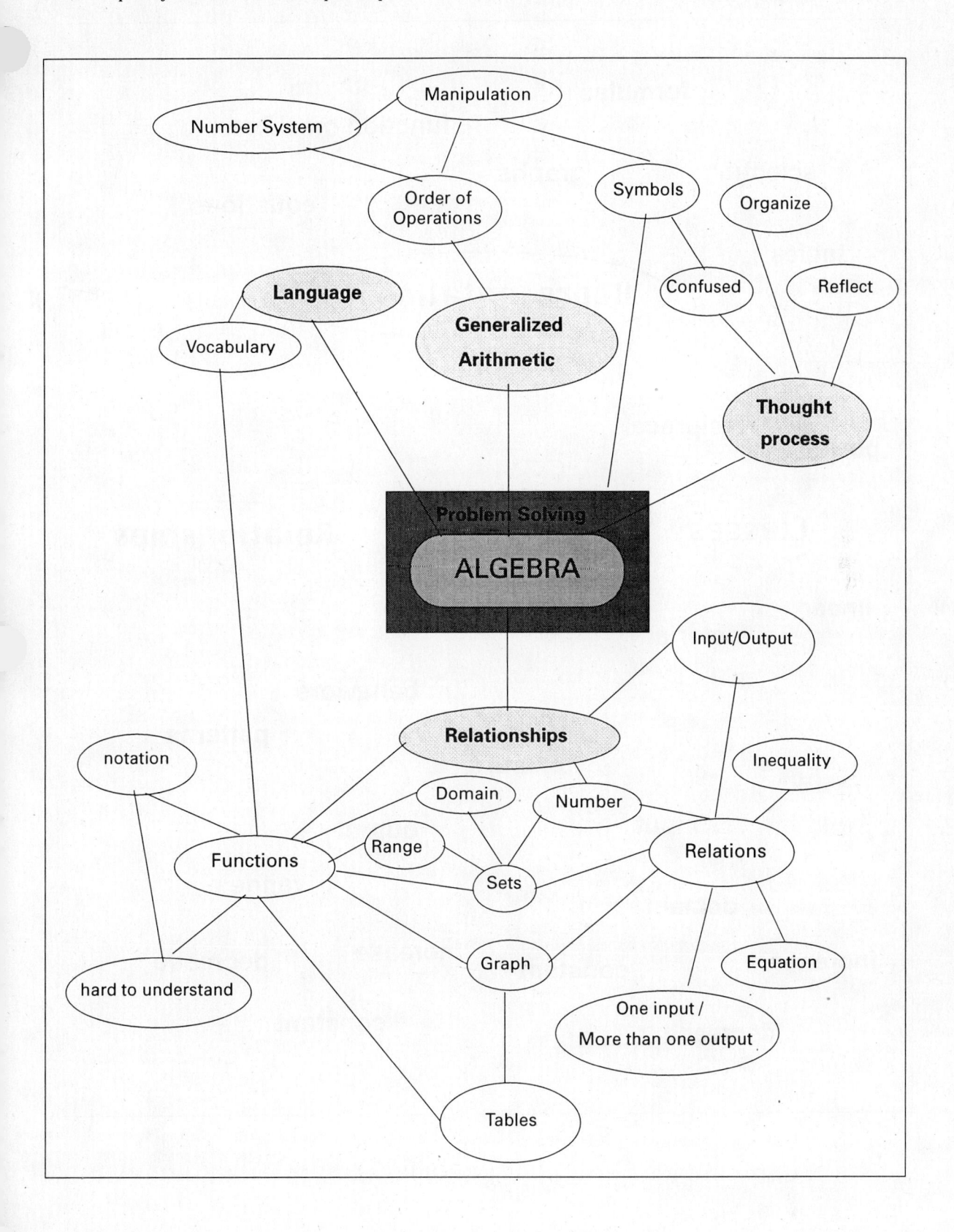

A Student's Concept Map on Function (after week 4)

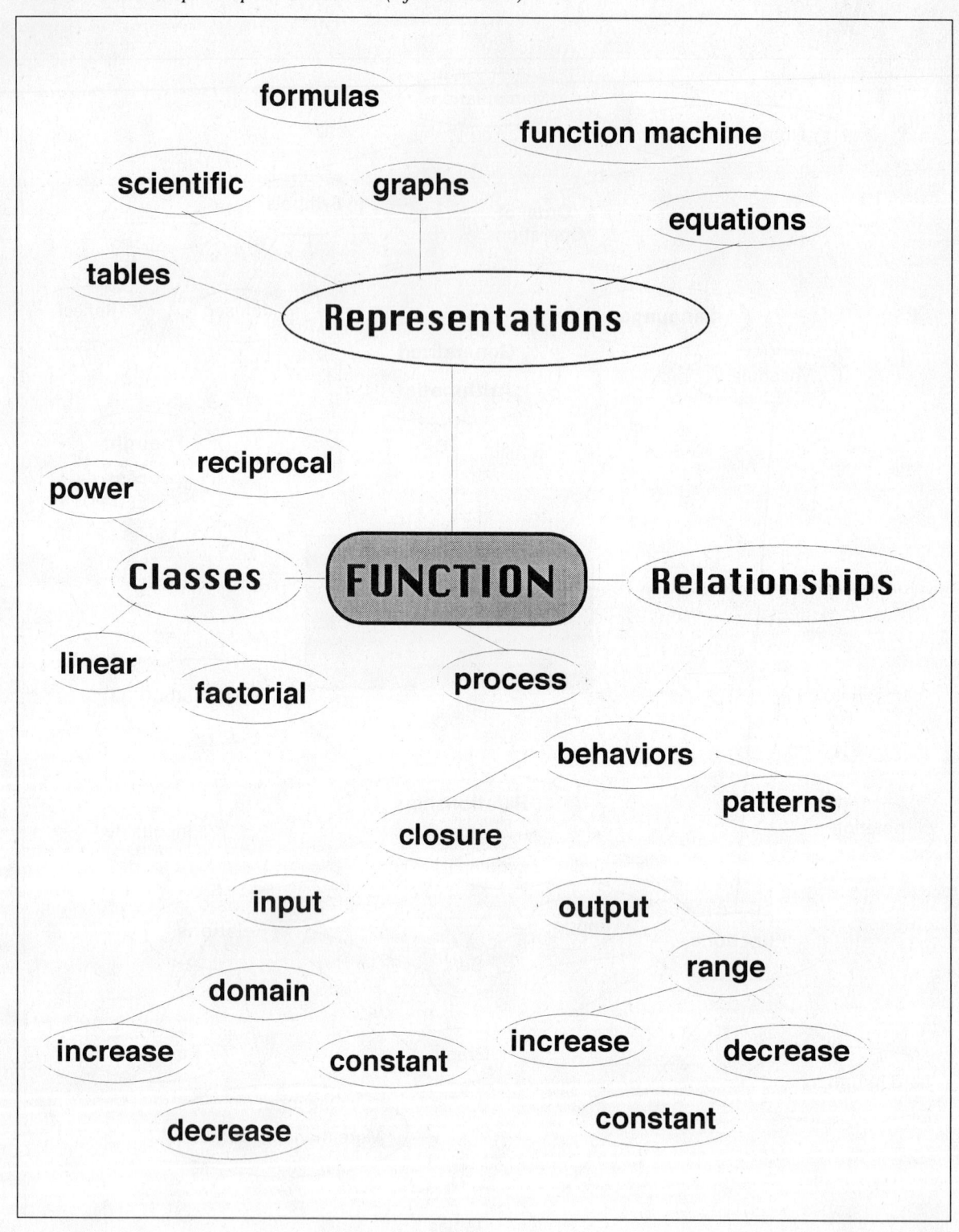

- formulas
- scientific
- graphs
- function machine
- equations
- tables

Representations

- power
- reciprocal

Classes — **FUNCTION** — **Relationships**

- linear
- factorial
- process
- behaviors
- patterns
- closure
- input
- output
- range
- domain
- increase
- constant
- increase
- decrease
- decrease
- constant

266

A Student's Concept Map of Function (after week 4)*

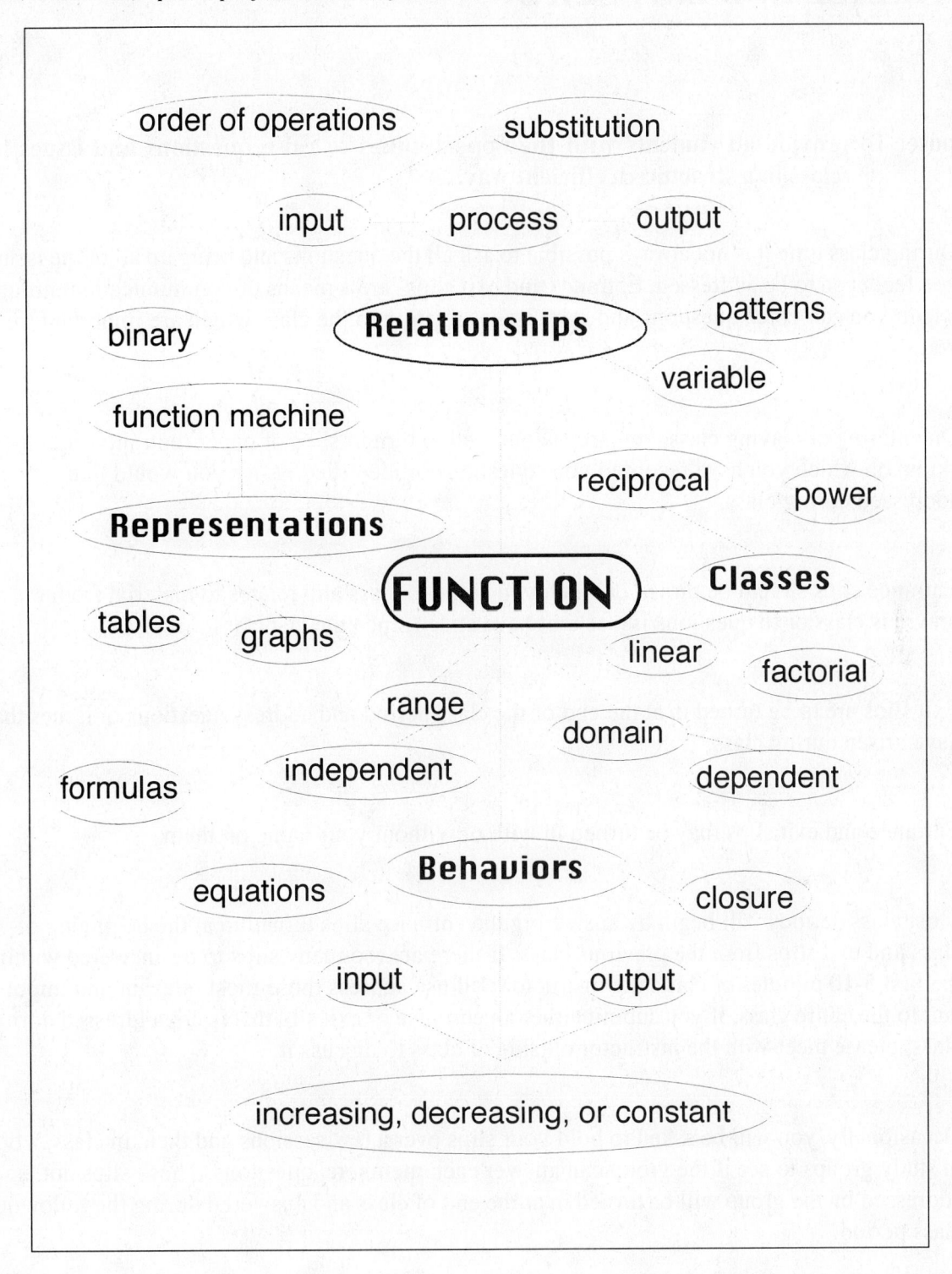

- order of operations
- substitution
- input
- process
- output
- **Relationships**
- patterns
- binary
- variable
- function machine
- reciprocal
- power
- **Representations**
- **FUNCTION**
- **Classes**
- tables
- graphs
- linear
- factorial
- range
- domain
- formulas
- independent
- dependent
- **Behaviors**
- equations
- closure
- input
- output
- increasing, decreasing, or constant

* Note the association of domain with dependent variable and range with independent variable.

ENTRANCE AND EXIT SLIPS

Purpose: To provide all students with the opportunities to raise questions and issues in class in a structured, efficient way.

During class time it is not always possible to ask all the questions and bring up all of the issues you feel need to be addressed. Entrance and exit slips[1] are a means of communication through which you can voice questions and raise issues relating to the class which are important to you.

On entering or leaving class, you are encouraged to turn in slips of paper (entrance or exit skips) on which you have recorded your questions or identified issues you would like addressed during class.

Entrance slips should be turned in when you come to class and relates to material from the previous class or to questions/issues that have arisen since the last class meeting.

Exit slips are to be turned in at the end of the class period and address questions or issues that have arisen during class.

Entrance and exit slips may be turned in with or without your name on them.

Most class periods will begin by answering the entrance slips turned in at the beginning of class and exit slips from the previous class. If there are too many slips to be answered within the first 5-10 minutes of class, the instructor will use address those most relevant and important to the entire class. If you submit either an entrance or exit slip that is not addressed during class, please meet with the instructor outside of class to discuss it.

Occasionally, you will be asked to hold your slips over a few sessions and then, in class, work in study groups to see if the group can answer each members' questions. Those slips not addressed by the group will be turned in at the end of class and answered during the following class period.

---

1. John Ingram, *Quality Teaching - Quality Learning;* Canterbury Educational Publishers, 1994

JOURNAL 1

Math Problem: State one way you worked within the S P C system and one way you worked outside the S P C system during your work in Section 1.2.

Journal Topic: Briefly describe your feelings about your competence in mathematics.

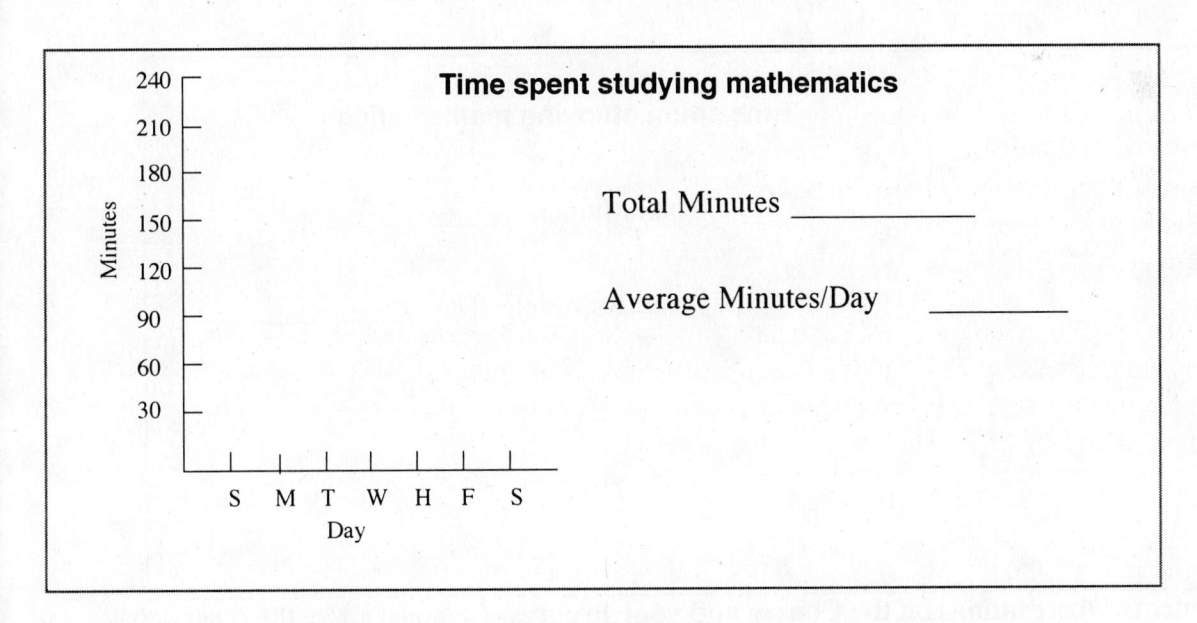

Time spent studying mathematics

Total Minutes _____

Average Minutes/Day _____

(Graph: Minutes (0–240) vs Day (S M T W H F S))

Comments/Observations on the Course and your Progress (optional): Use the other side.

JOURNAL 2

Math Problem: Identify what can be represented by the variable(s) when each of the following rules from the S P C system are applied to the path S P P C P P P C C P P.

1. Rule 1.

2. Rule 2.

3. Rule 3.

4. Rule 4.

Journal Topic: Write a paragraph stating your position on the following statement:

"Success in mathematics depends more on innate ability than on hard work."

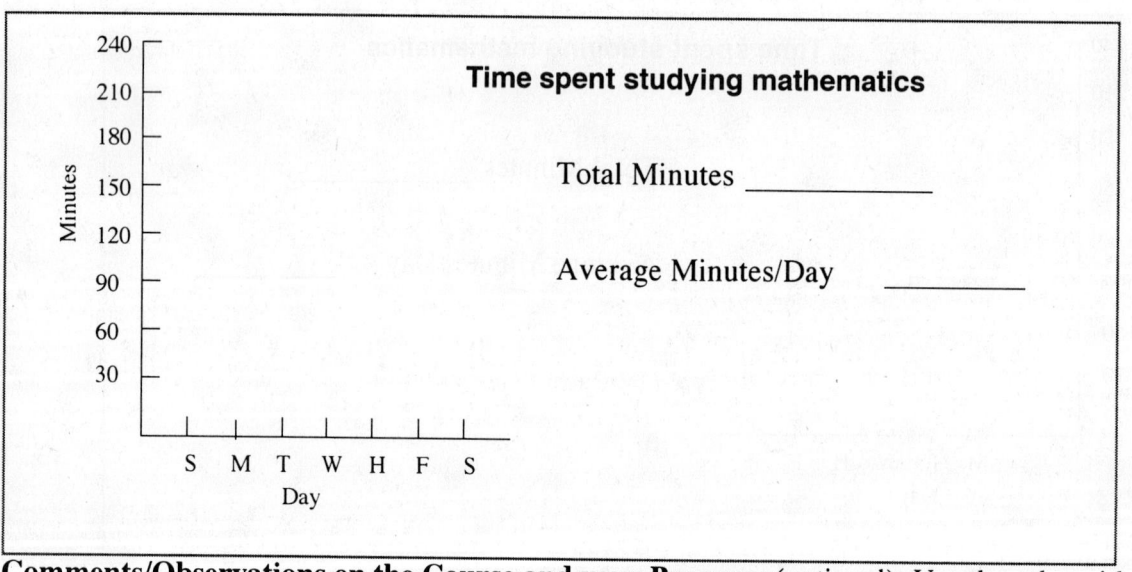

Time spent studying mathematics

Total Minutes _____

Average Minutes/Day _____

(graph: Minutes vs. Day, S M T W H F S, y-axis 30 to 240)

Comments/Observations on the Course and your Progress (optional): Use the other side.

JOURNAL 3

Math Problem: For each of the following, write an English statement clearly describing the order of operations.

1. $17 - 2(3 + 1)$

2. $7 + 3n$.

Journal Topic: Describe the best place and time of day for you to do math homework and explain why.

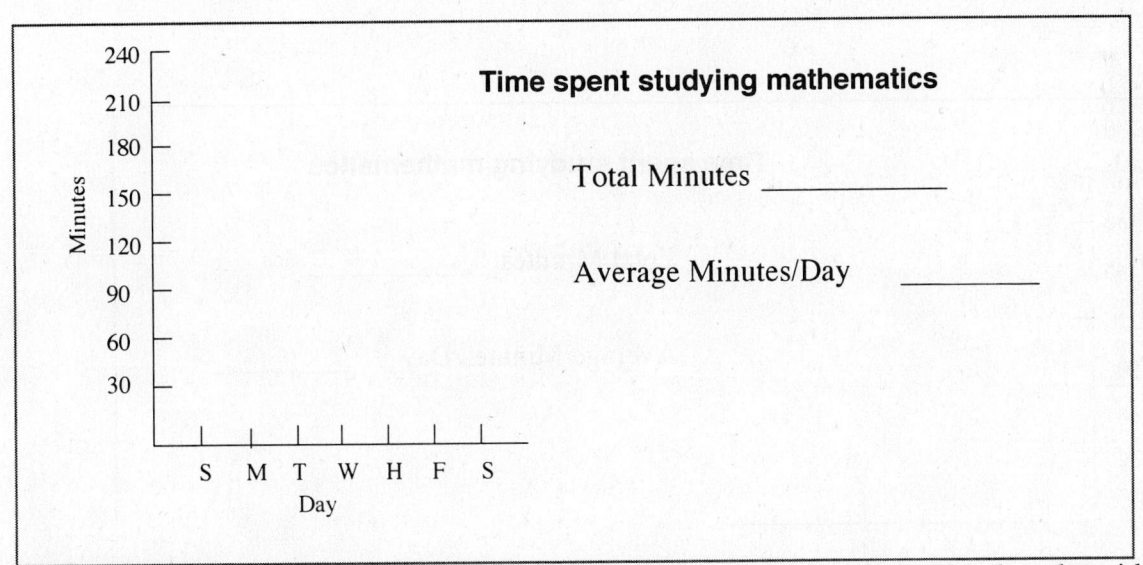

Time spent studying mathematics

Total Minutes _____

Average Minutes/Day _____

(Graph: Minutes (0–240, in increments of 30) vs. Day (S M T W H F S))

Comments/Observations on the Course and your Progress (optional): Use the other side.

JOURNAL 4

Math Problem: Draw a function machine representation for the expression $7t^2 - 3$.

Journal Topic: Write a paragraph stating your position on the following statement:

"To do mathematics is to calculate answers."

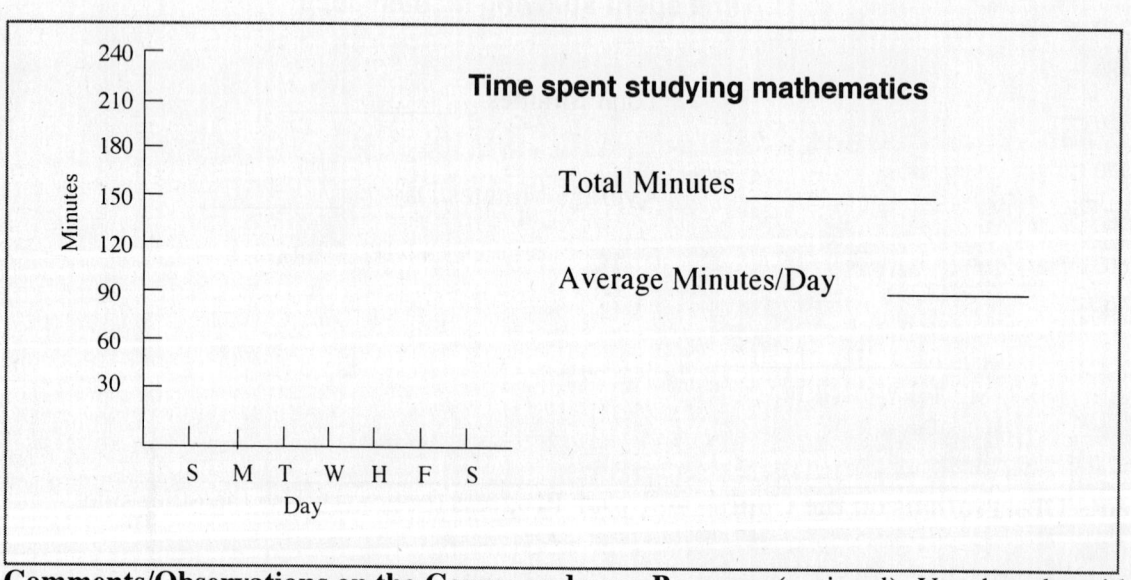

Time spent studying mathematics

Total Minutes _____

Average Minutes/Day _____

Comments/Observations on the Course and your Progress (optional): Use the other side.

JOURNAL 5

Math Problem: A restaurant estimates that it must have one employee on duty for every 13 customers.

1. If there are 11 employees on duty, how many customers can the restaurant handle. Show your work and write a complete sentence answering the question.

2. If C represents the number of customers and E represents the number of employees, write an equation that represents the relationship.

Journal Topic: Approximately how long did it take you to do the first problem set? Did you work alone or with a group? What was the easiest question? What was the hardest question? Why did you find this question difficult?

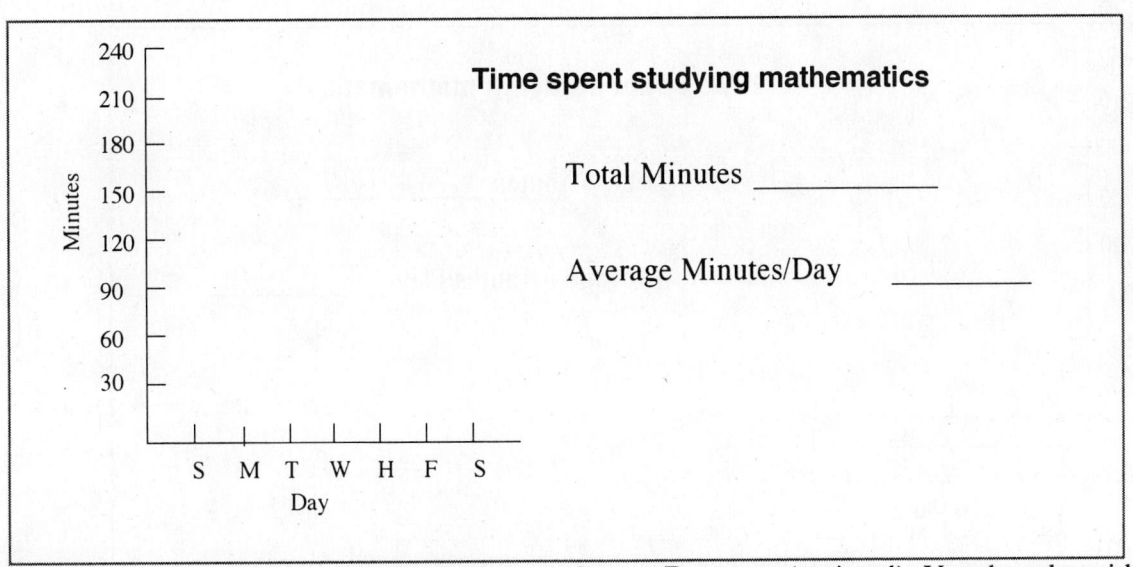

Time spent studying mathematics

Total Minutes _____

Average Minutes/Day _____

Comments/Observations on the Course and your Progress (optional): Use the other side.

JOURNAL 6

Math Problem: At We–Are–Cheap rental agency, it costs only $15 per day to rent a Yugo plus $0.20 per mile. Let the number of miles be the input and the total cost (not including taxes) be the output.

1. Create a table with 4 input/output pairs.

2. If m represents the number of miles driven and C represents the total cost, write an equation that represents the relationship.

Journal Topic: Write a paragraph describing a good student. Write a paragraph describing a good teacher. Use the back, if necessary.

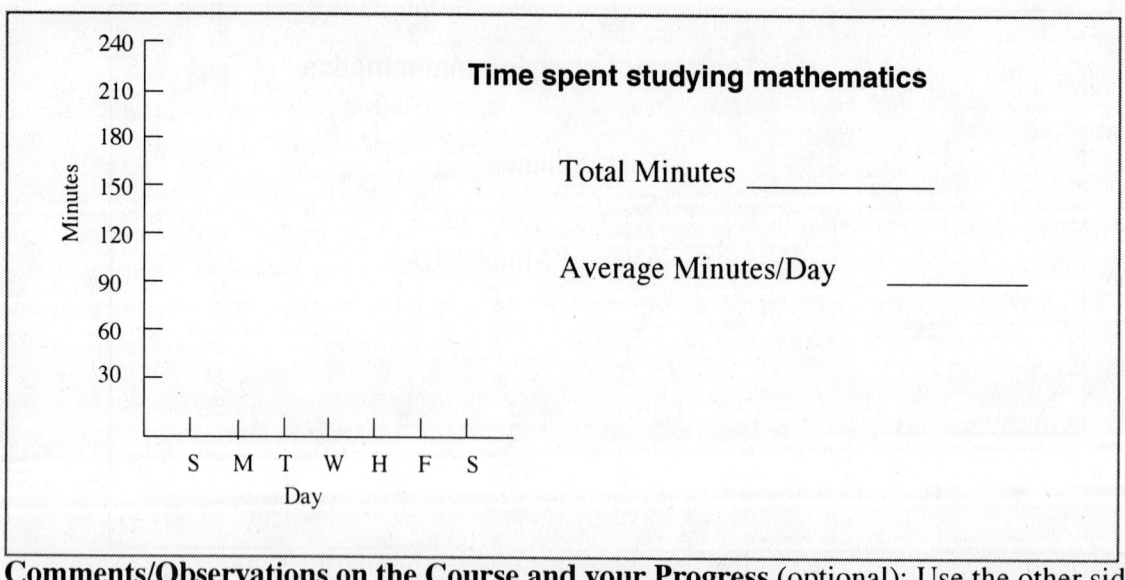

Time spent studying mathematics

Total Minutes _____

Average Minutes/Day _____

Comments/Observations on the Course and your Progress (optional): Use the other side.

JOURNAL 7

Math Problem: Given the ordered pairs (3, 61), (16, 347), (9, 845), (19, 490), (2, 126), (11, 760). write the boundaries (Xmin, Xmax, Xscl, Ymin, Ymax, Yscl) of a viewing window that would contain all the points corresponding to the ordered pairs.

Journal Topic: Describe the most stressful time related to your mathematics work this semester. How did you relieve some of the stress?

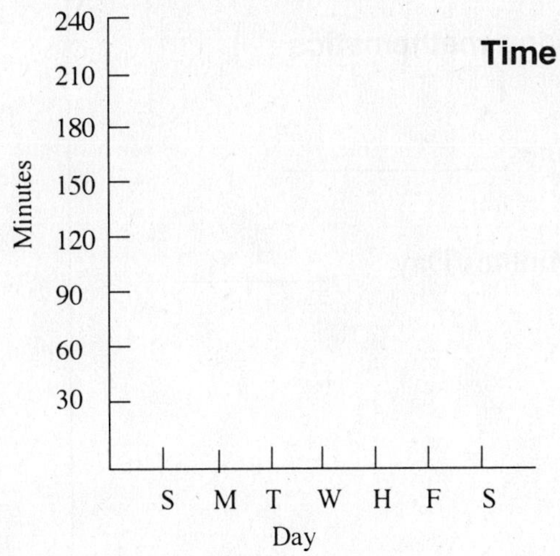

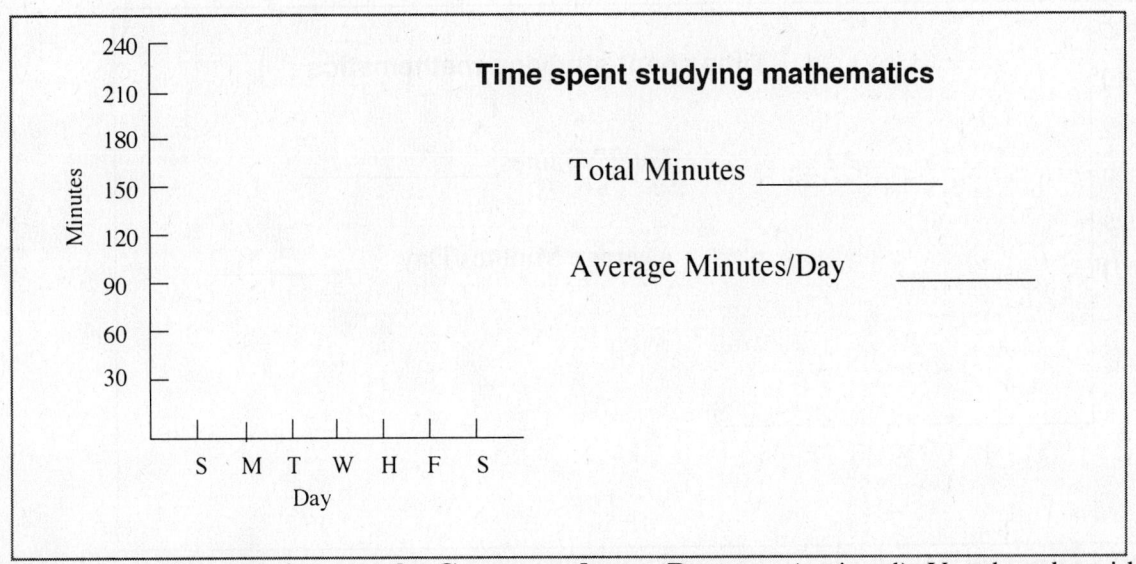

Time spent studying mathematics

Total Minutes ____________

Average Minutes/Day _________

Comments/Observations on the Course and your Progress (optional): Use the other side.

JOURNAL 8

Math Problem: Write a statement defending your answer for each of the following:

1. The opposite of $-2\frac{1}{3}$ is _____ .

2. The reciprocal of $-2\frac{1}{3}$ is _____ .

3. Write $(2x) \div (7a)$ as a multiplication: _____

Journal Topic: About one-third of the semester remains. Briefly describe where you stand in the course at this time. Where would you like to be by the end of the semester with respect to this course? What do you plan to do between now and the end of the semester to attain your goal in the course? (Use back if necessary.)

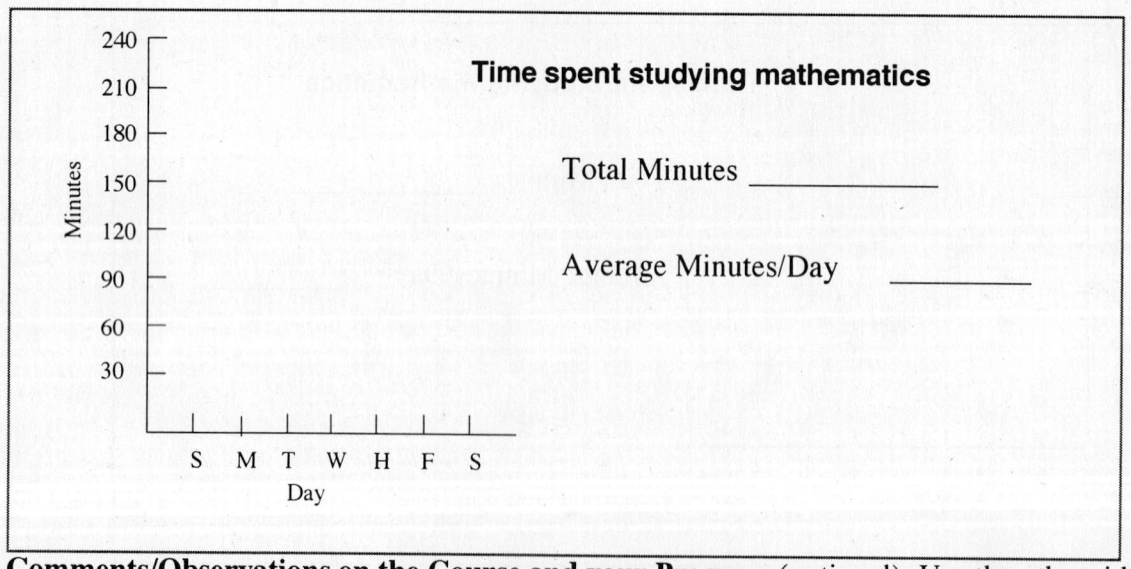

Time spent studying mathematics

Total Minutes _____

Average Minutes/Day _____

Comments/Observations on the Course and your Progress (optional): Use the other side.

JOURNAL 9

Math Problem: A line segment is formed by connecting the points $(-2, 6)$ and $(5, 11)$. What is the slope of the line between the two points? Briefly describe what you did.

Journal Topic: If there was one aspect of this course you could change, what would it be? Explain why and describe the change that you would make.

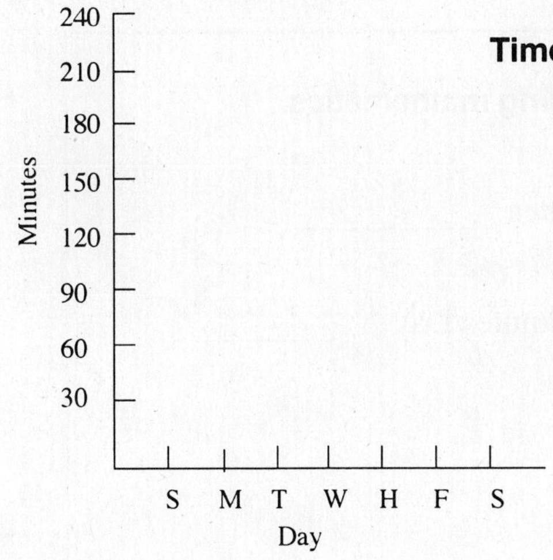

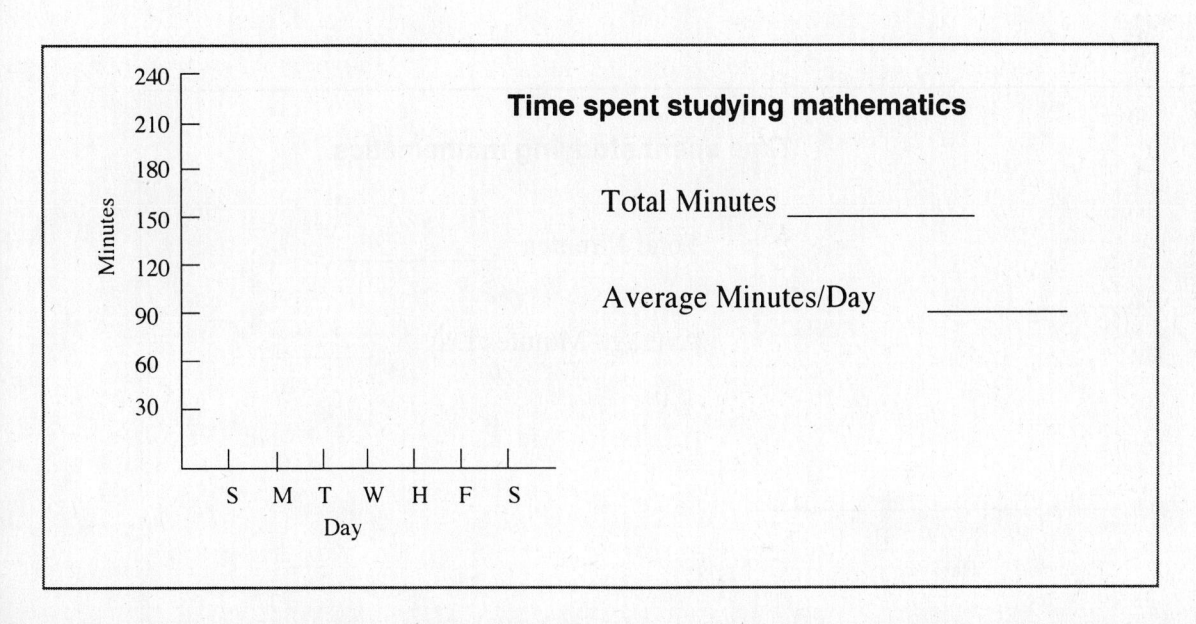

Time spent studying mathematics

Total Minutes ____________

Average Minutes/Day __________

Comments/Observations on the Course and your Progress (optional): Use the other side.

JOURNAL 10

Math Problem: Use an exponent property to rewrite each of the following. State what you did.

1. $(x^a)(x^b) = $ _____

2. $\dfrac{y^t}{y^r} = $ _____

3. $r^0 = $ _____

4. $p^{-x} = $ _____

Journal Topic: What activity do you believe makes the *best* use of class time. Why?

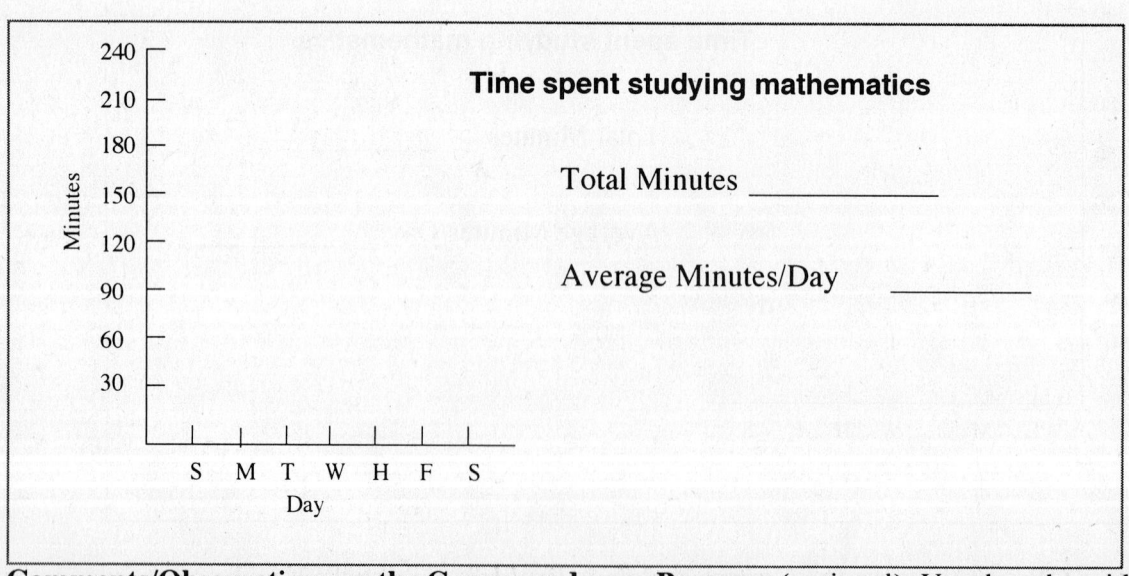

Time spent studying mathematics

Total Minutes _____

Average Minutes/Day _____

Comments/Observations on the Course and your Progress (optional): Use the other side.

JOURNAL 11

Math Problem: Consider the linear function $y(x) = -4x + 7$. Identify the slope, the vertical intercept, and the horizontal intercept. Briefly describe, in words, how you found each.

Journal Topic: Write a paragraph stating your position on the following statement:

"All useful mathematics was discovered long ago."

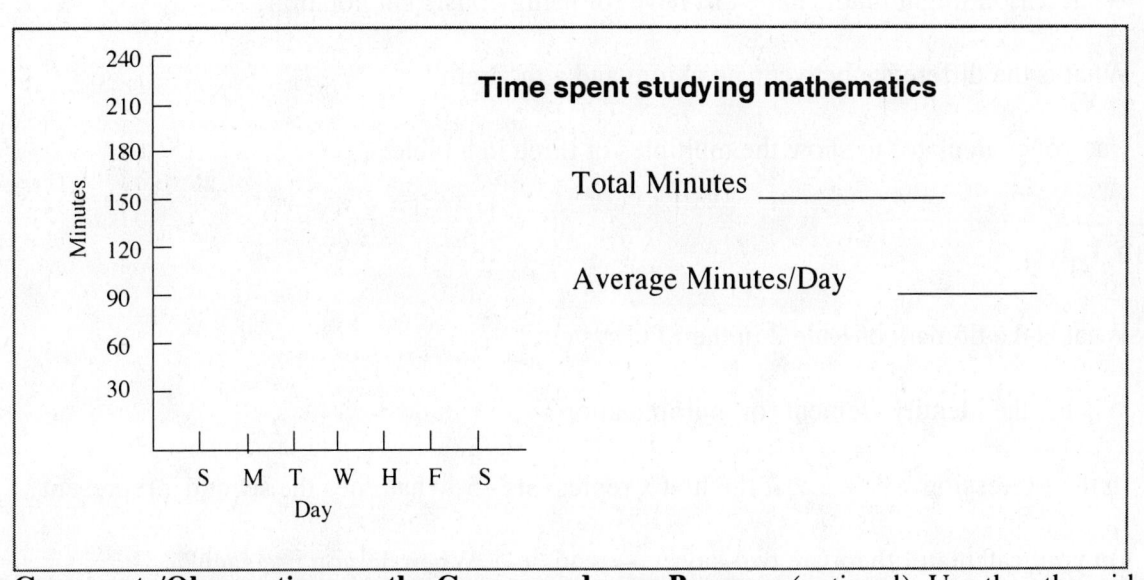

Time spent studying mathematics

Total Minutes _____

Average Minutes/Day _____

Comments/Observations on the Course and your Progress (optional): Use the other side.

MINUTE QUIZZES

There are several reasons to give these quizzes at the beginning of the period: they encourage promptness and attendance; they are a quick way to begin a discussion of the previous day's lesson; they provide a quick assessment of vocabulary, notation and concepts students have learned from the Investigations; and they send a message that what is being tested is worth learning since it is being tested. After collecting the papers, answer the question immediately--this often facilitates a summary for the last Investigation or a point of departure for the next Investigation. There are also basic technology questions included for some of the sections. Occasionally, there are some students who will, if not encouraged and shown that these skills are also to be learned, make little effort to learn to use the calculator.

The questions included here emphasize ideas which may be overlooked in the day-to-day activities. You may wish you hadn't asked some of these questions, since they often reveal how fragile the students' understandings really are.

Section 1.1

1. List four strategies for solving a problem. Which strategy might be appropriate to find the number of ways of paying for a $.25 candy bar?

2. List the criteria for determining a good solution process and explain one of them.

3. How do you turn your calculator on and adjust the contrast?

Section 1.2

1. What reason might mathematicians have for using consistent notation?

2. What is the difference between an axiom and a theorem?

3. Use your calculator to show the multiples of three in a table.

Section 1.3

1. What is the domain of Rule 2 in the SPC system?

2. What is the identity element for multiplication?

3. In the expression $3x^2 + 2x$, if the first x represents -5, what does the second x represent?

4. On your calculator there are two keys: $-$ and $(-)$. When do you use each?

Section 1.4

1. Tell how $5q - 7$ can be interpreted as a process. How is that different from the output?

2. When is the square of a number smaller than the number?

3. Use your calculator to display the powers of 5 in a table.

Section 2.1

1. What is the prime factorization of 23247?

2. List the digits. Is this list a finite or infinite set?

3. Julio works at a car wash. What set of numbers represents the domain for the variable representing the number of cars he has polished today?

Section 2.2

1. What is 5^0?

2. What is the order of the operations to be performed in $\dfrac{2^3 + 1}{7 \cdot \sqrt{16} - 1}$? Do this with pencil and paper and on the calculator.

3. List three symbols used for inclusion.

Section 2.3

1. What operation is implied $7x$?

2. Give an example of quadratic polynomial.

3. Evaluate $6x^3 - 4x + 1$ for $x = 2$.

Section 2.4

1. How does the distributive property allow us to disregard the rules for the order of operations?

2. List three different monomials which are like terms.

3. Expand $(3x - 1)^2$.

Section 3.1

1. Is the domain associated with input or output of a function?

2. How would finite differences help to determine the next number in the list: 3, 8, 13, 18,...?

3. Use the table feature on your calculator to list all the questions the first student answers in *The Nearsighted Professor* Problem.

Section 3.2

1. List three ways in which a function can be represented.

2. To solve we must find the _____ when the _____ is given. (Use input and output in the blanks.)

3. Write an equation to express the postal function which expresses the cost of postage in terms of the weight in ounces. (That is $.32 for the first ounce and $.23 for each additional ounce rounded up to the next ounce.)

Section 3.3

1. Use function notation to express the postal function which expresses the cost of postage in terms of the weight in ounces. (That is $.32 for the first ounce and $.23 for each additional ounce rounded up to the next ounce.)

2. In $f(x) = 2x + 1$, explain what $f(x)$ means.

Section 3.4

1. Plot these ordered pairs: (0, 5) (2, −1) (0, 0) (−3, −5) in the Cartesian Plane.

2. What are the coordinates of the origin?

3. Draw a vertical line two inches long on your paper.

4. Use your calculator to graph $f(x) = 2x + 1$.

Section 3.5

1. Explain how to find the sum of the first 30 counting numbers. Is your process efficient

enough to generalize to 3000 numbers?

2. How many dots lie on the side of the triangle if there are a total of 171 dots in the geometric representation of a triangular number?

Section 3.6

1. Write the power function of degree 5.

2. In a table of values, if the fourth finite differences of a power function are constant but not zero, what is the degree?

3. I have homework assignments in mathematics, biology and Spanish. If I complete one assignment before beginning the next, in how many different orders could I complete the assignments? List the possibilities.

Section 4.1

1. Explain why 5 acts as the arbitrary zero on a par 5 golf hole.

2. What is the opposite of –4?

3. What is the additive identity?

4. What is the unary operation which uses –? The binary operation? Give an example of both of them.

5. In the order of operations, do unary or binary operations take precedence? Give an example of each operation.

Section 4.2

1. Subtract –7 - 4 on your calculator.

2. The sum of a number and its opposite is _____.

3. Rewrite $7 - x$ as an addition problem. What happens to the size of the answer when x represents a negative number?

Section 4.3

1. Define the absolute value of x.

2. How is the domain important in simplifying $|2x - 7|$? (That is, in removing the absolute value signs.)

3. How do you enter a piecewise function into the calculator?

Section 4.4

1. Draw a coordinate system and label the axes and quadrants.

2. Give an ordered pair located in the second quadrant.

3. Give an ordered pair located on the vertical axis below the origin.

Section 4.5

1. In the viewing window, explain what Xmin represents.

2. What will happen if the difference between Ymax and Ymin is 100 and Yscl = 1?

3. Write the parametric equations needed to graph $y = 3x + 1$ so that x is just replaced by a whole number.

Section 4.6

1. Write the equation which represents the identity function.

2. Draw the function machine for the absolute value function.

3. Identity the function which is a line bisecting the second and fourth quadrants. (You may just want to show the graph.)

Section 5.1

1. Explain what the notation Δx means.

2. Why is slope called a "rate of change"?

3. Find the slope of the line thru (3, 8) and (7, 2).

Section 5.2

1. The product of reciprocals is _____.

2. What fraction is implied by "per cent"?

3. Write one-half as a percent. Write three as a percent.

4. A collector plate was touted as being a good investment because its value had soared to 87% of it purchase price. Explain if this was a good investment or not.

Section 5.3

1. If a and b are fractions between 0 and 1, then their sum is _____ than b. (Use greater than or less than in the blank.)

2. If a and b are fractions between 0 and 1, then their product is _____ than b. (Use greater than or less than in the blank.)

3. Draw two squares representing one-half and one-fourth. Explain what you are doing with the squares when you find a common denominator to add the fractions.

4. Enter 7/9 – 1/3 on your calculator. Display the answer as a fraction.

Section 5.4

1. In the first quadrant, is the graph of the reciprocal function increasing or decreasing? What about in the third quadrant?

2. Graph the postal function which expresses the cost of postage in terms of the weight in ounces. (That is $.32 for the first ounce and $.23 for each additional ounce rounded up to the next ounce.)

3. Identify the function graphed which passes thru (3, 27). (Show a cubic.)

Section 5.5

1. What does 6^7 mean?

2. What does 6^{-7} mean? Is the answer positive or negative?

3. Simplify $(3^5 \cdot 3^2)^4$. Show two ways this can be done.

Section 6.1

1. The decimal representation of a rational number either _____ or _____.

2. What is the period of the decimal representation of 1/7?

3. Sketch a diagram showing how real numbers, irrational numbers, rational numbers, whole numbers and integers are related.

Section 6.2

1. Is the square root of $\frac{4}{9}$ larger or smaller than the number?

2. What are the domains of $\sqrt{x}$ and $\sqrt{-x}$? Sketch the graphs.

3. What function has the same graph as $\sqrt{x^2}$? Why? When is x not the answer?

4. Find the distance between (−3, 2) and (4, 1). Explain why you gave your answer in the form which you chose.

Section 6.3

1. Write the function which relates the number assigned to each customer as input and the price paid for the daily $2.99 special. What kind of function is this?

2. Write the equation which represents the function defined by the table.

Input	1	0	5	−2/3	0.234	700
Output	−1	0	−5	2/3	−0.234	−700

Table 1: Mystery Function

3. Choose your favorite basic function. Name it. Write the equation. Draw a function machine. Make a table. Draw its graph. State its domain and range. (Okay so that one is not a minute question.)

4. Which basic function has no vertical intercept?

Section 6.4

1. Identify the slope and vertical intercept for the function $f(x) = 0.2x - 5$.

2. If the slope of a linear function is zero, the line is _____.

3. If the slope is -3, what are possible changes in x and y? If the change in x is 4, what is the change in y?

4. If the slope of a line is 3/7, what is the slope of a line perpendicular to it? Parallel to it?

Section 6.5

1. What is the vertex of a parabola?

2. What kind of function has a parabola as its graph?

3. What is the relation between the zeros of a quadratic function and its factors? Write the quadratic function with zeros at -2 and 0.

4. Write a quadratic function. Does it factor? How can you tell?

Section 7.1

1. Is 7 a solution to the equation $3x + 4 = 7$? Why or why not?

2. How many solutions does $2x + 1 > 12$ have? Write an expression which would include all the values that are a solution.

3. How do we read $\leq$?

4. Describe how to solve $3x + 4 = 7$ numerically, algebraically and geometrically or graphically. Should all of the solutions be the same?

Section 7.2

1. Write a system of linear equations in x and y.

2. When solving a system of equations by substitution, how do you decide which variable and equation to use?

3. Do all systems have a unique solution? Explain.

Section 7.3

1. Explain the relation between multiplication and factoring. Give an example.

2. Give an example of a prime polynomial over the rational numbers.

3. Many examples and cases were analyzed in this section. Tell what you are thinking as you factor $7x^2 + 28x + 21$.

4. How is factoring used in simplifying rational expressions?

Section 7.4

1. Factor $t^2 - 25$.

2. Write the general form of a quadratic function.

3. List the possible rational zeros of $7x^2 + 28x - 6$. How can you tell if this function really has rational zeros?

4. Use the calculator to solve for the zeros of $7x^2 + 28x - 6$.

PORTFOLIOS

Purpose: To provide a showcase for your work which demonstrates *evidence of your mathematical growth and knowledge* during the semester and serves as the basis for your evaluation.

Description: Each portfolio consists of your work in a three ring binder that contains:

1. Your Mid-term and Final Grade Evaluation and Rationale, completed, with the grade you believe your portfolio demonstrates. The rationale describes your mathematical growth, the difficulties you experienced and your reflections on your use of the graphing calculator. Specific examples and/or experiences are to be cited, not just general statements.

2. A concept map on **function** to be completed during the last two weeks of class. This final concept map is a significant element in your portfolio--another evidence of your growth during the semester.

3. One completely solved problem from each section's Explorations. The problem is selected by you from the assigned Exploration problems.

4. One journal accompanied by a rationale as to why you chose it.

5. The problem sets and skills tests *with corrections*.

6. Any additional work you have done during the course that demonstrates your mathematical growth (your choice).

Evaluation

Your portfolio serves as the primary source of evidence of your growth in mathematical understanding and your level of competence attained during the semester. The portfolio should include previously reviewed work you have selected (cited above), along with other items you've chosen that demonstrate your mathematical understanding, competence and growth.

At midterm and at the end of the semester, each of you will meet with me individually. Your portfolio will include a self-assessment in which you reflect on your performance in light of the goals and objectives of the course and will indicate the grade you feel your work justifies, with the evidence you feel supports that grade. Your semester grade is determined during discussion between you and me, based on the evidence of your completed portfolio. **Your portfolio is evaluated on completeness, correct and appropriate mathematics, and on demonstrated growth in mathematical understanding.** As you select work to include in your portfolio and prepare your self assessment and rationale for grade, realize that I value examples with more problem solving over those with straight computation and efficient algorithms over rote manipulation. I particularly value those assignments that demonstrate true mathematical understanding, fluent use of representations, and which demonstrate the effective use of mathematics to solve problems.

ASSESSMENT RUBRIC I

Advanced

The student's work goes well beyond expectations and demonstrates creativity, elegance, and logical reasoning. Work is complete, consistent and accurate. The choice of algorithm is appropriate and efficient. Responses are clear, coherent, and unambiguous. Supporting arguments are valid. Notation is used correctly. Creativity is demonstrated

Proficient

Student demonstrates consistent, appropriate application of knowledge and skills. Work is complete, consistent and accurate. There are a few minor errors in content knowledge and understanding. The student uses appropriate, efficient algorithms most of the time, with few instances of incorrect notation or algorithms. The student demonstrates, with minor exceptions, a grasp of mathematical ideas and processes and presents supporting arguments that may contain minor flaws.

Nearly proficient

The majority of work is complete, consistent and accurate, with some minor misconceptions, occasional use of inappropriate skills or concepts, and/or incomplete explanation. Notation is used incorrectly or inappropriately at times. Student responses are sometimes vague. Some responses indicate a misinterpretation of a problem situation or a failure to recognize all of the problem conditions. The student demonstrates, with a few exceptions, a grasp of mathematical ideas and processes but occasionally presents supporting arguments that are not valid.

Minimal

The student's work is incomplete and/or unorganized. The student demonstrates little, if any understanding of concepts or procedures. Only some of the essential conditions of the task have been addressed, with little evidence of understanding of the central ideas, concepts, and skills. There are some major flaws in content knowledge and responses are frequently not justified by work shown. Student communication is often vague, and there is use of inappropriate or incorrect notation.

Attempted

An attempt was made to do the work without meeting any of the essential criteria. A majority of the work is incomplete and not thoughtfully done. There are many major flaws in content knowledge. Most, if not all, of the student's responses are vague and fail to justify work. Student communication is always vague. The student shows little or no understanding of the mathematical ideas and processes.

Not scoreable

Nothing was turned in or the response is irrelevant and/or unreadable.

Evaluating your own work

In evaluating your work, ask yourself how complete and understandable to others your work is.

Specifically, does your work:

- reflect your initial thinking about the problem?
- include documentation of your thinking and any changes in your thinking or in the approaches you employed?
- and final response include a reflection on what, if any, are the new products of your learning---i.e., the new understandings you acquired as a result of your efforts to investiagte and solve the problem?
- demonstrate effective and efficient communication capabilities and representational fluency?

As you document your work at each succeeding stage of your investigation:

- does your work indicate the focus of your efforts?
- does your final response contain any necessary conditions and/or constraints on the appropriateness of the response?
- were you able to judge for yourself whether and when your response would be acceptable?
- is your solution generalizable--i.e., does your solution provide a useful prototype or model for interpreting other situations? Can it be used by yourself and others in other situations? Is it re-useable, modifiable, or shareable with others?

How do you know your work is acceptable? Were you able to

- explicitly identify your intuitive, informal initial understandings?
- identify any misconceptions/misunderstandings you became aware of during your investigation?
- extend, refine, or integrate your previous understandings and skills to develop new levels of more formal understandings?

Take a Risk – Input Effort and Output Success!

ASSESSMENT RUBRIC II

Noteworthy: achieves immediate objectives, with adequate conceptual and technical tools available; goes beyond solving the immediate problem. Your response:

- investigates a more general problem
- recognizes your own assumptions and limitations imposed on the problem.
- describes the accuracy of your solution and identifies sources of errors.
- offers alternataive solutions
- provides for "what if"
- makes connections within mathematics
- makes connections across disciplines
- makes new connections to your everyday life outside of the classroom.

Acceptable: achieves immediate objectives and contains everything necessary for a satisfactory solution. Your responses is complete, understandable, and well-grounded in the appropriate interpretation and use of mathematics notation, skills and concepts employed. Your work should:

- contain a satisfactory response to everything asked for.
- be clear, consistent, and well-organized.
- effectively represent your ideas/thinking with text, graphs, diagrams, tables, formulas, etc.
- use mathematics appropriately, reasonably and skillfully for the context of the problem.
- select, ignore, and use information appropriately.
- provide adequate support, justification, or explanation of your response(s).

Needs Minor Revision: With a few small changes it would be aceptable work. It closely resembles an acceptable response, but fails to provide everything that is necessary.. It may lack:

- a complete and accurate solution to all components of the problem asked for.
- adequate support, justification or explanation.
- a clear, consistent and well-organized response.
- use notation incorrectly and/or inappropriately.
- an effective representation of your ideas and thinking.

It may include:

- a few major or several minor mathematical misconceptions
- ignore or misinterpret important information
- lack adequate reasoning.
- reflect disorganized or incomplete thinking.
- be confusing to someone else because of poor organization, expression or representation.

Needs major revision: demonstrates a basic understanding of the problem, a reasonable start, and contains some ideas appropriate to the problem, but needs substantial work. It may lack important elements of an acceptable answer. It could:

- reflect disorganized or incomplete thinking.
- demonstrate poor organization, incomplete, inappropriate, or incorrect mathematical representation(s).
- employ incorrect or inappropriate procedures.
- reflect significant mathematical misconceptions.
- demonstrate an inability to effectively and efficiently interpret and use mathematical notation.
- reflect an inflexible approach and a failure to use various representations effectively.
- demonstrate a lack of adequate reasoning.

Needs redirection: response indicates a lack of understanding about the problem and/or a lack of the tools (skills and conceptual foundation) needed for an acceptable response. It could:

- just reiterate the information given in the problem.
- contain only an inappropriate response or one not asked for.
- indicate an appropriate initial attempt which is not developed.
- reflect significant mathematical misconceptions and/or lack of skills necessary to complete the task.
- reflect disorganized or incomplete thinking.

ASSESSMENT RUBRIC III

Excellent (E)

- Student's work is complete and thoughtfully done.
- No flaws in content knowledge
- Student responses are clear, coherent, and unambiguous.
- Student communicates effectively.
- Student demonstrates a firm grasp of mathematical ideas and processes.
- Student presents strong supporting arguments

Very Good (VG)

- Student's work is mostly complete and, for the most part, thoughtfully done.
- A few minor flaws in content knowledge.
- Most of the student's responses are clear and coherent.
- Student communication is clear, with minor exceptions.
- Student demonstrates a grasp, with minor exceptions, of mathematical ideas and processes.
- Student presents supporting arguments that may contain minor flaws.

Good(G)

- Majority of work is complete but often not thoughtfully done.
- Some flaws, mostly minor, in content knowledge.
- Student's responses are sometimes muddled and argumentation incomplete.
- Student communication is sometimes vague.
- Student demonstrates a partial grasp, with minor exceptions, of mathematical ideas and processes.
- Student presents supporting arguments that may contain major flaws.

Fair(F)

- Student's work is incomplete or not thoughtfully done.
- Some major flaws in content knowledge.
- Many of the student's responses are vague and muddled.
- Student communication is often vague.
- Student fails to show full understanding of the mathematical ideas and processes.
- Student seldom presents supporting arguments and they contain major flaws.

Poor (P)

- Majority of work is incomplete and not thoughtfully done.
- Many major flaws in content knowledge.
- Most, if not all, of the student's responses are vague and muddled.
- Student communication is always vague.
- Student shows little or no understanding of the mathematical ideas and processes.
- Student seldom, if ever, presents appropriate supporting arguments.

Blackline Masters

What Is Algebra?

- A study of structures

- The language of mathematics and science

- A study of relationships

- The study of procedures

- Generalized arithmetic

- A distinct symbol system with its own rules

Concept Maps:

- Provide a concise overview of a topic.

- Help us focus on relationships between concepts and their connections to each other.

- Support the examination and reflection about our ideas.

- Provide a visual picture of a topic or idea.

- Illustrate relationships between isolated pieces of knowledge.

- Focus discussion on alternative points of view.

My Favorite Parking Lot

All twenty parking spaces in my favorite parking lot are filled. Some are occupied by motorcycles and others by cars.

Some people count to ten when they get angry, but that wasn't nearly far enough. I counted wheels---sixty-six to be exact. How many cars and how many motorcycles had invaded my territory?

Problem-Solving Strategies

- *Guess and test*: guess a solution and test to see if the answer matches all the conditions.

- *Draw a picture*: can you visualize the parking lot? You don't have to draw the vehicles, just the wheels.

- *Use logic and arithmetic computation*: identify the numbers in the problem. Use only numbers given or implied to find the solution.

- *Make an organized list*: use a table to display information about the problem in an organized way.

More Strategies

- ***Write an equation***: use either the number of cars or the number of motorcycles as a variable to write an equation that represents the problem.

- ***Write a system of equations***: Use the number of cars as one variable and the number of motorcycles as a second variable. Write two equations that represent the relationships between the number of cars and the number of motorcycles.

Criteria for a Good Solution

- **Efficiency**—can I solve the problem with a minimal amount of effort? "No pain–no gain" applies more to physical exercise.
  In this problem a lucky guess might have been very efficient. For some it might have led to a lot of wasted effort.

- **Transferability**—can the process be generalized to solve a wide variety of problems like this, even when the answers are fractions or worse?

- **Comprehensibility**—can I understand what I am doing and why? Techniques that are not understood will be forgotten or remembered inaccurately so they result in incorrect answers.

Strategy – Draw a Picture

The number of wheels using twenty spaces

1 2 3 4 5 6 7 8 9 10

11 12 13 14 15 16 17 18 19 20

Strategy – Using a Table

Number of cars	Number of car wheels	Number of motor-cycles	Number of motorcycle wheels	Number of vehicles	Number of wheels

Using a Table

Number of cars	Number of car wheels	Number of motor-cycles	Number of motorcycle wheels	Number of vehicles	Number of wheels
20	80	0	0	20	80
19	76	1	2	20	78
18	72	2	4	20	76
17	68	3	6	20	74
16	64	4	8	20	72
15	60	5	10	20	70
14	56	6	12	20	68
13	**52**	**7**	**14**	**20**	**66**
0	0	20	40	20	40

Notice that an inspection of the list can reveal a pattern or many patterns that can be used to make a table that does not include every possible entry.

If there are multiple solutions, an organized table is more likely to reveal them than the guess and test method.

An Alternative Table

The process for the computations is displayed instead of the final answer. Patterns are more easily recognized if the processes are written out.

Number of cars	Car wheels	Motor-cycles	Mcycle wheels	Total Vehicles	Number of wheels
20	4(20)	20 – 20	2(0)	20 + (20 – 20)	4(20) + 2(20 – 20)
19	4(19)	20 – 19	2(1)	19 + (20 – 19)	4(19) + 2(20 – 19)
18	4(18)	20 – 18	2(2)	18 + (20 – 18)	4(18) + 2(20 – 18)
17	4(17)	20 – 17	2(3)	17 + (20 – 17)	4(17) + 2(20 – 17)
16	4(16)	20 – 16	2(4)	16 + (20 – 16)	4(16) + 2(20 – 16)
15	4(15)	20 – 15	2(5)	15 + (20 – 15)	4(15) + 2(20 – 15)
14	4(14)	20 – 14	2(6)	14 + (20 – 14)	4(14) + 2(20 – 14)
13	**4(13)**	**20 – 13**	**2(7)**	**13 + (20 – 13)**	**4(13) + 2(20 – 13)**
0	4(0)	20 – 0	2(20)	0 + (20 – 0)	4(0) + 2(20 – 0)
c	4(c)	20 – c	2(20 – c)	c + (20 – c)	4(c) + 2(20 – c)

Rules of the SPC System

Rule 1: Create a new production path by repeating the combination of processes, Ps, and quality control checks, Cs, following the initial start-up, S, of any production path.

For example:

$$SPCC \xrightarrow{Rule\ 1} SPCCPCC$$

Apply Rule 1, if possible, once to each of the following production paths.

a. S P P

b. S P P P

c. S C P

Rule 2: If a production path ends in a process P, add a quality control check, C, to make a new production path.

For example:

$$S P C C P \xrightarrow{\text{Rule 2}} S P C C P C$$

Apply Rule 2, if possible, once to each of the following production paths.

a. S P P

b. S P P P P

c. S C P

d. S P C

Rule 3: If a production path contains three consecutive processes, replace the three Ps with a quality control check, C, to make a new production path.

For example:

$$SPPPCCP \xrightarrow{\text{Rule 3}} SCCCP$$

a. Given the path S P P P P P P P, use Rule 3 once.

b. List all possible new paths that can be derived using with only one application of Rule 3.

Rule 4: If a production path contains two consecutive quality control checks, remove the two Cs to make a new path.

For example:

$$SPCCP \xrightarrow{Rule\ 4} SPP$$

Apply Rule 4, if possible, once to each of the following paths.

a. S P C C

b. S C C P P P C

c. S P P C P P

Summary: SPC Rules

Rule 1: Create a new production path by repeating the combination of processes, Ps, and quality control checks, Cs, following the initial start-up, S, of any production path.

Rule 2: If a production path ends in a process P, add a quality control check, C, to make a new production path.

Rule 3: If a production path contains three consecutive processes, replace the three Ps with a quality control check, C, to make a new production path.

Rule 4: If a production path contains two consecutive quality control checks, remove the two Cs to make a new path.

P-Counts

In the first column, record ten valid paths you produced in Investigation 8 that have **different** P–counts. Record the P–count for each valid path in the second column.

Valid Paths	P–Counts

Assumptions about Variables

- Every time a variable is used in a given problem it represents the same thing.

- The variable probably represents something different when we work a new problem.

- Variables make it possible to describe a few specific examples as a generalized statement.

- A generalized statement must always be true. In order to guarantee this, the use of restrictions, or conditions, may be required. For example:

    $\frac{n}{d}$ represents any fraction only if $d \neq 0$.

- Variables allow us to save a lot of manual effort since a single variable can represent several words or a tedious repetition.

- The cost of using a variable to simplify our work is increased mental effort needed to deal with the abstraction.

Finding a Pattern

1. Use your calculator to determine the answers.

$$1^2 =$$
$$11^2 =$$
$$111^2 =$$

2. Predict and check

$$1111^2 =$$
$$11111^2 =$$

3. Since the answer to 111111^2 is displayed in scientific notation, all the digits you would get if you multiplied this by hand will not be visible. Predict exact answers to the following.

$$111111^2 =$$
$$1111111^2 =$$
$$111111111111^2 =$$

4. Check the last answer on a scientific or graphing calculator. Does your generalization approximate the calculator's answer?

Rule 1 and P-Counts

Begin with the path **S P** and repeatedly apply only Rule 1. Record your results.

Number of times Rule 1 is applied	P–count
0	1
1	
2	
3	
4	
5	
6	
7	
8	

P-Counts with Rules 1 and 3

Apply Rule 1 five times to the path S P.

Then analyze the new production path and complete the table below.

Number of times Rule 3 is applied	Total number of Ps removed from theorem	Number of Ps remaining in theorem
0	0	32
1	3	
2		
3		
4		
5		
6		

Setting a Viewing Window

On the TI–82, set the viewing window boundaries as follows:

$$xmin = -22 \quad ymin = -20$$
$$xmax = 25 \quad ymax = 70$$
$$xscl = 5 \quad yscl = 10$$

Graph the family of lines for

$$n = 6, 5, 4, \text{ and } 3$$

Note: On the TI-81, set *xmin* to -22.5.

Use the TRACE feature to investigate the P–counts on your graphs.

Record P–counts for each value of *n*.

The Parking Lot Revisited

In this investigation, the parking lot (20 spaces) is not necessarily full. Instead of considering the number of cars or the number of motorcycles, use one variable, w, to represent the number of wheels parked in the lot at any given time.

If only cars could park in the lot:

- list the domain for variable w, the number of wheels, that might be in the parking lot at any given time.

- Describe the set of numbers listed as an answer for the domain.

If only motorcycles could park in the lot:

- list the domain for w, the number of wheels, that might be in the parking lot at any given time.

- Describe the set of numbers listed as an answer for the domain.

If both cars and motorcycles are allowed to park in the lot:

- list the domain for w, the number of wheels, that might be in the parking lot at any given time.

Investigating Prime Numbers

A *prime number sieve* is a means of identifying prime numbers. Complete the table below for the numbers up to and including 103 to locate primes.

2	3	4	5	6	7
8	9	10	11	12	13
14	15	16	17	18	19
20	21	22	23	24	25
26	27	28	29	30	31
32	33	34	35	36	37
38	39	40	41	42	43
44	45	46	47	48	49
50	51	52	53	54	55
56	57	58	59	60	61
62	63	64	65	66	67
68	69	70	71	72	732
74	75	76	77	78	79
80	81	82	83	84	85
86	87	88	89	90	81
92	93	94	95	96	97
98	99	100	101	102	103

Two Lottery Tickets

John borrowed $12 from his brother. He spent $2 to buy two lottery tickets—the first quadrupled his money and the second added six dollars to his total. John then split the money he had with his brother, Mike. How much money did each man get?

1. Indicate the computations you did to solve this problem.

2. Write *one* expression that contains all the computations needed to solve this problem.

3. Record the answer to the problem.

Jennifer had $12. She bought two paperback books that cost $4 each and cashed a check for six dollars. Jennifer combined the money she had with the money from the cashed check and divided it into two equal parts. She put one part in a secret compartment in her wallet. How much money did she hide?

4. Indicate the computations you did to solve this problem.

5. Write *one* expression that contains all the computations needed to solve this problem.

6. Record the answer to the problem.

7. How are the problems in Investigations 1 and 2 the same?

8. How are they different?

The Near-sighted Professor

Consider the following seating arrangement:

	2	3	4	5
9	8	7	6	
	10	11	12	13
17	16	15	14	

Determine the column in which the following numbers appear:

    a. 100   b. 1000

    c. 1835   d. 9993

    e. 99,997

Recording Differences

Record each *difference* to the right on the line provided.

```
          2     3     4     5
         ___   ___   ___

   9     8     7     6    ___
         ___   ___   ___

  ___   10    11    12    13
         ___   ___   ___

  17    16    15    14    ___
         ___   ___   ___

  ___   18    19    20    21
         ___   ___   ___

  25    24    23    22
```

Remainders

Divide each number in Table 1 by 8. Record only the remainders in place of the original numbers.

A Function Machine

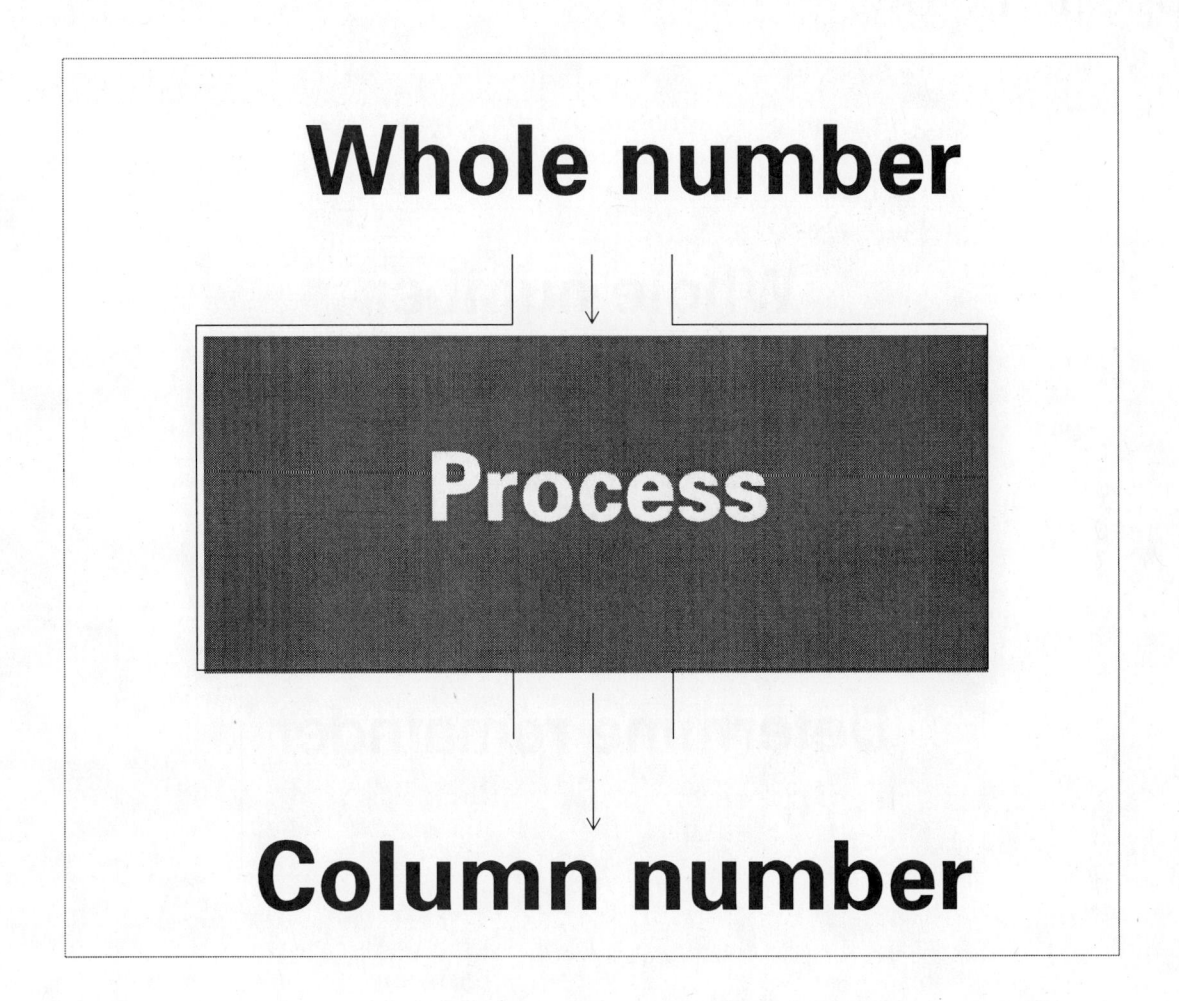

Whole number

Process

Column number

A function machine for the process of using a whole number as input and receiving its remainder when the whole number is divided by 8 appears below.

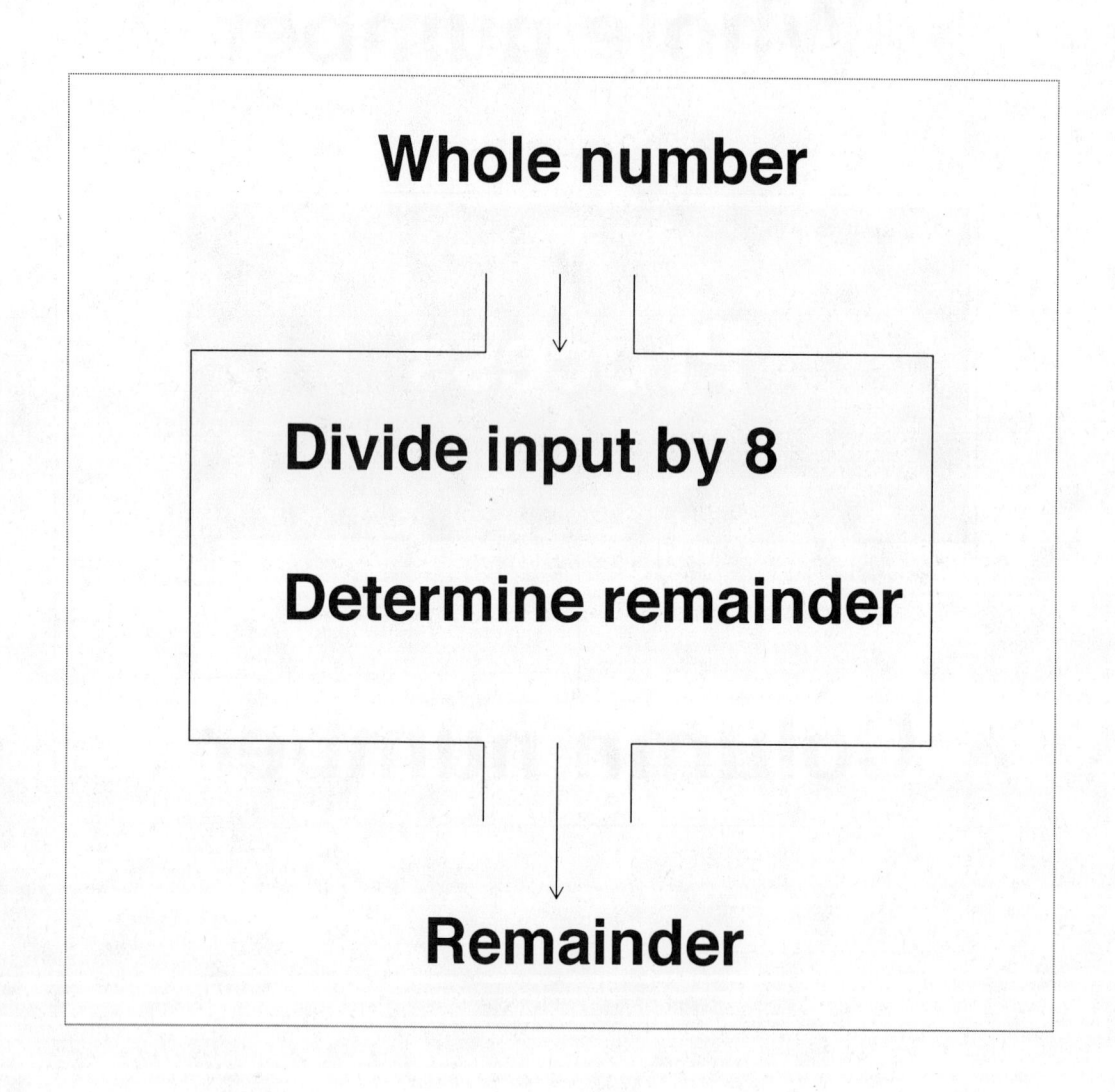

Whole number

Divide input by 8

Determine remainder

Remainder

A function machine using a remainder from division by 8 to determine the column a number would appear in.

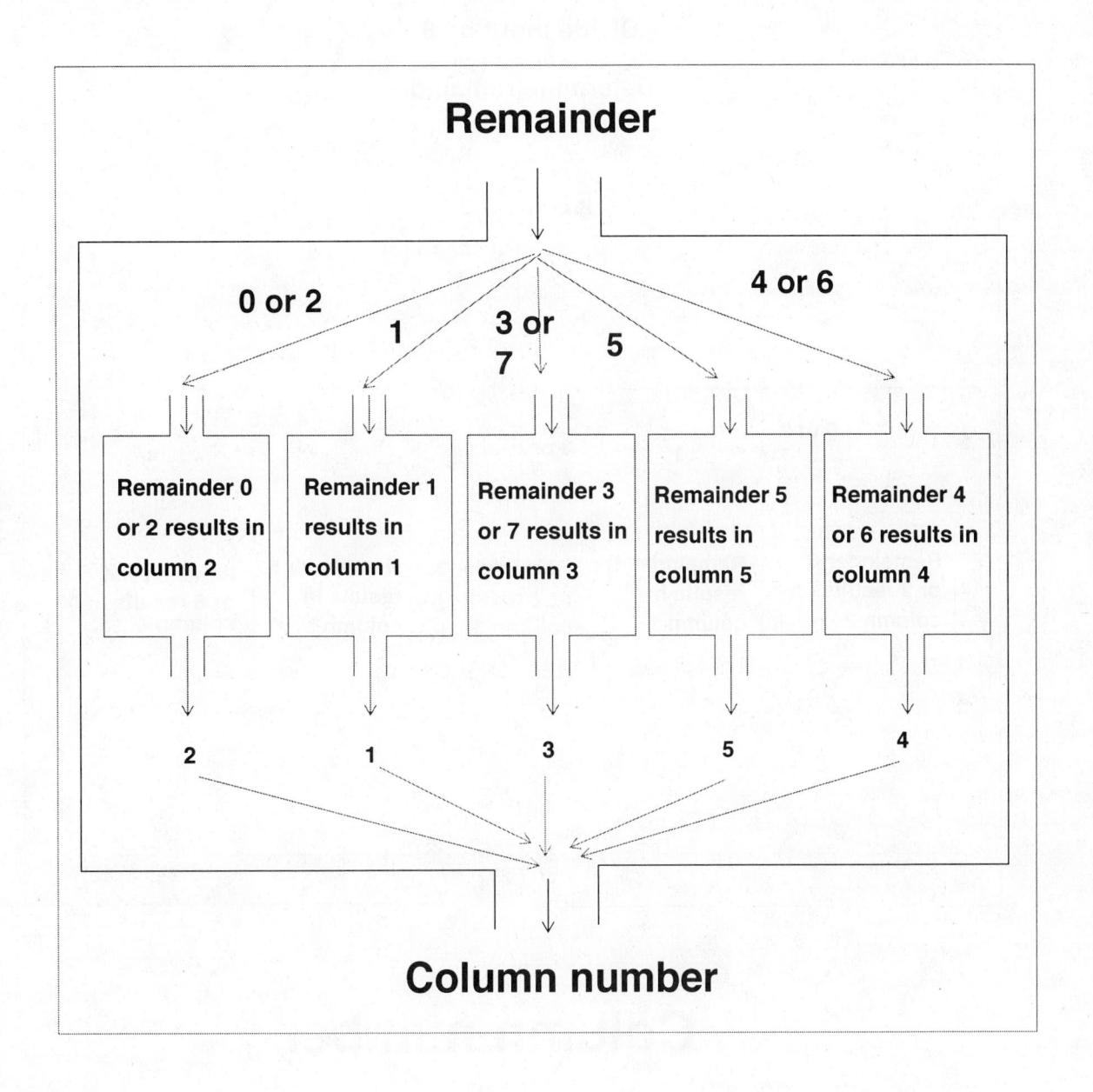

Remainder

- **0 or 2** → Remainder 0 or 2 results in column 2 → 2
- **1** → Remainder 1 results in column 1 → 1
- **3 or 7** → Remainder 3 or 7 results in column 3 → 3
- **5** → Remainder 5 results in column 5 → 5
- **4 or 6** → Remainder 4 or 6 results in column 4 → 4

Column number

Whole number

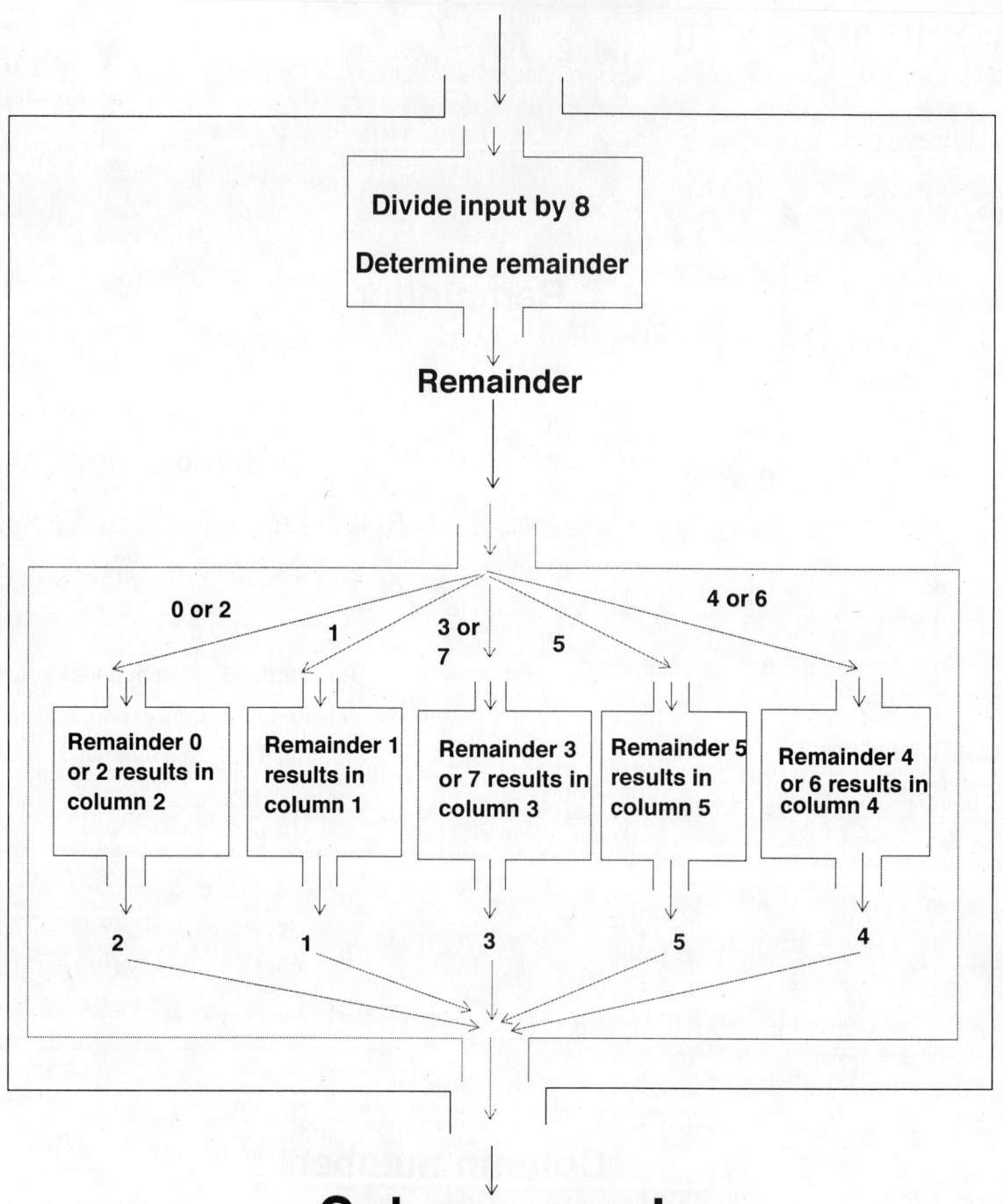

```
Divide input by 8
Determine remainder
```

Remainder

0 or 2	1	3 or 7	5	4 or 6
Remainder 0 or 2 results in column 2	Remainder 1 results in column 1	Remainder 3 or 7 results in column 3	Remainder 5 results in column 5	Remainder 4 or 6 results in column 4
2	1	3	5	4

Column number

Investigating Car Rentals

My car is in the shop so I must rent a car for only one day. If I rent a car for one day the charge is $35 plus 17¢ per mile.

1. What are the constants?

2. What are the two variables? Identify the dependent variable. Identify the independent variable.

3. Suppose I drive 53 miles on this day. What will my cost be? Show your work and explain what you are doing.

4. Describe in words how to find the total charge for the car given the number of miles driven.

5. Draw a function machine that demonstrates the relationship.

Search Plan & Reference Map

Reference Map

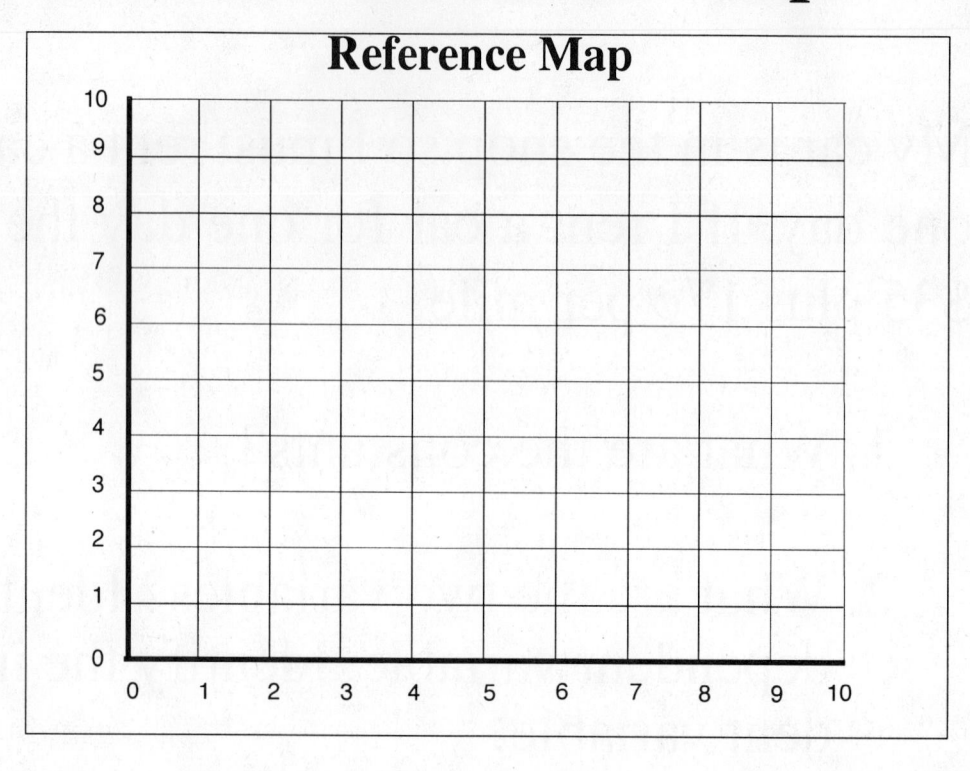

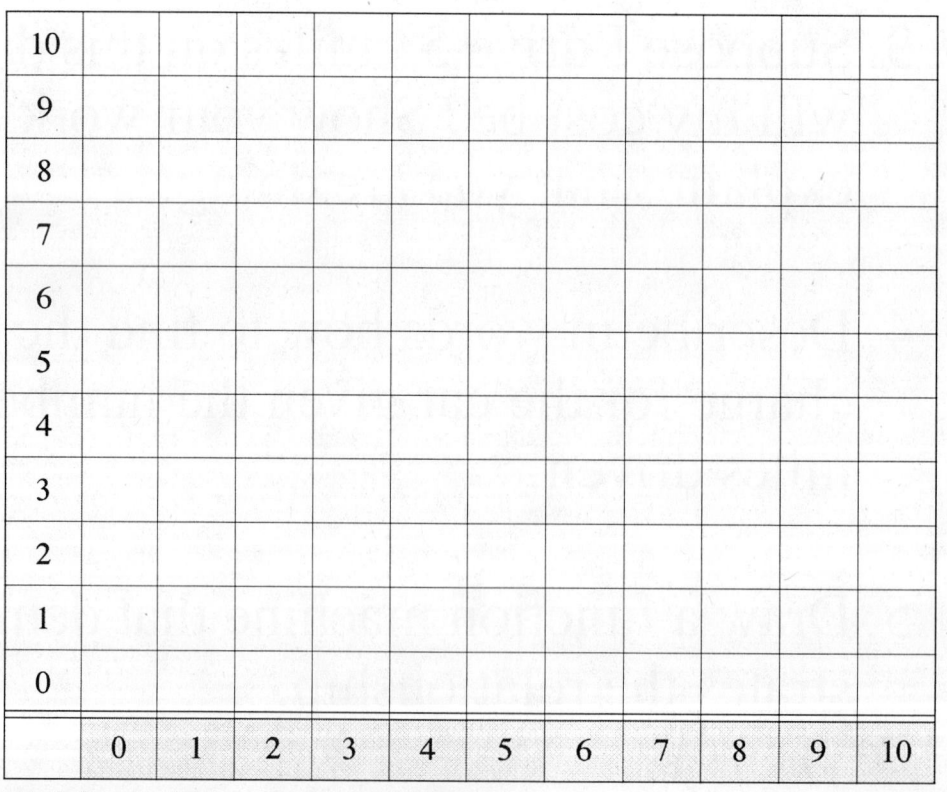

Table 1: The Search Plan

Doubling Your Earnings

Suppose you are offered a job in which you will be paid $1 the first day, $2 the second day, $3 the third day, etc. Naturally, you want to keep track of how much money you will have at the end of two days, nine days, or a year. You also want to compare this way of being paid with other, more traditional ways of being paid. Is this a good pay plan?

Under this pay plan, how much money will you earn:

  a) after two days on the job?

  b) after nine days of the job?

  c) after a year on the job?

What strategies make sense for exploring this problem?

Visually representing the problem can be an efficient and effective way to understand the problem. Let each dot represent a dollar, each row represent one day's wages, and each grouping of dots represent the money earned to date.

1. The pattern of four triangles in Figure 1 visually displays the total amount of money earned after one, two, three, and four days.

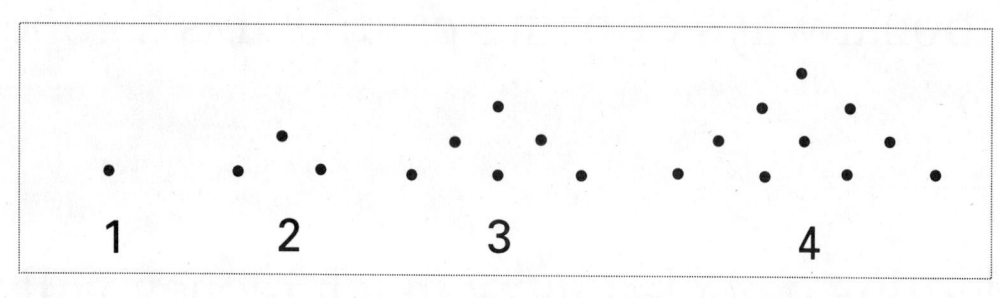

```
                    .           .
             .    . .      .  . . .
        .   . .  . . .   . .  . . . .
    .   . . . . . . . . . . . . . . .
    1    2      3            4
```

Figure 1

2. Draw the next two triangles in the **sequence**.

3. How many dots were added to the fourth triangle to generate the fifth triangle? How many dots were added to the fifth triangle to generate the sixth triangle? How does this relate to the amount paid each day?

Doubling Your Wages

Day	Day's wages	Total wages to date expressed as a sum	Total wages to date
1	1	1	1
2	2	1 + 2	3
3	3	1 + 2 + 3	
4			
5			
6			
7			
8			
9			
10			

Calculator Golf

In any type of golf game (regular, miniature, frisbee, etc.), the score on a hole is often measured by its signed distance from par—the "expected" score on a hole.

For example, a "birdie" represents one shot below par (–1) and a "bogey" represents one shot over par (+1). Par is represented by 0. The person with the lowest score has performed the best.

Let's investigate positive and negative numbers by playing "calculator golf."

You will play a nine–hole game. Par for each hole is given in the table below.

Hole	1	2	3	4	5	6	7	8	9
Par	4	5	4	3	5	4	4	3	4

Generating Random Numbers

Use the random number generator on your calculator to "play" each hole. You will generate a random number between one and eight, inclusive, which will be used as a score on a hole, since some of us are "duffers."

On the Home Screen enter the expression

$$Int(Rand \times 8) + 1$$

Both the *Int* and *Rand* functions are accessed from the **MATH** menu.

- Press **MATH**

- Use the ▷ to highlight the **PRB** menu.

- Enter **1** to select the random seed generator.

- Press **ENTER** to display your score.

Tuition Policy

The tuition policy at a large university allows students to take twelve or more semester hours for a flat charge of $1425. The charge is $98 per semester hour if a student takes less than twelve credit hours and student fees and perks are not charged.

Complete the table for the tuition problem situation.

Semester hours	Tuition ($)
6	
9	
11	
12	
15	
18	

Variable Rates of Change

One morning, exactly at sunrise, a monk began to climb a mountain. The graph below displays the relationship between the horizontal distance he travelled and the corresponding change in altitude. There is a third distance to consider–the length of the path, which we will consider in a later section.

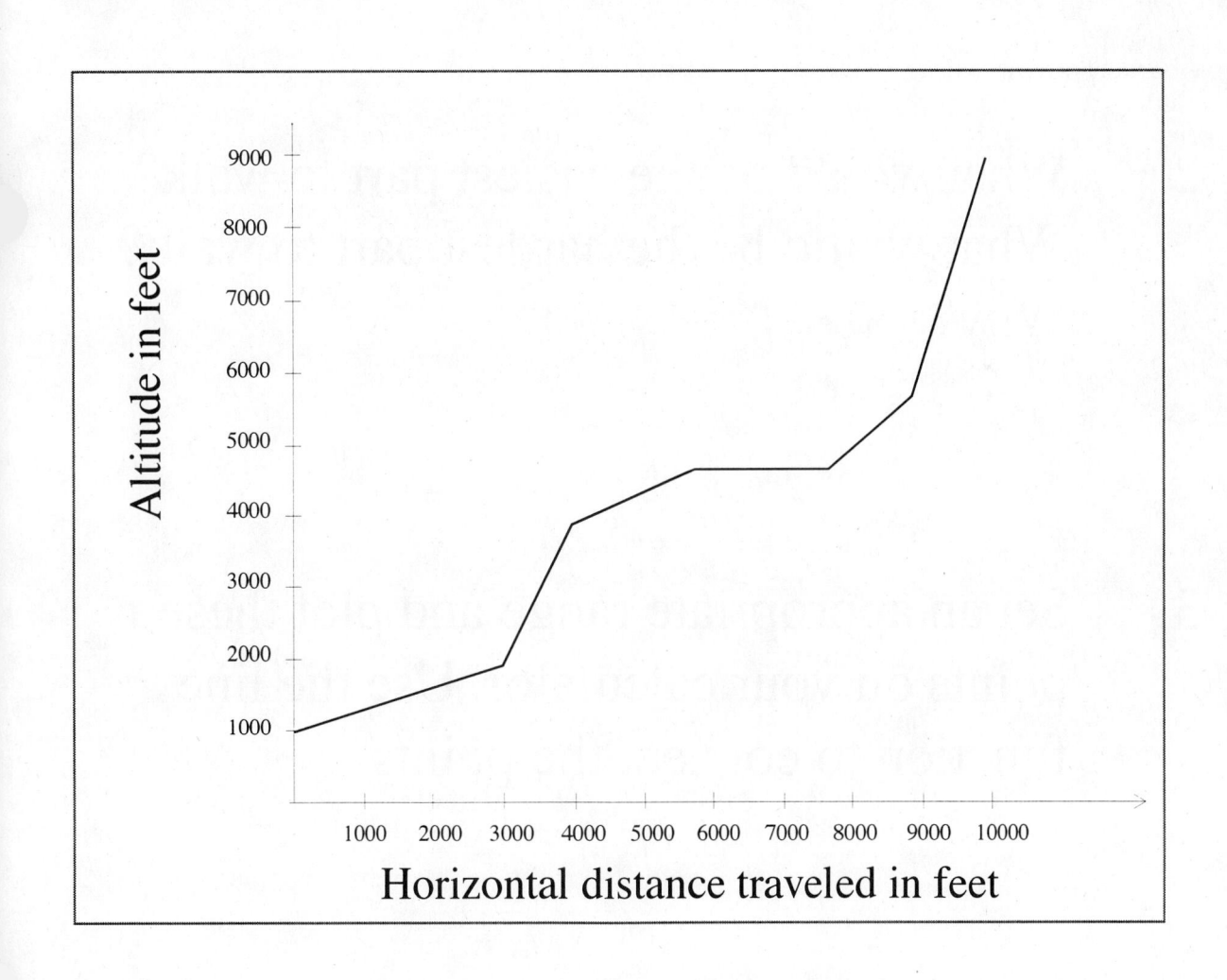

Variable Rates of Change

1. What points would you choose as interesting points to describe this hill? Label the points using A, B, C, . . . and write each as an ordered pair. Defend why you chose these points.

2. What would be the easiest part to walk? What would be the hardest part to walk? Why?

3. Set an appropriate range and plot these points on your calculator. Use the line function to connect the points.

4. How would you find a mathematical expression or number to compare parts of the climb?

5. Describe why walking the first 3000 feet in distance might be easier than walking the distance from 3000 feet to 4000 feet.

6. Describe why going a distance of 1000 feet horizontally might be *easier* than walking 1000 feet vertically.

7. Describe why going a distance of 1000 feet horizontally might be *harder* than walking 1000 feet vertically.

Linear Investigations

Consider the problem: Two business majors, Sue and Tom, sell cookies during breaks between classes. They spend $30 on supplies and sell cookies for 25¢ each.

Investigate the relationship between the number of cookies sold (input) and the net profit (output) by completing Table 1.

Number of cookies sold	Net profit (dollars)
0	
50	
100	
150	
200	

A Monk's Journey

One morning, exactly at sunrise, a monk began to climb a tall mountain. A narrow path, no more than a foot or two wide, spiraled around the mountain to a glittering temple at the summit. The monk ascended at a varying rate of speed, stopping many times along the way to rest and eat dried fruit he carried with him.

He reached the temple shortly before sunset. After many days of fasting and meditation he began his journey back along the same path, starting again at sunrise and walking at variable speeds with many pauses along the way.

His average speed descending was, of course, greater than his climbing speed. Prove that there is a spot along the path that the monk will occupy on both trips at precisely the same time of day.

A Monk's Journey

1. Make a list of the unnecessary information included in this problem.

2. Identify the relevant information needed to describe the problem situation.

3. How many variables are there? Name them.

4. Can you represent the problem situation by writing an equation? Why or why not? If you can, write an equation to describe the problem situation.

TI-82
Graphing Calculator
Reference Manual

TI-82 REFERENCE MANUAL

Turn the calculator on and off

Press **ON** to turn the calculator on. If nothing appears on the screen, adjust the display contrast (see below). If you still see nothing, check the batteries.

Press **2nd** **ON** to turn the calculator off.

Keyboard organization

The keys are grouped by both color and location for ease of use.

The keys are color–coordinated.

To choose a command that is in blue and appearing above a key, press the blue **2nd** key first. By convention, any key that is in blue will appear with black, rather than white, letters in this manual.

To choose a command in gray and appearing above a key, press the gray **ALPHA** key first.

Keys by row beginning at the top of the calculator:

Row 1: Graphing and table keys.

| Y= | WINDOW | ZOOM | TRACE | GRAPH |

Rows 2 and 3: Editing keys.

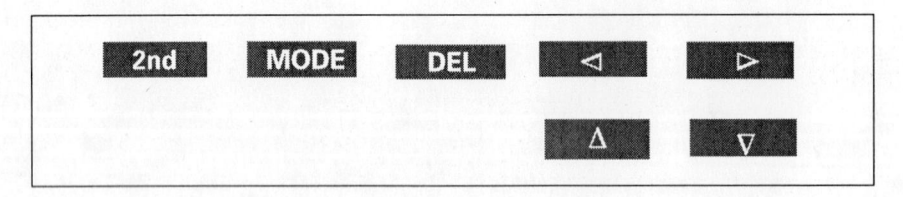

| 2nd | MODE | DEL | ◁ | ▷ |
| | | | △ | ▽ |

Rows 3 and 4: Advanced functions accessible through pull-down menus.

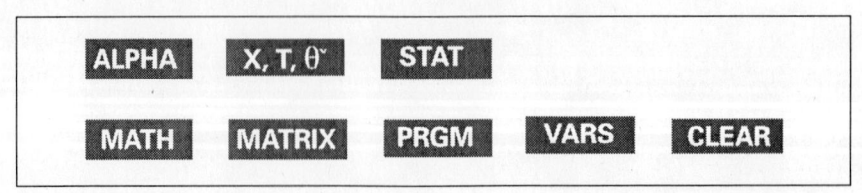

| ALPHA | X, T, θ | STAT |
| MATH | MATRIX | PRGM | VARS | CLEAR |

344

Rows 5–10: Scientific calculator keys.

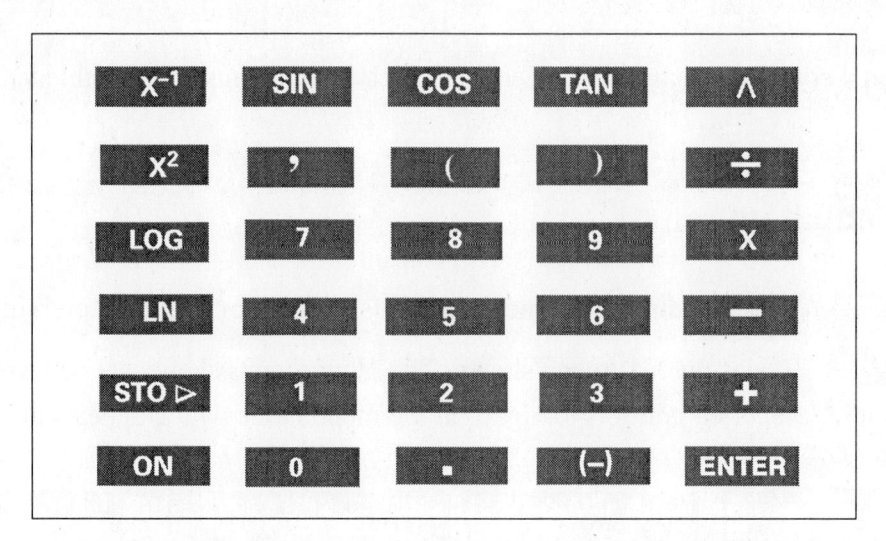

Adjust the display contrast (darken or lighten screen)

Press and release the [2nd] key.

Press and hold [▲] to increase the contrast (darken the screen).

Press and hold [▼] to decrease the contrast (lighten the screen).

The calculator can be returned to factory settings by pressing [MEM]. Since MEM is in blue above the "+" key, access MEM by pressing [2nd] [+].

```
MEMORY
1:Check RAM...
2:Delete...
3:Reset...
```

Change the mode (the Mode Screen)

Press [MODE] to customize the calculator settings. Activated settings are highlighted.

```
Normal Sci Eng
Float 0123456789
Radian Degree
Func Par Pol Seq
Connected Dot
Sequential Simul
Fullscreen Split
```

Change a setting by moving the cursor to the desired setting using the arrow keys.

Press [ENTER] to select a setting.

345

Menus

Rows 3 and 4 contain keys that allow access to both math menus and variable values.

Select a menu item

Press the key(s) corresponding to the menu to be displayed. For example, pressing the **MATH** opens the first menu shown. The other menus shown are accessed by using the right or left cursor controls to highlight the menu name which appears at the top of the screen.

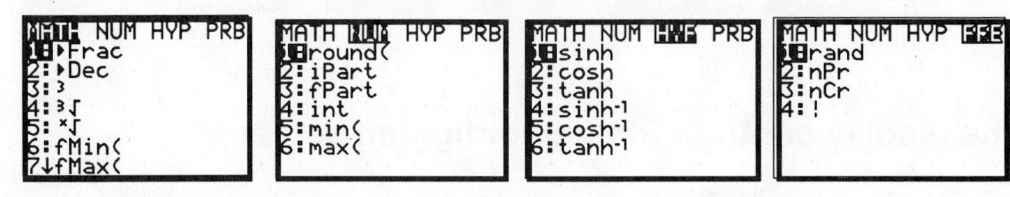

Select a menu item by either using the arrow keys to highlight the desired choice and then press **ENTER** ; or

if a number is associated with a menu item, press the number next to the menu item.

Home screen

The home screen is used for calculations.

You may return to the home screen at any time by executing the **QUIT** command. This is done by pressing **2nd** **MODE**.

Graphics screen

The graphics screen is used to display graphs.

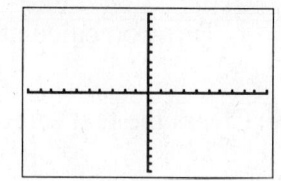

The adjustments to the graphics screen are set using the window screen to be discussed later in the manual. To show the graphics screen press **GRAPH**.

Menu screens

There are several menu screens which display menus of selections which allow you to control the functionality of the calculator and display values of variables.

Table screen

This screen displays a table of inputs and outputs for each active user-defined function.

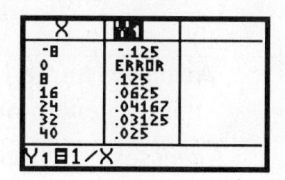

Choose **TABLE** to display this screen by pressing **2nd** **GRAPH**. Notice the word TABLE appears above the GRAPH key printed in blue.

List screen

This screen displays up to six lists where users may enter data.

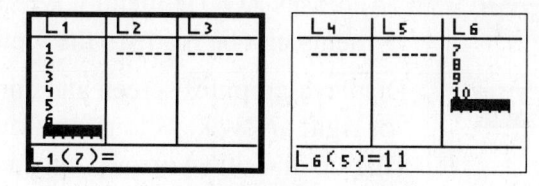

Press **STAT** and then press **1** to choose the Edit item on the menu. This shows the list screen and allows you to edit the list. Press the right cursor control to show more list.

347

Important keys and their function

Keystroke	Function Description
ON	Turns on calculator; interrupts graph being sketched.
CLEAR	Clears the text screen; deletes functions when in the Y= screen.
ENTER	Executes a command; moves cursor to next line.
(−)	The unary operation of opposing (not used for subtraction).
MODE	Displays current operating mode.
DEL	Deletes character at the cursor.
∧	Symbol used for exponentiation
Y =	Accesses screen where up to eight functions may be entered and stored.
X, T, θ	Enters the variable x in function mode; enters the variable t in parametric mode; enters Θ in polar mode.
MATH	Accesses numerical representations; random numer generator and combinatoric entries.
2nd	Accesses functions and menus printed in blue.
ALPHA	Accesses letters and symbols printed in grey to the right above many keys.
STO ▷	Used to assign a numeric value to a variable
GRAPH	Displays the graphical representation of entered functions.
WINDOW	Accesses menus for setting the viewing window and selecting the window format.
ZOOM	Accesses ZOOM menu where various adjustments can be made to the viewing window.
TRACE	Displays graphics screen and cursor. Use the left or right arrow keys to trace along a graph. Use the up and down arrow keys to move the trace cursor from one function to another function.

Access functions using the **2nd** key (commands written in blue, above and left of center of the keys).

Editing Functions Accessed with the **2nd** key.

Function	Keystrokes	Description
OFF	2nd ON	Turns the calculator off.
ENTRY	2nd ENTER	Allows the last line entered to be edited and re-executed.
INS	2nd DEL	Allows insertion of character(s) to the left of the cursor.
LINK	2nd STAT	Allows transfer of data or programs between two TI-82 calculators.
QUIT	2nd MODE	Exits menus; returns calculator to home screen.

Table Functions Accessed with the **2nd** key.

Function	Keystrokes	Description
TABLE	2nd GRAPH	Displays a table of x-values and corresponding output values functions stored in the **Y=** menu.
TblSet	2nd WINDOW	Accesses the TABLE SETUP menu where the minimum table value and the table increment are set.
STATPLOT	2nd Y=	Accesses the STAT PLOT menu where statistical data may be graphically displayed using a histogram, point plot, line plot, or box-and-whisker plot.

Calculations

Calculations are done in the home screen. If you are not in the home screen, choose `QUIT` which is accessed by pressing `2nd` `MODE`.

Enter expressions as you would write them. Observe the algebraic order of operations.

After you complete typing an expression, press `ENTER` to obtain the answer.

If you make a mistake during entry, use the arrow keys, `INS` (for insert), or `DEL` (for delete) to edit your expressions.

Raise a number to a power `^`

Enter an exponent first entering the base and then pressing the `^` key followed by the exponent. For example, 4^5 appear on the home screen as shown. Press `ENTER` to display the value of the expression.

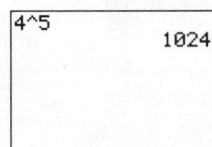

```
4^5
       1024
```

Negative numbers `(-)`

Enter a negative number using the *gray* key `(-)` (located to the left of the `ENTER` key), not the blue key, which is reserved for the operation of subtraction.

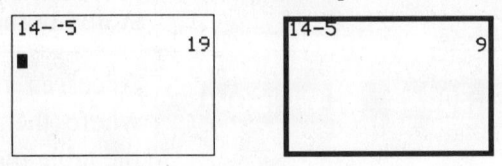

```
14--5
       19
■
```

```
14-5
        9
```

Scientific notation (3.46E11 means 3.46×10^{11})

This notation is uses powers of ten to represent very large and very small numbers. Numbers are written as a decimal times a power of 10.

To enter on the calculator, replace the **X 10** with **EE**. The exponent appears immediately to the right of the E in the display. The display, 3.46E11, actually represents 3.46×10^{11}.

Enter **EE** by pressing [2nd] [,]. Only one **E** appears on the screen to indicate that the next number entered will be accepted as a power of ten.

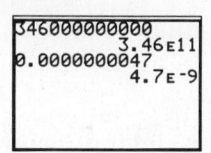

```
346000000000
           3.46E11
0.0000000047
           4.7E-9
```

Edit a previously-entered expression [2nd] [ENTER]

The last entered in the calculator can be displayed on the screen and edited. Execute the **Entry** command to display the last expression entered by pressing the following keys.

Choose [ENTRY] by pressing [2nd] [ENTER].

A copy of the previously entered expression should appear on the screen. This expression may be edited. The new expression may be executed by pressing [ENTER].

Reciprocal function [x⁻¹]

To find the reciprocal of a number first enter the number whose reciprocal you wish to find on the home screen. If the number is a fraction it must be enclosed in parentheses.

Press [x⁻¹].

Press [ENTER] to display the reciprocal of the original number in decimal form. To display the answer as a fraction choose item number one from the math menu by pressing [MATH] then [1].

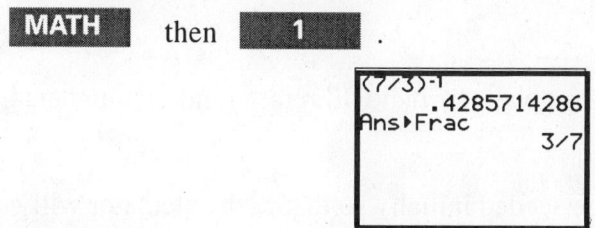

```
(7/3)⁻¹
         .4285714286
Ans▶Frac
                3/7
```

INT function

The greatest integer function (**int**) returns the largest integer less than or equal to a number, expression, list or matrix. For nonnegative numbers and negative integers, the **int** value is the same as **iPart**. For negative noninteger numbers **int** returns a value one integer less than **iPart**.

Press [MATH]

351

Use the ▷ key to highlight the **NUM** menu.

Enter **4** or use the ▽ to highlight **4** and press **ENTER** to select the greatest integer function *4: int*.

The calculator returns to previous screen and inserts **int** function at cursor. Follow the function name with an input value in parentheses. The function returns an output which is the largest integer that does not exceed the input.

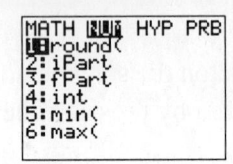

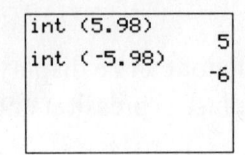

```
MATH NUM HYP PRB        int (5.98)
1:round(                          5
2:iPart                 int (-5.98)
3:fPart                          -6
4:int
5:min(
6:max(
```

Obtain fractional results

If a computation results in a decimal output, the result can often be converted to a fraction by using the **Frac** command.

Press **MATH** following a computation.

Press **1** to select **Frac**. The calculator returns to previous screen.

Press **ENTER** and the calculator outputs a fraction.

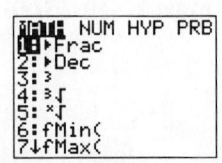

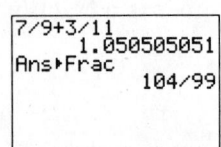

```
MATH NUM HYP PRB        7/9+3/11
1:▶Frac                      1.050505051
2:▶Dec                  Ans▶Frac
3:³                              104/99
4:³√
5:ˣ√
6:fMin(
7↓fMax(
```

RAND function

The calculator contains a random number generator that will return random numbers greater than zero and less than one.

The random number generator should be seeded initially so that each calculator will generate a different sequence of numbers. To control a random number sequence, first store an integer seed value in **rand**. The factory-set seed value is used whenever you reset the TI-82 or when you store **0** to **rand**.

Start in the home screen.

Type a favorite whole number. This number acts as the seed.

Press **STO▷**. This is the key used for storing a value in the calculator.

Press **MATH**.

Use the **◁** key to highlight the **PRB** menu.

Press **1** to select **rand**.
The calculator returns to the home screen and inserts the **rand** function.

Press **ENTER** to complete the seeding process.

Recall **rand** by pressing **MATH**, arrowing to **PRB**, and selecting **1**.

Press **ENTER** to see the first random number.

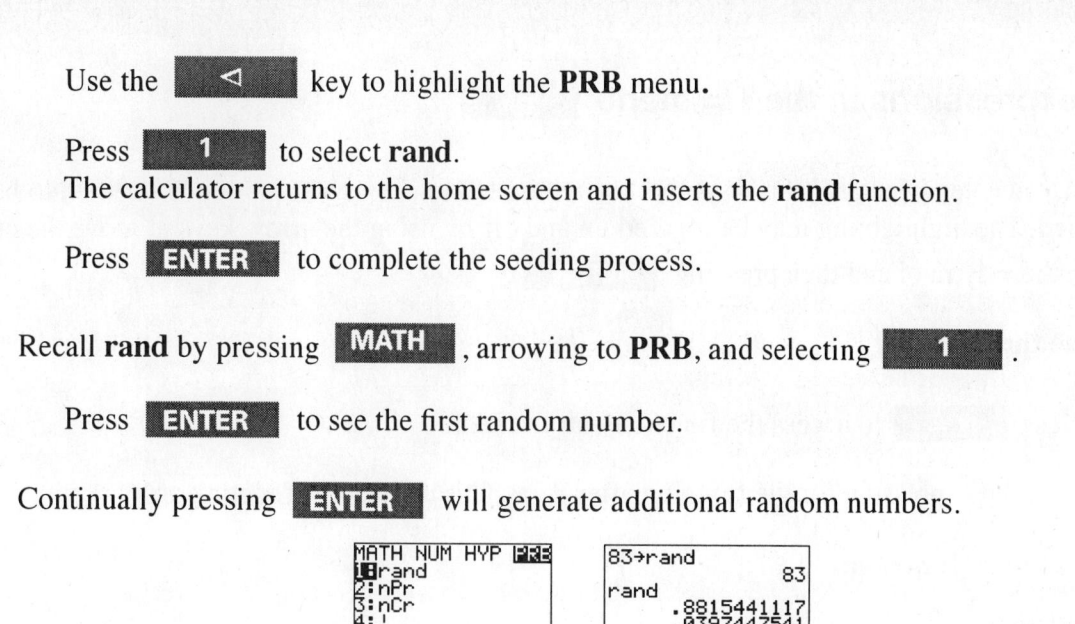

Continually pressing **ENTER** will generate additional random numbers.

Factorial function

The factorial of a whole number is the product of all whole numbers less than or equal to the given whole number.

Type a whole number.

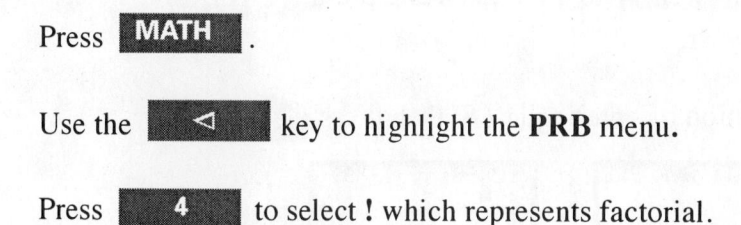

Press **MATH**.

Use the **◁** key to highlight the **PRB** menu.

Press **4** to select ! which represents factorial.

The calculator returns to the home screen and inserts the ! after the number you typed.

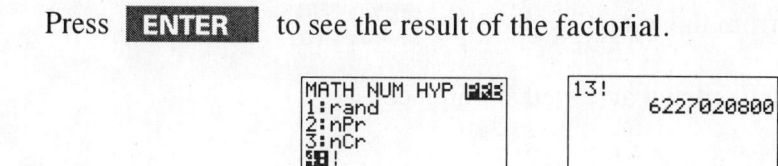

Press **ENTER** to see the result of the factorial.

353

Enter expressions in the Y = menu [Y=]

When you enter a function the = symbol is highlighted to show the function is selected to be graphed. The highlighting may be toggled on and off by using the arrow keys to move the cursor to the = symbol and then pressing [ENTER].

Define the function:

Press [Y=] to access the function entry screen.

Type the expression for the function after $Y_1 =$. This defines the function as Y_1.

Select the *int* function:

Example: Enter the function $f(x) = x - 8int(x/8)$ as Y_1:

Press [Y=] to access the function entry screen.

Enter the function expression $x - 8$ at the cursor.

Press [MATH].

Use the [▷] key to highlight the **NUM** menu.

Press the **4** key or use the [▽] to highlight **4** and press [ENTER] to select the greatest integer function *4: int*.

Complete the entry of the function by entering $(x / 8)$ then press [ENTER].

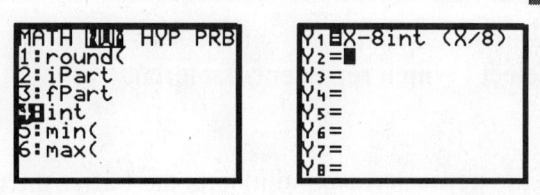

Functions may be removed from this menu by pressing [CLEAR].

Up to ten functions can be defined and accessed simultaneously.

Display a table of values from a function in the Y= menu

The TI–82 will display a table of values for a function defined under the Y = menu.

Example: Display a table of values for the function $y(x) = x - 8\,int\left(\dfrac{x}{8}\right)$

First Define the function:

Press **Y=** to access the function entry screen and enter *x - 8 int (x/8)* as Y_1.

Press **2nd** **WINDOW** to access **TblSet**, the table set up menu.

Set the beginning table value and the increment:

Enter a beginning table value for **TblMin** at the flashing cursor.

Press the **▽** to place the cursor on the number next to **ΔTbl =** to change the increment value.

ΔTbl represents the increment on the input values. The default value is **1**. If you wish a different increment, enter that value.

Check to see that **Auto** is highlighted for both Indpnt and Depend.

If not, use arrow keys to place cursor on **Auto** and press **ENTER** to select it.

To view the table:

Press **2nd** **GRAPH** to select the **TABLE** command which displays the table screen.

By using the up and down arrows, you can scroll through the table, either up or down.

Storing a value for a variable

In the home screen enter the input value and store it in the variable x.

Type the number you wish to store.

Press [STO ▷] . This is the key used to store values.

When pressed, an arrow pointing right appears.

Press [X, T, Θ] . This will display the variable x.

```
3→X█
```

Press [ENTER] . this causes the storage to take place.

A common mistake is to forget to press [ENTER] .

Example: Store the value of 3 in x.

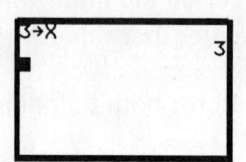

```
3→X
          3
█
```

Function evaluation: storing input

Example: Evaluate the function $y(x) = -x^2 + 8x + 19$ for $x = 3$.

Once a function is defined using the **Y =** menu, you can evaluate the function for a given input value by storing the value in the variable x.

Define the function:

Press `Y=` to access the function entry screen.

Type the expression $-x^2 + 8x + 19$ for the function after $Y_1 =$.
This defines the function as Y_1.

Return to the Home Screen:

Return to the Home Screen by pressing `2nd` `MODE`.

Enter the **input value** followed by `STO ▷` `X, T, Θ`.
This will have the effect of storing a value for variable x.

Writing more than one command on the same line (chaining commands):

Enter a colon by pressing `2nd` `■`. The colon is used to chain two or more commands together on the same line.

Select the function you wish evaluated:

Select Y_1 by pressing `2nd` `VARS` `1` `1`.

Evaluate the function:

Press `ENTER`. The function value is displayed.

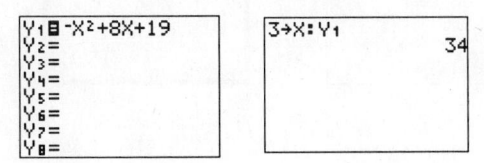

```
Y1■-X²+8X+19
Y2=
Y3=
Y4=
Y5=
Y6=
Y7=
Y8=
```

```
3→X:Y1
              34
```

Function evaluation using function notation

To evaluate a function using function notation you must first define the function in the Y = menu as one of the Y variables. Then on the Home Screen enter the Y variable with the input value in parentheses next to that Y variable.

Example: Evaluate the function $y(x) = -x^2 + 8x + 19$ at $x = 3$.

First, define the function:

Press **Y=** to access the function entry screen.

Type the expression $-x^2 + 8x + 19$ for the function after $Y_1 =$. This assumes that Y_1 is clear. This defines the function as Y_1.

Second, return to the Home Screen:

Return to the home screen by pressing **2nd** **MODE** .

Third, select the Y variable you wish evaluated:

Type **2nd** **VARS** **1** **1** to access Y_1 in the Y variables menu. Y_1 appears in the home screen.

Finally, complete the process by giving the input value in parentheses:

Immediately after Y_1,

Type **(3)** followed by **ENTER** .

This returns the output of the function stored in Y_1 at an input value of 3.

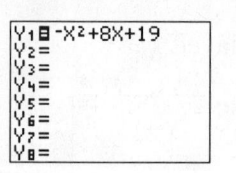

```
Y1■-X²+8X+19
Y2=
Y3=
Y4=
Y5=
Y6=
Y7=
Y8=
```

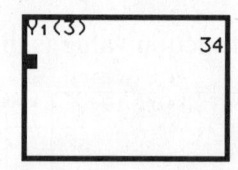

```
Y1(3)
         34
```

Function evaluation using Ask mode with the table screen

You can define a function and, using the table with the independent variable set to Ask mode and the dependent variable set to Auto mode, enter values for the independent variable and by pressing ENTER after each entry, displaying the value of the function for the input entered.

Evaluate the function $y(x) = -x^2 + 8x + 19$ at $x = 3$.

Define the function:

Press **Y=** to access the function entry screen.

Type the expression $-x^2 + 8x + 19$ for the function after $Y_1 =$. This assumes Y_1 is clear. This defines the function as Y_1.

Select the table setup TblSet :

Press **2nd** **WINDOW** to access **TblSet** , the table set up screen.

Arrow down to **Indpnt** and over to **Ask**.

Press **ENTER** to select the **Ask** mode for the independent variable. The dependent variable should still be in **Auto** mode.

Press **2nd** **GRAPH** to access **TABLE** , and display the table screen. A blank table appears.

Enter the input value:

Type the input followed by **ENTER** . The output for this input appears.

You may continue to enter inputs and the calculator will return the corresponding outputs.

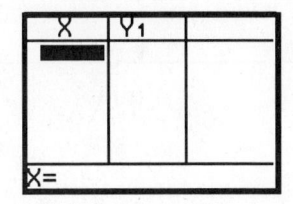

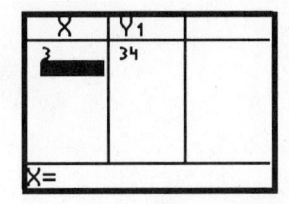

Graph a function

Once a function is defined in the Y = menu, its graph can be displayed on the graphics screen. It will be necessary to adjust the viewing rectangle which controls the portion of the graph shown.

> **Example:** Graph the function $f(x) = -x^2 + 8x + 19$.

Define the function:

Press **Y =** to access the function entry screen.

Type the expression $-x^2 + 8x + 19$ for the function after $Y_1 =$.
This defines the function as Y_1.

Choose a viewing window:

A *viewing window* is a rectangular portion of the coordinate plane.

Xmin puts the left edge of the viewing window at this x-coordinate.

Xmax puts the right edge of the viewing window at this x-coordinate.

Xscl defines the distance between tick marks on the horizontal axis.

Ymin puts the lower edge of the viewing window at this y-coordinate.

Ymax puts the upper edge of the viewing window to this y-coordinate.

Yscl defines the distance between tick marks on the vertical axis.

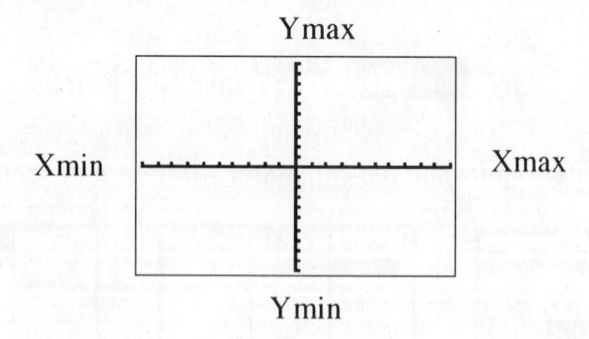

Standard viewing window

The *standard viewing window* is defined as:

Xmin = –10	Ymin = –10
Xmax = 10	Ymax = 10
Xscl = 1	Yscl = 1

10 (Ymax)

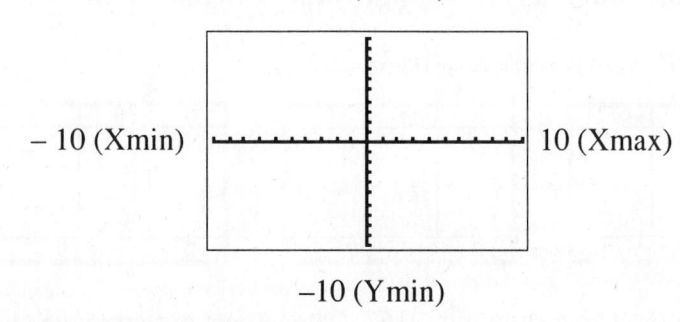

– 10 (Xmin)   10 (Xmax)

–10 (Ymin)

Set the view window:

Press **WINDOW** and enter the viewing window values.

Press the **GRAPH** key to display the graphics screen with a portion of the graph shown.

```
WINDOW FORMAT
Xmin=-10
Xmax=10
Xscl=1
Ymin=-10
Ymax=10
Yscl=1
```

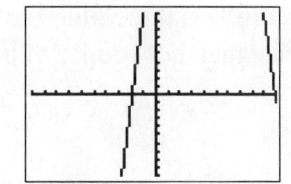

To select the standard viewing window using the Zoom menu:

Press **ZOOM** **6**.

```
ZOOM MEMORY
1:ZBox
2:Zoom In
3:Zoom Out
4:ZDecimal
5:ZSquare
6:ZStandard
7↓ZTrig
```

Change the viewing window

The previous graph is only a portion of the complete graph. To see more of the graph, change the viewing window. This is done by pressing the **WINDOW** key which displays the window screen. First decide how to change the window. The table feature helps determine a better viewing window.

Press **2nd** **GRAPH** to display the table.

Determine the smallest and largest output value between Xmin (–10) and Xmax (10).

Examine the table values for x = –10 and x = 10.

X	Y1
-10	-161
-9	-134
-8	-109
-7	-86
-6	-65
-5	-46
-4	-29

Y1 = -X²+8X+19

X	Y1
-3	-14
-2	-1
-1	10
0	19
1	26
2	31
3	34

Y1 = -X²+8X+19

X	Y1
4	35
5	34
6	31
7	26
8	19
9	10
10	-1

Y1 = -X²+8X+19

The smallest output value shown is –161. The largest output value shown is 35.

Press **WINDOW**.

Enter a value for Ymin which is smaller than the smallest output –161. –170 would be a good choice.

Enter a value for Ymax which is larger than the largest output value 35. Say, 40.

For Yscl, choose a reasonable value for the distance between tick marks. Approximately one-tenth of the distance between Ymin and Ymax is a reasonable choice. Set Yscl = 20.

Graph the function:

Press **GRAPH**. A more complete graph of the function appears.

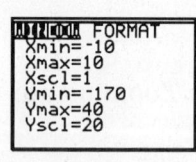

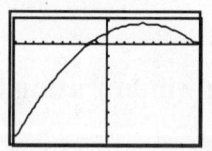

```
WINDOW FORMAT
Xmin=-10
Xmax=10
Xscl=1
Ymin=-170
Ymax=40
Yscl=20
```

Change Xmin and Xmax to see more of the graph to the left or to the right. For instance one might increase the Xmax value to 20. It would then be wise to increase the Xscl value so there won't be an excessive number of tick marks on the x-axis.

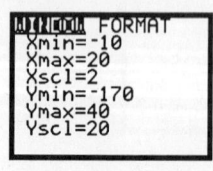

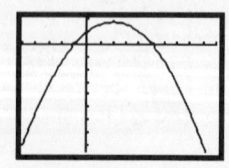

```
WINDOW FORMAT
Xmin=-10
Xmax=20
Xscl=2
Ymin=-170
Ymax=40
Yscl=20
```

Display coordinates of points on a graph using TRACE

The cursor can be moved from one plotted point to the next along a defined function. The coordinates of the point located by the cursor are displayed at the bottom of the screen.

Example: Trace along the graph of $f(x) = -x^2 + 8x + 19$.

Define the function:

Enter the expression $f(x) = -x^2 + 8x + 19$ as Y_1 at the Y = menu.

Verify the viewing window settings:

If the viewing window is not in standard setting, enter the following:

Xmin = –10	Ymin = –170
Xmax = 10	Ymax = 40
Xscl = 1	Yscl = 20

Graph the function:

Press **GRAPH** . The graph of the function is displayed.

Press **TRACE** .

A flashing cursor appears on the graph along with a value for input and output at the bottom of the screen.

Press the left arrow key to move the cursor toward smaller inputs. Moving to the left of Xmin causes the graph to scroll by automatically decreasing Xmin.

Press the right arrow key to move the cursor toward larger inputs. Moving to the right of Xmax causes the graph to scroll by automatically increasing Xmax.

As the cursor moves, the values of the x and y-coordinates at the cursor appear at the bottom of the screen. .

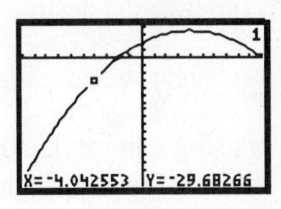

Note that the choices for the x-coordinates are determined by the programing of the graphing utility. The coordinates which are displayed can be controlled. as is explained in the next section of this manual.

Friendly viewing windows to control the Trace Display

A "friendly" viewing window displays values for x with at most one decimal position. There are 95 pixels across the screen. The distance between Xmin and Xmax is divided by 94 to determine the size of the increment in x when moving the cursor in the graphics window.

To obtain a "friendly" window, Xmin and Xmax are selected so that when the difference Xmax − Xmin is divided by 94, the quotient terminates by the tenths position.

Tracing by x increments of 0.1: Zoom Menu Item Number 4:ZDecimal

Press `ZOOM` `4` to automatically set the "friendly" viewing window to

Xmin = −4.7   Ymin = −3.1

Xmax = 4.7   Ymax = 3.1

Xscl = 1   Yscl = 10

Notice that Xmax − Xmin = 9.4. Divide this difference by 94, to determine the x-increment displayed. The quotient is 0.1. The x values displayed by the Trace feature will increment by 0.1 in this window.

ZOOM options to adjust the viewing window

Use the `ZOOM` key to open the following menu of options for altering the viewing window.

1:ZBox: creates a new viewing window based on a rectangle that you draw.

Press `ZOOM` `1` to return to the graphics window.

Use the arrow keys to move the cursor to one corner of the desired viewing window.

Press `ENTER` to set one corner.

Move the cursor to the opposite corner of the desired viewing window.

Press `ENTER` and the box (rectangle) becomes the new viewing window.

2:Zoom In: Use this item to zoom in using a preset factor to decrease the size of the viewing window. The factors can be set by selecting the Memory submenu from the Zoom screen and selecting item number **4:SetFactors** in that submenu.

Press `ZOOM` `2` which returns you to the graphics window.

Use the arrow keys to position the cursor at the desired center of the new window.

Press `ENTER`.

You can now repeatedly zoom in by moving the cursor to the center of the desired window and pressing `ENTER`.

3:Zoom Out: Use this item to zoom out using a preset factor to increase the size of the viewing window. The factors can be set by selecting the Memory submenu from the Zoom screen and selecting item number **4:SetFactors** in that submenu.

Press `ZOOM` `3` which returns you to the graphics window.

Use arrow keys to position the cursor at the desired center of the new window.

Press `ENTER`.

You can now repeatedly zoom out by moving the cursor to the center of the desired window and pressing `ENTER`.

4:Decimal: This item sets a viewing window so that the x values will increment by 0.1 as the cursor moves across the window or when using the trace feature. These are "friendly" values for analyzing the graph.

Press `ZOOM` `4` to change the window so that Xmin = -4.7, Xmax = 4.7, Ymin = -3.1, and Ymax = 3.1. Both Xscl and Yscl are 1.

5:Square: sets the viewing window so that the aspect ratio is 1. Graphs, such as circles, will appear distorted if the aspect ratio is not one.

Press `ZOOM` `5`.

6:Standard: returns graph to the standard viewing window.

Press `ZOOM` `6` to graph in the standard viewing window.

7:Trig: sets a "friendly" viewing window for graphing trigonometric functions.

Press `ZOOM` `7` to change the window.

8:Integer: sets the viewing window so that the x values will increment by one as the cursor moves across the window. These are "friendly" values for analyzing the graph.

Press `ZOOM` `8` to change the window.

9:ZoomStat: This item resets the viewing window so that all the points in the active StatPlots are displayed. Press `ZOOM` `9` to change the window.

Connected Mode for continuous graphs

By choosing the MODE setting on the fifth line to the MODE screen, you can display either a connected graph, where the points plotted are connected to make the graph look relatively smooth, or a discrete graph of a function, where only the plotted points are shown. The calculator comes from the factory set in **Connected** mode. This is the default setting. This setting must be changed to **Dot** to view a discrete graph.

As an example the graph of the function $y(x) = 2x - 1$ is shown with both mode settings using the "friendly" viewing window set by choosing item number 4:ZDecimal in the Zoom menu.

Connected mode

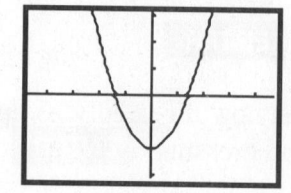

Dot Mode for discrete graphs

Dot mode

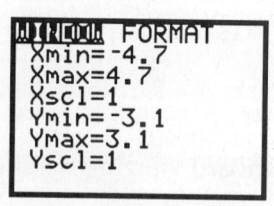

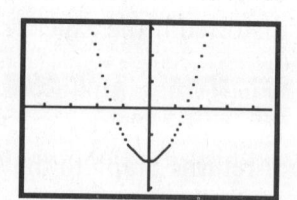

Parametric graphing

Parametric equations consist of two equations, representing the *x*-component and the *y*-component, each expressed in terms of a third independent variable, *T*. Parametric equation may be used to control the points which are plotted on the graph of a function.

Before entering the two parametric equations, you must change from function mode to parametric mode and adjust your viewing window.

Press **MODE** and use the **▽** key to highlight **Par**.

Press **ENTER** .

Highlight **Dot**.

Press **ENTER** . The mode is now set appropriately for parametric plotting.

Example: Suppose we wish to plot the function $y(x) = 10x - 1000$ but only show the points on the graph whose x-coordinates are 0, 5, 10,... 200. First we select Parametric and Dot in the Mode window.

Now we open the Y= menu and define $x = t$ and $y = 10t - 1000$.

Press **Y =** and enter *t* for X_{1T} and $10t - 1000$ for Y_{1T}.

Press **WINDOW** and set the viewing window appropriately. (Shown below.)

The value of Tmin will determine the smallest *t* and hence *x*- coordinate to be plotted and the Tmax will determine the largest value of t and therefore *x* to be plotted. The size of the Tstep will determine the distance between the *x*-coordinates of the points which are plotted. The larger Tstep, the fewer points will be plotted.

Press **GRAPH** when all this is set up.

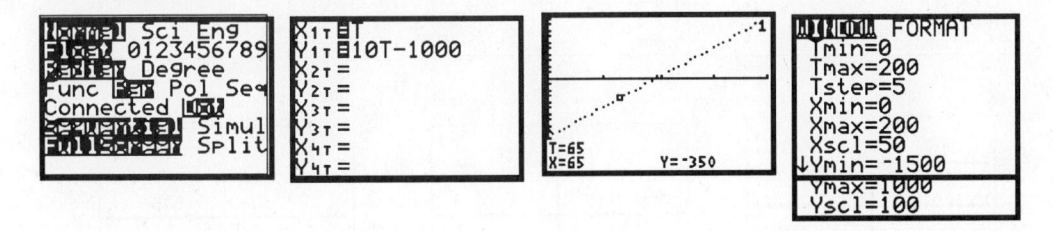

When you use the trace feature you will see the points on the graph highlighted by the cursor and the *t*, *x* and *y* coordinates will be displayed at the bottom of the screen.

Graph the solution interval to inequalities in one variable

Inequalities in one variable, such as $2x + 3 < 5$, can be entered directly in the Y= menu. The calculator will graph the solution set by drawing a horizontal line at $y = 1$ for all x values that make the inequality true. The "1" is the calculator's way of saying "true" whereas "0" is what the calculator uses for "false".

Set the calculator to **Dot** mode:

Press MODE , arrow down and over to **Dot**, and press ENTER .

Press Y= .

Type the left hand side of the inequality.

Open the **TEST** menu by pressing 2nd MATH .

Select the desired inequality relation by typing the number next to the relation, in this case **5**. The calculator returns to the Y= screen.

Type the right hand side of the inequality.

Set an appropriate viewing window. It is important that $y = 0$ and $y = 1$ be clearly visible on the graphics screen. It would be appropriate to use Ymin = –2 and Ymax = 2.

Press GRAPH . A horizontal line segment at $y = 1$ appears for x values that satisfy the inequality.

You can trace on the graph. An output of 1 indicates the corresponding x is in the solution set to the inequality. An output of 0 indicates the corresponding x makes the inequality false.

Example: Graph the solution set to $2x + 3 < 5$.

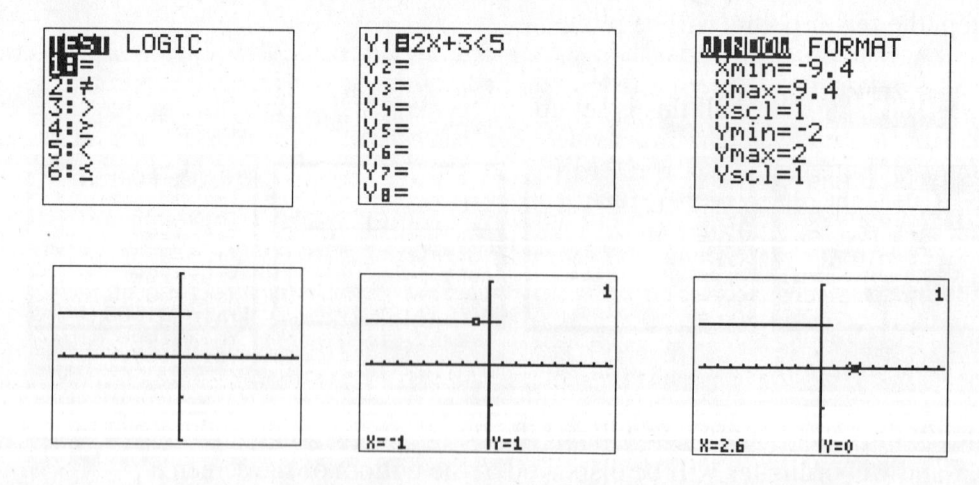

Find zeros of a function using the CALCULATE menu item 2:root

Enter the function in the Y= menu(the mode setting should be **Func**) and graph the function in a window so that the points where the graph crosses the horizontal axis are visible in the graphics screen.

Press **2nd** **TRACE** to display the **CALCULATE** menu, abbreviated as CALC in blue above the Trace key.

Type **2** to select **2:root** from the menu.

The calculator returns to the graphics window and prompts you to select a **Lower Bound** for one of the roots.

Use the **left arrow key** to move the cursor until the value for x at the cursor is smaller than the desired zero or root.

Press **ENTER** and the x value of the cursor is selected as the lower bound.

Now the calculator prompts you to select an **Upper Bound** for the root.

Press the **right arrow key** at least twice to move the cursor until the value for x at the cursor is larger than the desired zero.

Press **ENTER** and the x value of the cursor is selected as the upper bound.

The calculator prompts you to select a **Guess** for the root.

Press the **left arrow key** to move the cursor close to the zero which is where the graph intersects the x-axis. Be careful not to move the cursor past the lower bound. The guess must be between the lower and upper bounds. When the cursor is close to the root, press **ENTER**.

The calculator begins approximating the zero. A moving vertical line segment is displayed in the upper right corner of the screen to indicate that the calculator is computing. When finished, the calculator displays the approximate zero within the interval defined by the lower and upper bounds. This solution is correct with an error of no more than 0.00001.

Example: Find the largest zero of the function $y = x^2 + x - 3$.

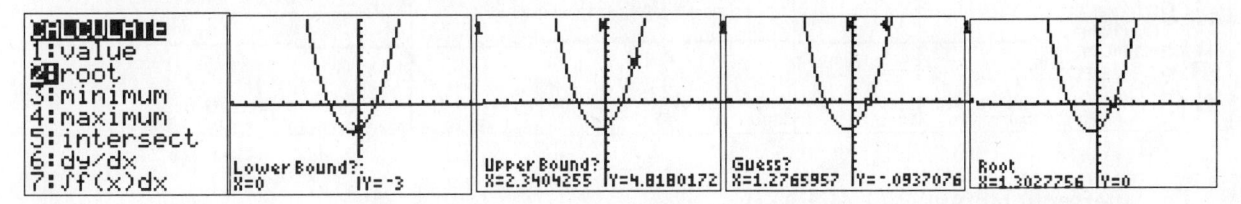

The largest zero or root of the function is 1.3027756, approximately.

Find intersection points using the CALCULATE menu

Enter the two functions in the Y= menu as Y_1 and Y_2. (You may use any of the other y variables if you do not wish to erase the functions you have entered for Y_1 and Y_2.)

Graph both functions in a window so that the desired intersection point is visible on the graphics screen.

Press **2nd** **TRACE** to display the **CALCULATE** menu.

Type **5** to select **5:intersect**.

The calculator returns to the graphics screen. The calculator prompts you to select the **First curve**, meaning the graph of one of the functions. The cursor is on the graph of Y_1.

Press **ENTER** to select Y_1 as the first curve.

Now the calculator prompts you to select the **Second curve**. The cursor is on the graph of Y_2. (If there were more than two graphs on the graphics screen you could move the cursor to one of the other graphs by using the up or down cursor control.)

Press **ENTER** to select Y_2 as the second curve.

The calculator prompts you to supply a **Guess** for the intersection point.

Do this by using the arrow keys to move the cursor near the intersection point.

Press **ENTER** when the cursor is close to the intersection point.

The calculator begins approximating the intersection point. A moving vertical line segment is displayed in the upper right corner of the screen to indicate that he calculator is computing. When finished, the calculator displays the approximate ordered pair for the intersection point. The solution is correct with an error of no more than 0.00001.

Example: Find the intersection point of the functions $y = 2x - 1$ and $y = -3x + 7$. The graph appears in the standard viewing window.

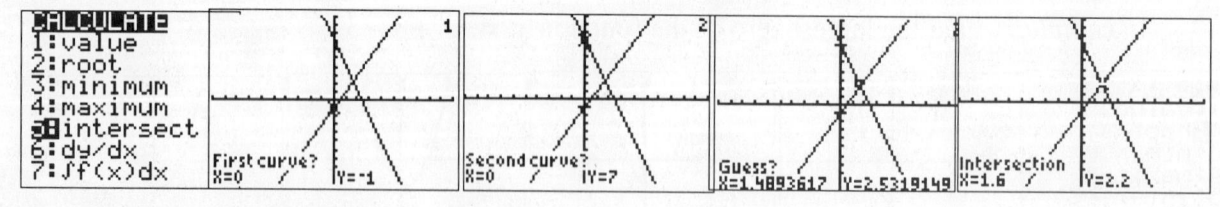

The intersection point is $(1.6, 2.2)$.

Find a function's maximum using the CALCULATE menu

Enter the function in the Y= menu.

Graph the function in a window so that the graph displays a highest point.

Press **2nd** **TRACE** to open the **CALCULATE** menu.

Type **4** to select **4:maximum**.

The calculator returns to the graphics window. The calculator prompts the user to select a **Lower Bound** for the interval containing the function's maximum.

> Use the **left arrow key.** Move the cursor until it is to the left of the function's maximum.
>
> Press **ENTER** to select this x value as the lower bound.

Now the calculator prompts the user to select an **Upper Bound** for the interval containing the function's maximum.

> Press the **right arrow key** at least twice to move the cursor until the cursor is to the right of the function's maximum.
>
> Press **ENTER** to select this x value as the upper bound.

The calculator prompts the user to select a **Guess** for the function's maximum.

> Press the **left arrow key** to move the cursor close to the function's maximum.

Be careful not to move the cursor past the lower bound. The guess must be between the lower and upper bounds.

> Press **ENTER** when the cursor is close to the maximum.

The calculator begins approximating the function's maximum. A moving vertical line segment is displayed in the upper right corner of the screen to indicate that the calculator is computing. When finished, the calculator displays the approximate maximum of the function within the interval defined by the lower and upper bounds. This solution is correct with an error of no more than 0.00001.

> *Example:* Find the vertex of $y = -x^2 + x + 4$. For this function the vertex is the highest point on the graph. The graph appears in the standard viewing window.

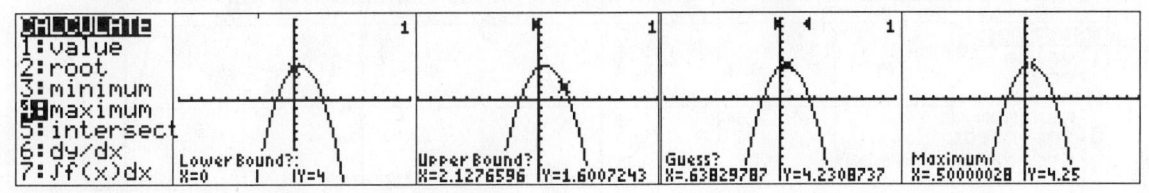

The vertex is actually the point (0.5, 4.25).

371

Find a function's minimum using CALC

Enter the function in the Y= menu.

Graph the function in a window so that the graph displays a lowest point.

Press **2nd** **TRACE** to open the **CALCULATE** menu.

Type **3** to select **3:minimum**.

The calculator returns to the graphics window. The calculator waits for you to select a **Lower Bound** for the interval containing the function's minimum.

Use the **left arrow key** to move the cursor until it is to the left of the function's minimum.

Press **ENTER** and this x value becomes the lower bound.

The calculator waits for you to select an **Upper Bound** for the interval containing the function's minimum.

Press the **right arrow key** at least twice to move the cursor until it is to the right of the function's minimum.

Press **ENTER** and this x value becomes the upper bound.

The calculator waits for you to select a **Guess** for the function's minimum.

Press the **left arrow key** to move the cursor close to the function's minimum.

Be careful not to move the cursor past the lower bound. The guess must be between the lower and upper bounds.

Press **ENTER** when the cursor is close to the minimum.

The calculator begins approximating the function's minimum. A moving vertical line segment is displayed in the upper right corner of the screen to indicate that the calculator is computing. When finished, the calculator displays the approximate minimum of the function within the interval defined by the lower and upper bounds. This solution is correct with an error of no more than 0.00001.

Example: Find the vertex of $y = x^2 - 3x - 2$. The vertex of this function is the lowest point on the graph. The graph appears in the standard viewing window.

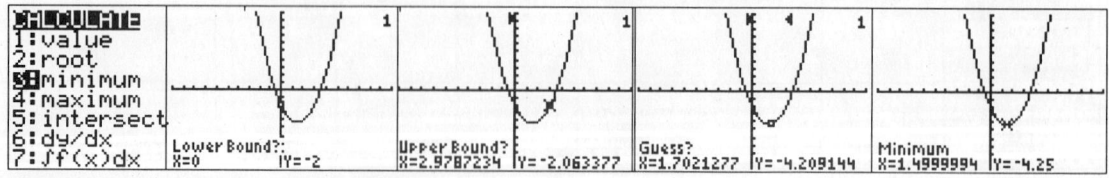

The vertex is actually (1.5, −4.25).

Enter data into a list

A list of data can be entered, displayed, copied to another list, stored, sorted, used to graph families of curves, and in mathematical expressions. Data that has been entered in lists can be used to plot statistical data. The types of plots available include scatter plots, box-and-whisker plots, x-y plots and histograms. Calculations to fit the data entered in lists to one or more models is also possible. The resulting equations can then be stored in the Y = menu, graphed and traced.

From the Home Screen:

Press **STAT** **ENTER** to select **1:Edit** form the **STATEDIT** menu and to display the first three lists, L_1, L_2, and L_3.

With the lists displayed:

Position the cursor in the column with heading L_1 using the arrow keys. At the bottom of the screen you should see $L_1(1) =$. The 1 in the parentheses indicates that the cursor is at the first position in the list.

Enter each of the data elements.

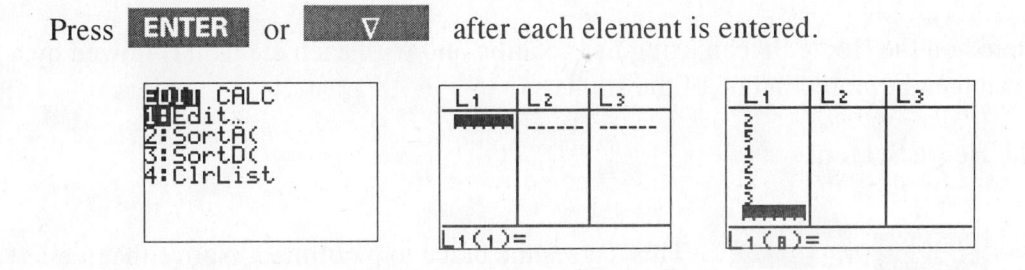

Press **ENTER** or **▽** after each element is entered.

Clear data from a list

If data is already entered in a list you wish to use and you would like to erase the entire list:

Use the **△**, **▷** or **◁** keys and position the cursor in the column heading over the list name.

Press **CLEAR** **ENTER** to remove the existing data elements and clear the list.

The empty list is ready for entry of new data elements.

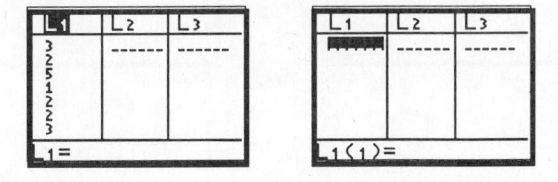

373

Change numbers in an existing list

A number in a list can also be changed by positioning the cursor over that number and typing in the desired new number, followed by **ENTER** .

Insert new numbers into an existing list

If you wish to insert a new number into an existing list without removing any of the numbers from the list first position the cursor where you wish to place the new number. Then put the calculator in **insert** mode by pressing **2nd** **DEL** . The abbreviation for **insert** appears above the **DEL** key in blue. A zero appears in the list where you wish to insert a new number. Type the new number and press the enter key.

Enter data into a list from the Home Screen

Data entered on the Home Screen using list notation and with each element followed by a comma can then be stored in one of the six lists.

From the Home Screen:

Press **2nd** **(** . This causes the brace to be printed. Notice that the left brace appears above the left parenthesis key in blue.

Enter the data elements separating each element with a comma.

Press **2nd** **)** after the last data element is entered. This prints the right brace which appears above the right parenthesis key in blue.

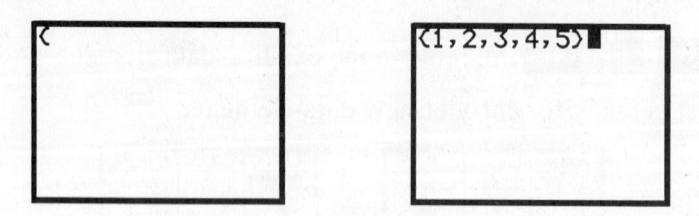

```
{                    {1,2,3,4,5}
```

374

Store and view the newly-generated list:

After entering the list in braces on the home screen press `STO ▷` `2nd` `1` , then press `ENTER` to store the list in L_1. Notice that L_1 appears above the key with 1 on it, printed in blue.

Press `STAT` `ENTER` to select **1:Edit** from the **STATEDIT** menu and view the newly-generated list on the list screen.

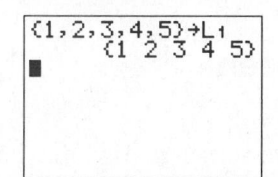

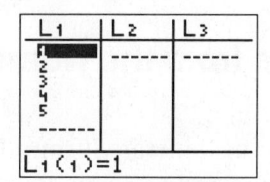

Pressing `STO` `2nd` `2`, and then `ENTER` would store the list in L_2.

Press `STAT` `ENTER` to select **1:Edit** from the **STATEDIT** menu and view the newly-generated list on the list screen.

Apply arithmetic operations to an existing list to create new list

A new list can be created either on the home screen and then stored in a blank list or on the list screen.

Generate new data on the list screen:

Press `STAT` `ENTER` to select **1:Edit** from the **STATEDIT** menu and to display L_1, L_2, and L_3.

Position the cursor on the column heading L_2 using the arrow keys.

Create a new list L_2, by adding three to each element of L_1:

Type $L_1 + 3$ and press `ENTER`. The calculator adds 3 to each element of L_1 and stores this new list in L_2.

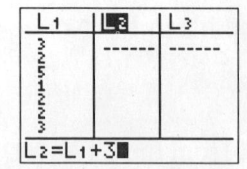

Generate new data from the Home Screen

A new list of data can be generated from the home screen using an existing list name. The data generated can be stored as a new list and viewed on the list screen.

From the Home Screen:

Press `STAT` `ENTER` to select the **EDIT** menu and to display L_1, L_2, and L_3.

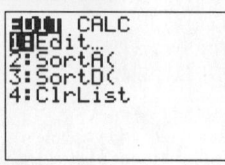

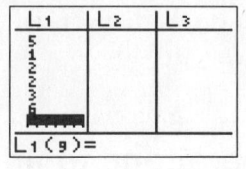

Create a new list L2, adding three to each element of an existing list L1:

Press `2nd` `QUIT` to return to the Home Screen.

Type $L_1 + 3$

Store and view the newly-generated list into L_2.

Press `STO` `2nd` `2`, then press `ENTER` to store the list in L_2.

Press `STAT` `ENTER` to select **1:Edit** from the **STATEDIT** menu and view the newly-generated list.

Use right and left arrows to view additional elements of the list on the Home Screen.

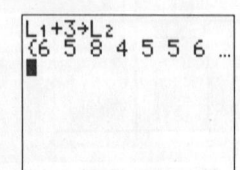

Generate a list of data using the sequence function in the LIST OPS menu

From the Home Screen access the **LIST OPS** menu:

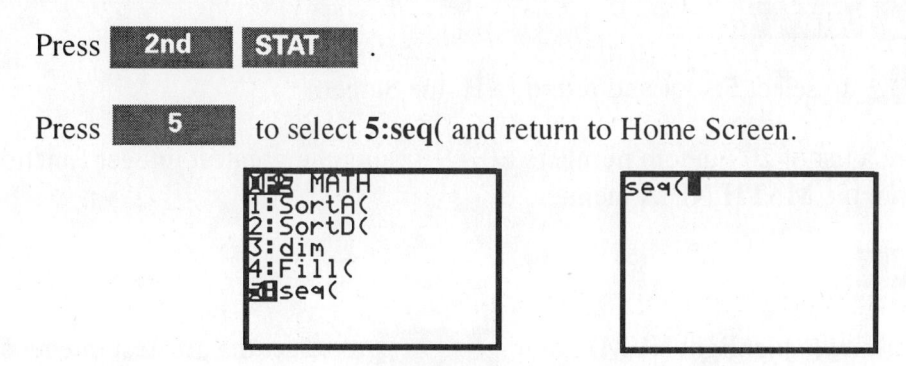

Press `2nd` `STAT` .

Press `5` to select **5:seq(** and return to Home Screen.

```
OPS MATH
1:SortA(
2:SortD(
3:dim
4:Fill(
5:seq(
```

```
seq(
```

Type $x^2, x, 1, 10, 1)$ to generate the squares of the numbers 1–10.

Press `ENTER` to display the list of elements on the Home Screen. Use left and right arrows to view additional list elements on the Home Screen.

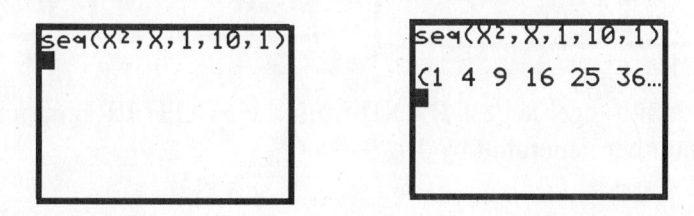

```
seq(X²,X,1,10,1)
```

```
seq(X²,X,1,10,1)
{1 4 9 16 25 36…
```

Note: The general format for entry of a sequence is:

seq (generating rule, variable used in generating rule, initial value of the variable, ending value of the variable, increment of the variable)

Commas are necessary and must be included after each input for the sequence function.

Store and view the newly-generated list

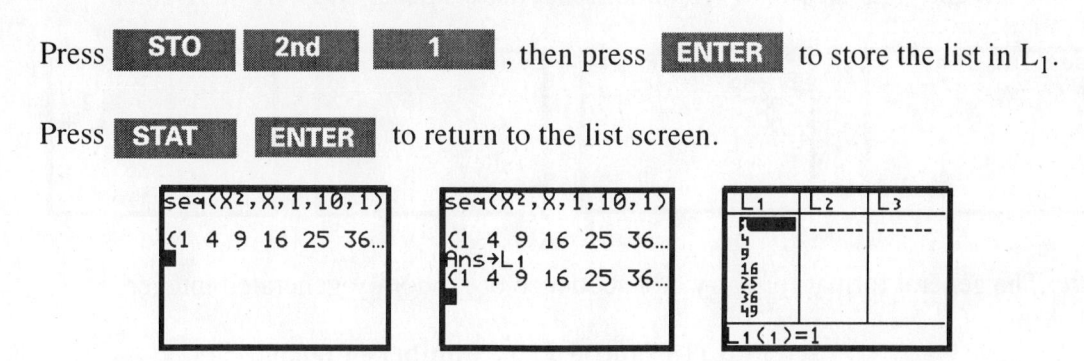

Press `STO` `2nd` `1` , then press `ENTER` to store the list in L_1.

Press `STAT` `ENTER` to return to the list screen.

```
seq(X²,X,1,10,1)
{1 4 9 16 25 36…
```

```
seq(X²,X,1,10,1)
{1 4 9 16 25 36…
Ans→L1
{1 4 9 16 25 36…
```

```
L1    L2    L3
1     ---   ---
4
9
16
25
36
49
L1(1)=1
```

Generate a list of random numbers using sequences

From the Home Screen access the **LIST OPS** menu:

Press `2nd` `STAT`.

Press `5` to select **5:seq(** and return to Home Screen.

Generate and store a list of 20 random numbers (0 – 9) using the greatest integer function, **int**, accessed from the **MATH NUM** menu.

Press `MATH`.

Use the right arrow to highlight **NUM**. Enter `4` to select the greatest integer function *4: int*.

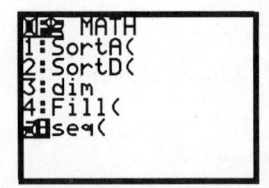

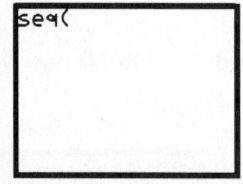

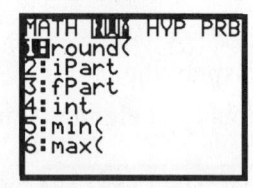

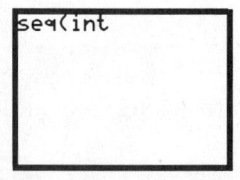

To generate a random integer access **RAND** from the **MATH PRB** menu and multiply the random decimal number generated by 10:

Enter **(10**

Press `MATH`.

Use the right arrow to highlight **PRB.**

Press `ENTER` to select the random seed generator.

Complete the sequence **), x, 1, 20, 1)**. The screen should look like the one below.

Press `ENTER` to display the list of elements on the Home Screen.

Use left and right arrows to view additional list elements on the Home Screen.

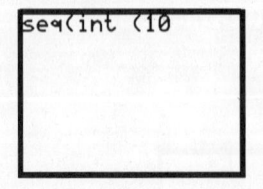

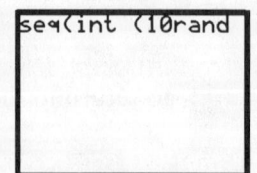

Note: The general format for entry of a sequence of randomly-generated integers is:

seq (int (10 rand), x, 1, number of terms, 1)

Store the newly-generated list in L₁:

Press **STO** **2nd** **1**, then press **ENTER** to store the list in L₁.

Press **STAT** **ENTER** to view the list on the list screen.

Use up and down arrows to view additional list elements on the list screen.

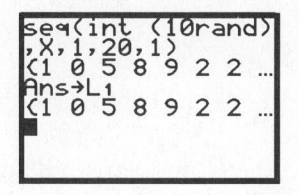

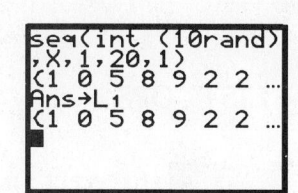

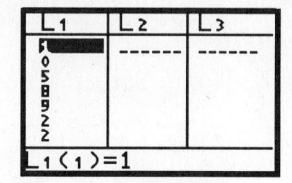

Add a random number to each element of a list:

With the lists screen displayed:

Position the cursor on column heading **L₂**. At the bottom of the screen is displayed **L₂ =**

Complete the right-hand side by entering **L₁ + int(10 RAND)**.

Press **ENTER** to display the list.

Use up and down arrows to view additional list elements.

NOTES:

- **RAND** generates and returns a random number between zero and 1 (0 < rand < 1).

- To generate more random number sequences, have individuals each store a different **integer** seed value in **RAND** first.

- If using a new set of TI-82s, have each person store an integer (social security numbers are a good choice) to generate random numbers.

- When you reset the TI-82 or store **0** to **RAND**, the factory-set seed value is restored.

Generate data using a sequence and an existing list (from the Home Screen)

A new list of data can be generated using an existing list with a sequence command from the Home Screen or with the lists displayed.

Using the list of twenty numbers randomly generated, return to the Home Screen.

Press **2nd** **QUIT**.

From the Home Screen access the **LIST OPS** menu:

Press **2nd** **STAT** **5** to select **5:seq(**. The Home Screen displays **seq(**.

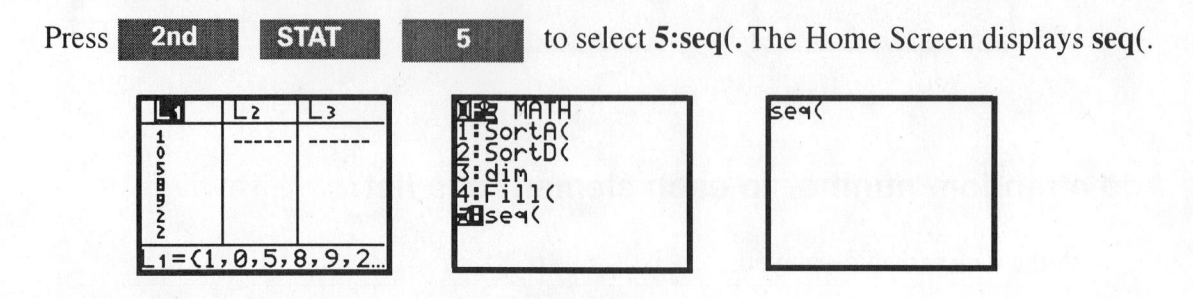

Type $L_1(x) + 3, x, 1, 20, 1)$ then press **ENTER** to display the list.

Press **STO** **2nd** **2**, then press **ENTER** to store the list in L_2.

Press **STAT** **ENTER** to return to the list screen.

Use up and down arrows to view additional elements of the list.

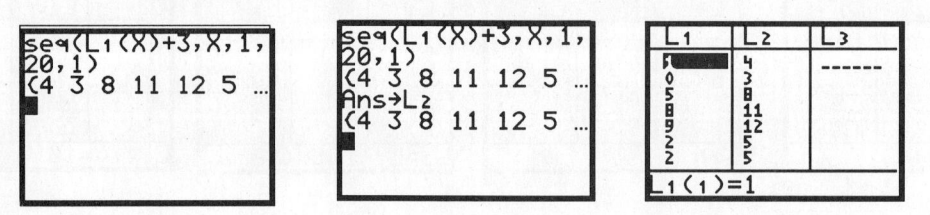

Note: The general format for entry of a sequence using an existing list is:

seq (L_{old} (x) with operation(s), x, 1, number of elements in the list,1)

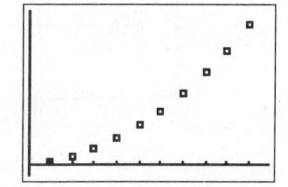

Construct a scatter plot of data

A scatter plot is a two dimensional graph of ordered pairs and requires two lists of data values, one for the first coordinates and the second for the second coordinates. Scatter plots plot the data points from two lists, Xlist and Ylist, as ordered pairs. You can select one of three options for highlighting the points: as a box (□), a cross (+) or a dot (•). The number of data points in the Xlist and in the Ylist must be the same. The frequency of occurrence of each data value does not apply to scatter plots.

From the Home Screen:

Press **Y =** and turn off all selected Y= equations.

Press **STAT** **ENTER** to view the list screen and display L_1, L_2, and L_3.

If data has been previously entered in L_1, L_2, and L_3, clear existing data.

Enter the first set of data in L_1 and the second set of data in L_2.

Press **2nd** **Y =** to access the **STAT PLOTS** menu.

Select **1** for **Plot1.** Highlight **On** and press **ENTER**.

Use the **▽** and **▷** to highlight the first icon under the **Type:** options.

Press **ENTER**. This darkens the first icon which is the **scatter plot** icon.

Use the **▽** and **▷** keys to highlight L_1 in **Xlist,** L_2 in **Ylist,** and the square in the **Mark:** options.

Press **ENTER** after each selection.

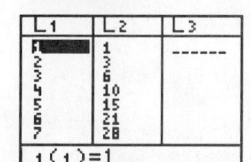

Set a viewing window that contains all the data points and then display the scatter plot on the graphics screen.

Press **ZOOM** **9** to select **9:ZoomStat** from the **ZOOM** menu, set a viewing window and display the scatter plot.

Construct a line plot of data

Enter the data in lists L_1 and L_2.

Press `2nd` `Y=` to access the **STAT PLOTS** menu.

Select `2` for **Plot2**. Highlight **On** and press `ENTER`.

Use the `▽` and `▷` to highlight the second icon in the **Type:** options.

Press `ENTER`. This selects the second type which is the **line plot** option.

Use the `▽` and `▷` keys to highlight L_1 in **Xlist,** and L_2 in **Ylist**.

Press `ENTER` after each selection

Select **ZoomStat** by pressing `ZOOM` `9`. This command sets a viewing window that contains all the data points and then displays the scatter plot.

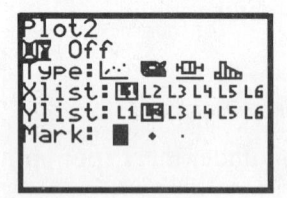

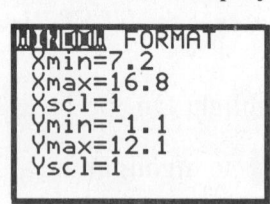

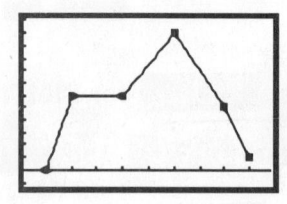

Turn on/off Stat Plots

When finished with them, it is a good idea to turn off the statistical plots. Not doing so often results in confusing graphs and error messages at a later time when you may be graphing functions entered in the Y= menu.

Press `2nd` `Y=` to access the **STAT PLOTS** menu.

Select `4` for **PlotsOff**. The calculator returns to the home screen.

Press `ENTER`. All statistical plots are now off.

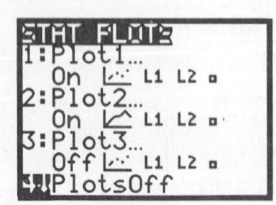

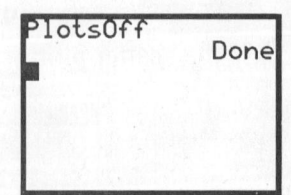

Finite differences and ratios using the sequence function

Finite differences can be calculated using the sequence function for data entered in a list. If the data consists of ordered pairs, enter the x-coordinates in the first list and the y-coordinates in the second list.

Example: Enter the following table into two list on the list screen and calculate the finite differences for the values in the second column:

X	S(x)
1	35,000.00
2	37,000.00
3	39,000.00
4	41,000.00
5	43,000.00
6	45,000.00
7	47,000.00
8	49,000.00
9	51,000.00
10	53,000.00

If the function is known and entered in the Y= menu, then the ordered pairs can be entered into two lists element by element or by using the function rule with the sequence command. In the table above, the left column is a list of sequential numbers (years) from 1 to 10; the right column is a list of the first ten years of salaries for a fictitious company named CD-R-Us, given an initial starting salary of $35,000, with annual pay raises of $2000 each subsequent year.

Enter the column of sequential years, X, as the first list, L_1, using the sequence function:

Press **STAT** **ENTER** to view the list screen and to display L_1, L_2, and L_3.

Position the cursor on the column heading **L_1** using the arrow keys.

Press **2nd** **STAT** **5** to select **5:seq(**.

Type $x + 1, x, 0, 9, 1)$ to generate the list of numbers 1 – 10. Press **ENTER**.

383

Enter the column of salaries, S(X), as the second list, L$_2$:

Position the cursor on the column heading **L$_2$** using the arrow keys.

Press **2nd** **STAT** **5** to select **5:seq(**

Type: *35000 + 2000(x-1), x, 1, 10, 1)* to generate the list of salaries. Press **ENTER**.

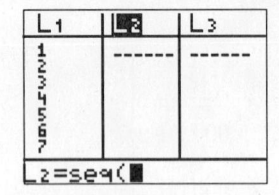

Generate the first finite differences of L$_2$ and store in L$_3$:

Position the cursor on the column heading **L$_3$** using the arrow keys.

Press **2nd** **STAT** **5** to select **5:seq(**.

Type **L$_2$ (x + 1) - L$_2$ (x), x, 1, 9, 1)** to generate the list of first finite differences.

Press **ENTER**.

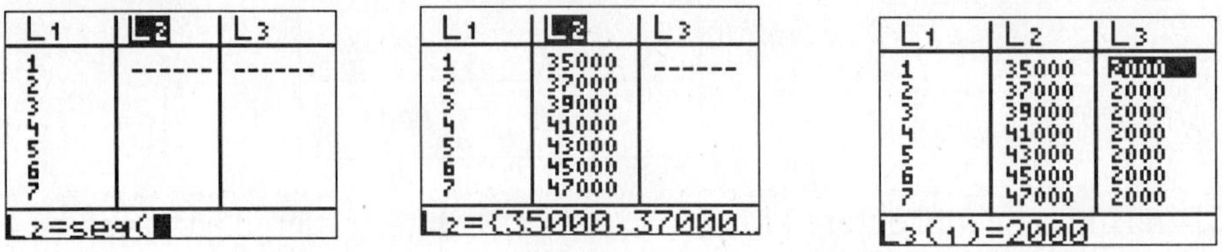

Generate the finite ratios of L$_2$ and store in L$_4$:

Position the cursor on the column heading **L$_4$** using the arrow keys.

Press **2nd** **STAT** **5** to select **5:seq(**.

Type **L$_2$ (x + 1) / L$_1$ (x), x, 1, 9, 1)** to generate the list of finite ratios.

Press **ENTER**.

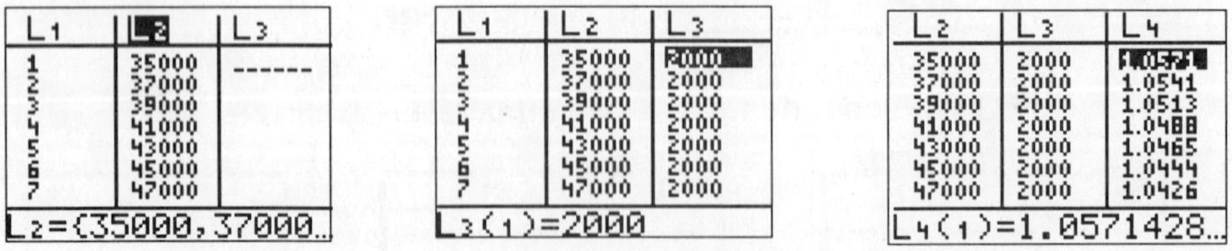

Note: The first finite difference is constant, an indication that the function is linear.

Finite differences used to select an appropriate regression equation

Data generated by an unknown sequence can be investigated using finite differences in order to select an appropriate regression equation and determine the function.

Example: Use finite differences to investigate the sequence

$$1, 3, 6, 10, 15, 21, 28, 36, 45, 55, \ldots$$

The first list consists of the positions of the known terms in the sequence; the second list consists of the terms that are given of the unknown sequence.

Position: 1, 2, 3, 4, 5, 6, 7, 8, 9, 10, ...
Seq term: 1, 3, 6, 10, 15, 21, 28, 36, 45, 55, ...

Enter the position numbers as the first list, using a sequence.

Press **STAT** **ENTER** to select the **EDIT** menu and to display L_1, L_2, and L_3.

Position the cursor on the column heading L_1 using the arrow keys.

Press **2nd** **STAT** **5** to select **5:seq(**.

Type $x + 1, x, 0, 9, 1)$ to generate the list of numbers 1 – 10.

Press **ENTER**.

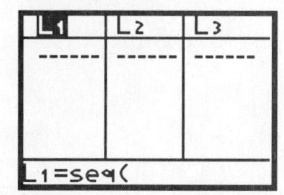

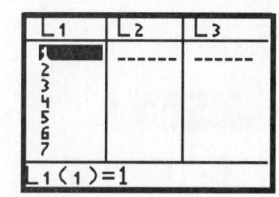

Enter the terms of the sequence as the second list, L_2, element by element.

Position the cursor in L_2 for the first entry.

Enter the terms of the sequence, element by element. Press **ENTER** after each term.

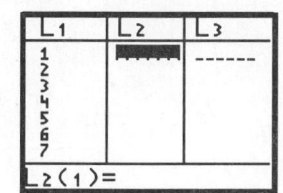

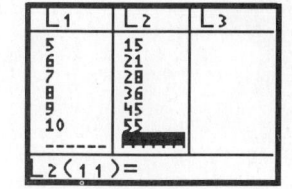

Generate the first finite different as L₃:

Position the cursor on the column heading **L₃** using the arrow keys.

Press `2nd` `STAT` `5` to select **5:seq(**.

Type **L₂ (x + 1) - L₂ (x), x, 1, 9, 1)** to generate the list of first finite differences.

Press `ENTER`.

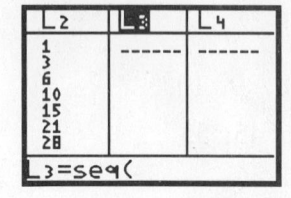

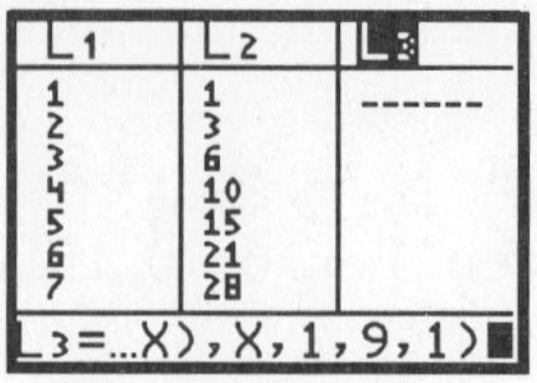

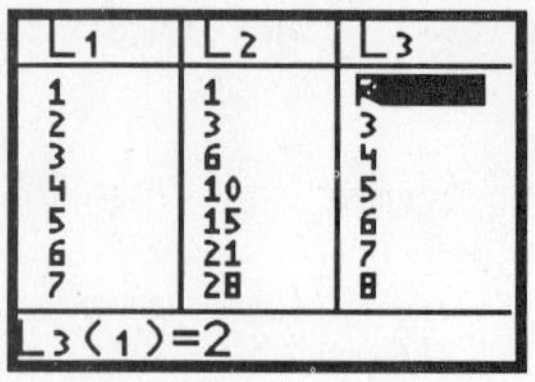

A linear model is indicated by the first finite difference being constant. If, however, the first finite difference is not constant, the function is not linear and you should continue to take finite differences.

Generate the second finite difference as L₄:

Position the cursor on the column heading **L₄** using the arrow keys.

Press `2nd` `STAT` `5` to select **5:seq(**.

Type **L₃ (x + 1) - L₃ (x), x, 1, 8, 1)** to generate the list of second finite differences.

Press `ENTER`.

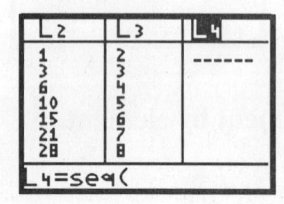

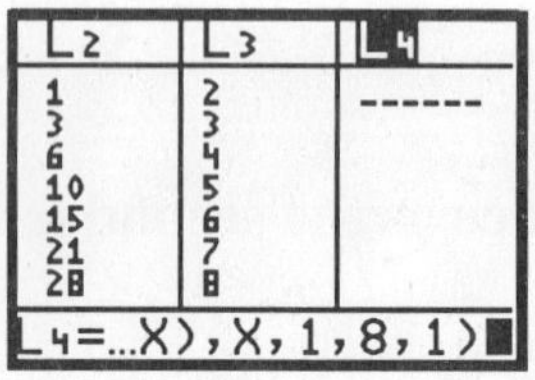

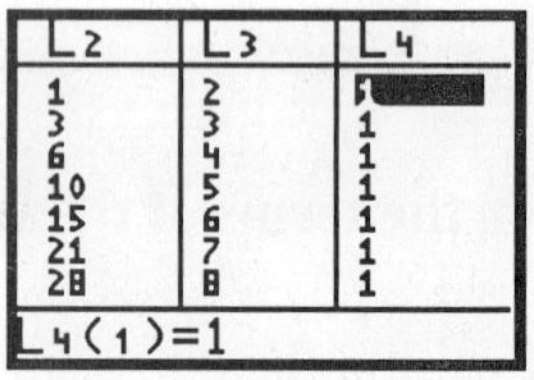

Note: The second finite difference is constant, an indication that the function is quadratic. The function can be determined using the quadratic regression model.

Create a linear regression model

The TI–82 has the capability of creating the equation of a line that best fits a set of paired data. Such a line is called a linear regression model for the data.

Example: Enter the following paired data as lists L_1 and L_2.

x	1	2	7	9	12
y	4	11	23	32	45

Enter the paired data in two lists, L_1 and L_2.

Construct a scatter plot of the data (see index of procedures).

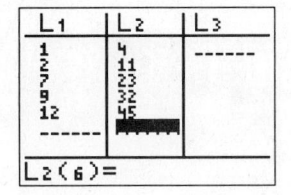

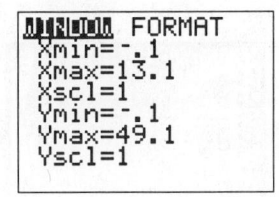

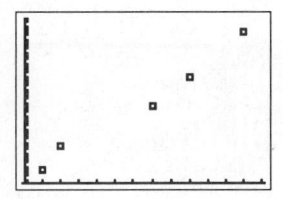

Press **STAT** **▷** to select the **CALC** menu.

Press **5** **ENTER** to select **LinReg**. The calculator returns to the home screen.

Press **2nd** **1** **,** **2nd** **2** **ENTER** to enter the names of the two lists, separated by a comma. The best–fit linear model is displayed

The value of *a* is the slope of the line.

The value of *b* is the vertical intercept of the line.

The value of *r* measures how well the line fits the data. The closer the absolute value of *r* is to one, the better the fit.

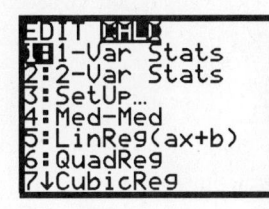

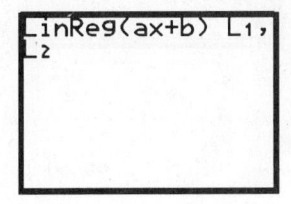

387

Store the regression model as one of the functions Y$_1$ to Y$_0$

Use "Creating a linear regression model for data" to create the best-fit linear equation.

Press **Y =** and place cursor on the function that will be assigned the regression model.

Press **VARS**.

Press **1** to select **STATISTICS**.

Press **7** to select **RegEQ**.

The regression model appears in the **Y=** menu.

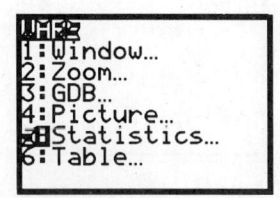

```
VARS              X/Y Σ EQ BOX PTS    Y1■3.47926267281
1:Window...       1:a                 11X+1.4285714285
2:Zoom...         2:b                 72■
3:GDB...          3:c                 Y2=
4:Picture...      4:d                 Y3=
5:Statistics...   5:e                 Y4=
6:Table...        6:r                 Y5=
                  7:RegEQ             Y6=
```

Graph the regression model

Store the regression model as a function using the **Y =** menu.

Adjust the viewing window if necessary.

Press **GRAPH** to display the regression model.

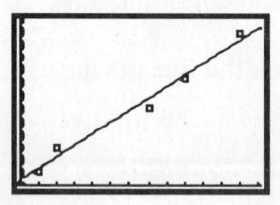

Create a quadratic regression model

The TI–82 has the capability of creating the equation of a line that best fits a set of paired data. Such a line is called a regression model for the data.

The first step in creating a regression model is to construct a scatter plot of the data. A graph can be constructed using the data entered into two lists and the statistical plot capability of the calculator.

Construct a scatter plot of data

Enter the data in lists L_1 and L_2 as indicated below.

Press **2nd** **Y =** to access **STAT PLOT**.

Select **1** for **Plot1**.

Highlight **On** and press **ENTER**.

Use the arrow keys to move down to **Type** and highlight the first figure (scatter plot).

Press **ENTER** to turn on the **scatter plot.**

Repeat the process of using arrow keys to highlight L_1 under **Xlist**, and L_2 under **Ylist.**

Press **ENTER** after each selection is highlighted.

Set a viewing window that contains all the data points and then display the scatter plot.

Press **ZOOM** **9** to select **ZoomStat** and set a viewing window.

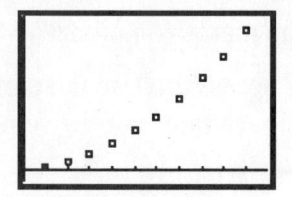

Select an appropriate regression model.

The data points displayed on the scatter plot give the appearance of being quadratic. A good first choice would be the quadratic regression model.

Press **STAT** ▷ to select the **CALC** menu.

Press **6** **ENTER** to select **QuadReg**. The Home Screen is displayed.

Press **2nd** **1** **,** **2nd** **2** **ENTER** to enter the names of the two lists, separated by a comma. The best–fit quadratic model is displayed.

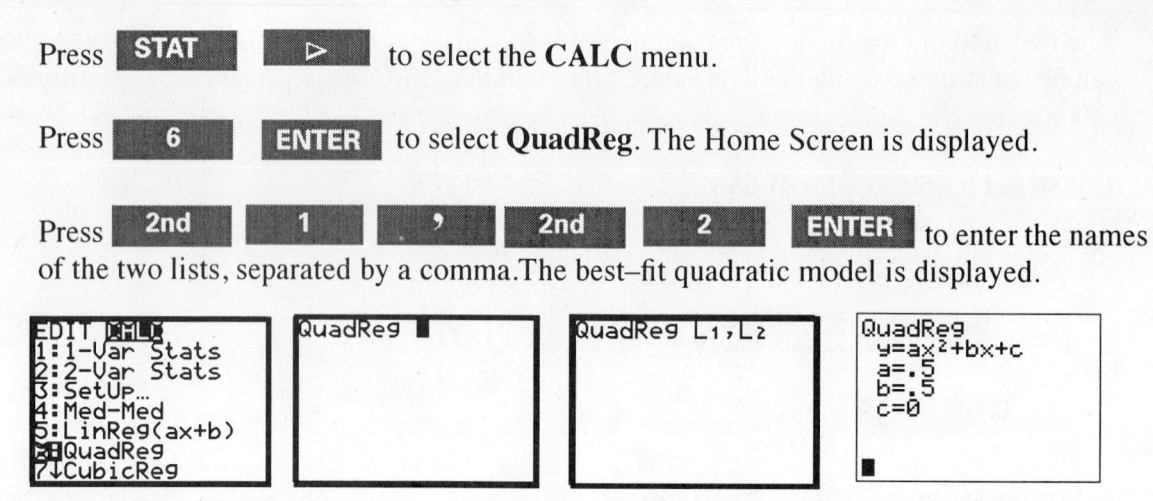

The value of a is the coefficient of the quadratic term. The value of b is the coefficient of the linear term and the value of c is the vertical intercept and constant term.

Store the regression model as one of the functions Y_1 to Y_0.

Press **Y=** and place cursor on the function assigned the regression model.

Press **VARS** **5** to select **STATISTICS**.

Press **7** **ENTER** to select **RegEq**.

The regression model appears in the **Y=** menu.

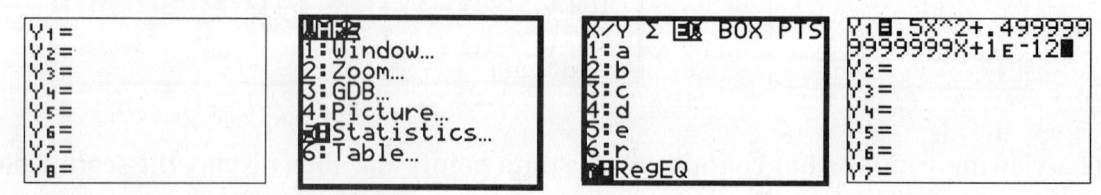

Graph the regression model.

Press **GRAPH** to display the regression model.

Adjust the viewing window if necessary. Use the table generated by the equation entered in the Y = menu to view additional values of the sequence.

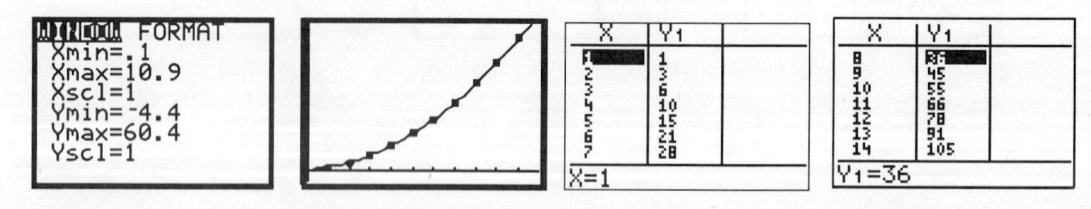

Transfer data between calculators: using LINK

Connect the two calculators using the connection cable.

Person receiving the data:

Press `2nd` `X, T, θ` to select **LINK**.

Press `STAT` `▷` to select **RECEIVE** and press `ENTER`. Now just wait.

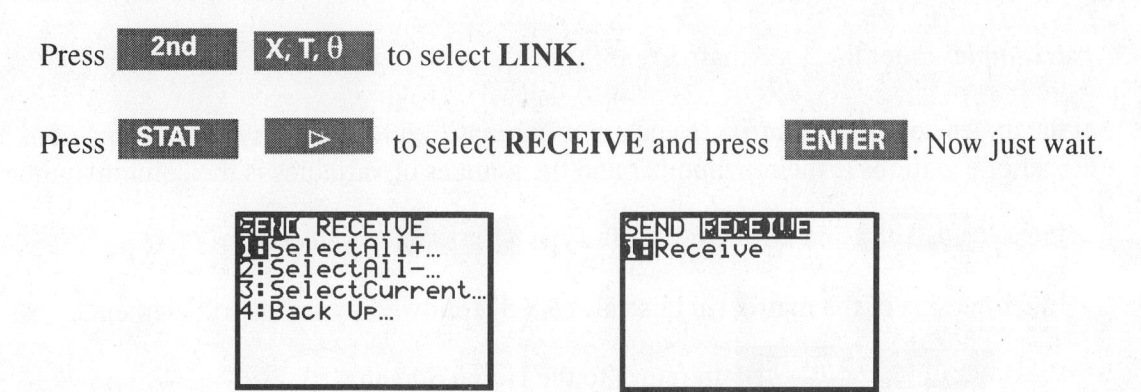

```
SEND RECEIVE
1:SelectAll+…
2:SelectAll-…
3:SelectCurrent…
4:Back Up…
```

```
SEND RECEIVE
1:Receive
```

Person sending the data:

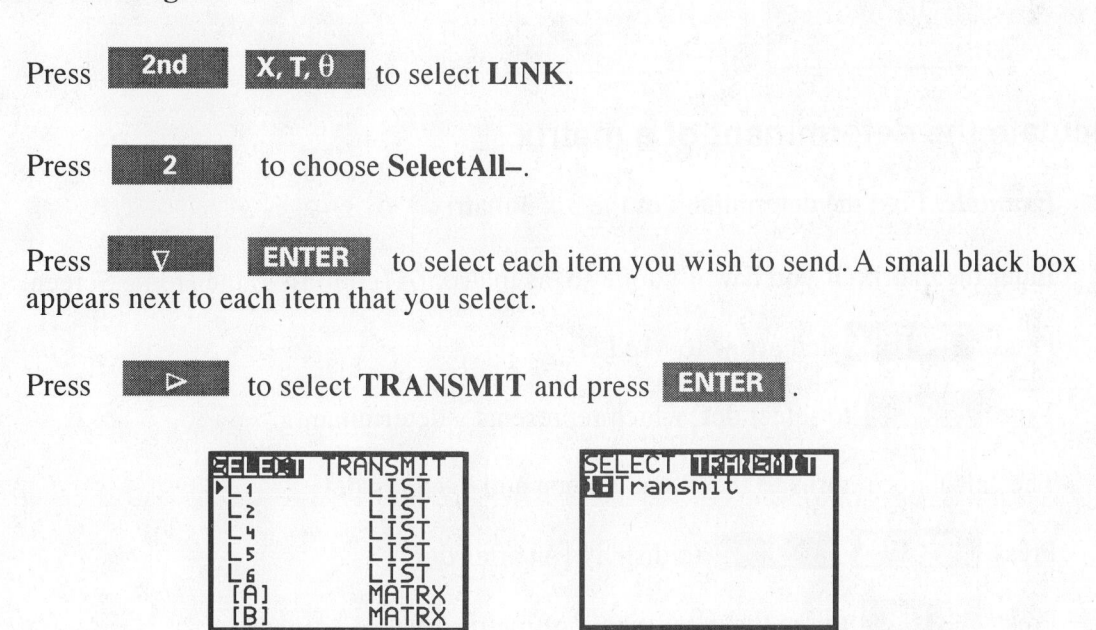

Press `2nd` `X, T, θ` to select **LINK**.

Press `2` to choose **SelectAll–**.

Press `▽` `ENTER` to select each item you wish to send. A small black box appears next to each item that you select.

Press `▷` to select **TRANSMIT** and press `ENTER`.

```
SELECT TRANSMIT
▶L1      LIST
 L2      LIST
 L4      LIST
 L5      LIST
 L6      LIST
 [A]     MATRX
 [B]     MATRX
```

```
SELECT TRANSMIT
1:Transmit
```

Disconnect calculators.

Enter and display a matrix

A matrix is a two-dimensional array of rows and columns. The dimensions of a matrix, the number of rows and the number of columns, is designated with the row number preceding the column number. The TI-82 stores as many as five matrices using the matrix variables [A], [B], [C], [D], and [E]. You can display, enter, or edit a matrix in the matrix editor.

Example: Enter the 3 x 3 matrix $\begin{bmatrix} 2 & -5 & 3 \\ 0 & 1 & -2 \\ -4 & 7 & 6 \end{bmatrix}$.

Set the dimension of the matrix. If you are using matrices to solve a system of equations, the number of equations is the row number and the number of variables is the column number.

Press **MATRIX** and arrow to **EDIT**. Type **1** to select matrix [A].

Enter the size of the matrix (in this case, 3 x 3) followed by the matrix elements.

Press **2nd** **MODE** to return to the Home Screen.

Press **MATRIX** **1** **ENTER** to display matrix [A] in the Home Screen.

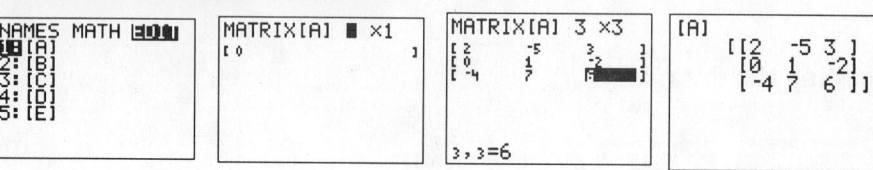

Calculate the determinant of a matrix

Example: Find the determinant of the 3 x 3 matrix $\begin{bmatrix} 2 & -5 & 3 \\ 0 & 1 & -2 \\ -4 & 7 & 6 \end{bmatrix}$.

Enter the matrix, if you haven't done so, as matrix [A]. Return to the Home Screen.

Press **MATRIX** and arrow to **MATH**.

Type **1** to select **det**, which represents a determinant.

The calculator returns to the Home Screen and displays **det**.

Press **MATRIX** **1** to display [A] after **det**.

Press **ENTER** to see the determinant of matrix [A].

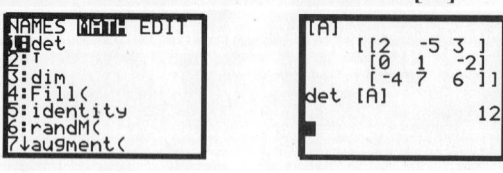

Delete a matrix from memory

Press **2nd** **+** to access the **MEM** menu.

Press **2** to display the 2: Delete options.

Press **4** to display the matrix deletion options. Select the matrix you wish to delete.

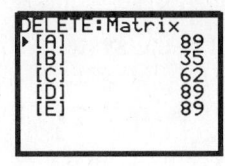

```
MEMORY
1:Check RAM…
2:Delete…
3:Reset…
```

```
DELETE FROM…
1:All…
2:Real…
3:List…
4:Matrix…
5:Y-Vars…
6:Prgm…
7↓Pic…
```

```
DELETE:Matrix
▶[A]      89
 [B]      35
 [C]      62
 [D]      89
 [E]      89
```

Press **ENTER** to delete the selected matrix.

Find the inverse of a square matrix

Example: Find the inverse of the 3 x 3 matrix $\begin{bmatrix} 2 & -5 & 3 \\ 0 & 1 & -2 \\ -4 & 7 & 6 \end{bmatrix}$.

Enter the matrix, if you haven't done so, as matrix [A].

Return to the Home Screen.

Press **MATRIX** **1** **X⁻¹** **ENTER** to display the inverse of matrix [A] in decimal form.

The left and right arrow keys can be used to see all elements of the 3 x 3 inverse matrix.

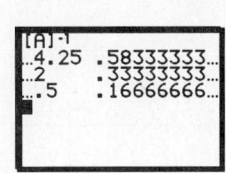

```
[A]⁻¹
[[1.666666667 4…
 [.6666666667 2…
 [.3333333333  .…
```

```
[A]⁻¹
…4.25  .58333333…
…2     .33333333…
….5    .16666666…
```

Press **MATH** **1** **ENTER** to display the inverse matrix in fractional form.

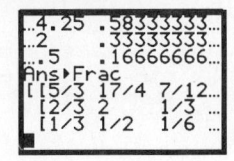

```
…4.25  .58333333…
…2     .33333333…
….5    .16666666…
Ans▶Frac
[[5/3  17/4  7/12…
 [2/3   2    1/3 …
 [1/3  1/2   1/6 …
```

Solve linear systems using matrix inverses

A system of linear equations can be solved by entering the coefficients as elements in an augmented matrix, then finding the solution using matrix row operations. If the coefficient matrix is square (same number of rows and columns) and the determinant is not equal to zero, you may solve the system by finding the product of the inverse of the coefficient matrix and the constant matrix.

Example: Solve the system
$$\begin{aligned} 2x - 5y + 3z &= 7 \\ y - 2z &= -3 \\ -4x + 7y + 6z &= 1 \end{aligned}$$

The coefficient matrix is $[A] = \begin{bmatrix} 2 & -5 & 3 \\ 0 & 1 & -2 \\ -4 & 7 & 6 \end{bmatrix}$ and the constant matrix is $[C] = \begin{bmatrix} 7 \\ -3 \\ 1 \end{bmatrix}$.

Enter the coefficient matrix as matrix[A], if you haven't done so.

Press **MATRIX** and arrow to **EDIT**.

Type **3** to select matrix [C].

Enter the size of the matrix (in this case, 3 x 1) followed by the matrix elements.

Press **2nd** **MODE** to return to the Home Screen.

The determinant of [A] is 12. The system has one solution since the determinant of the coefficient matrix is not zero.

If the linear system has a unique solution, the computation $[A]^{-1}[C]$ will return a column matrix that contains the solution to the system.

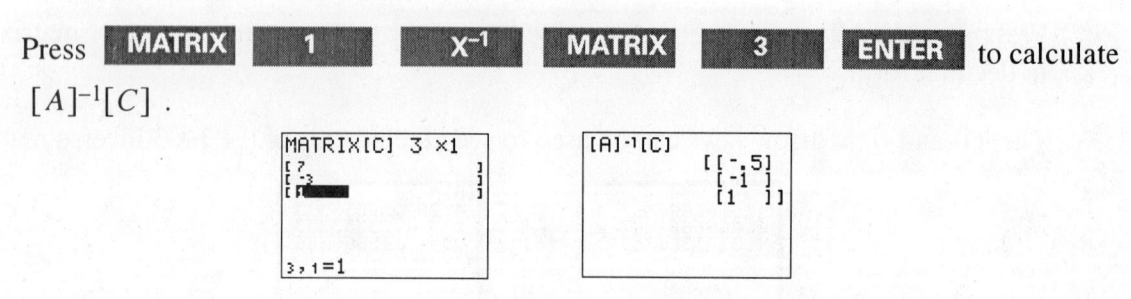

Press **MATRIX** **1** **X⁻¹** **MATRIX** **3** **ENTER** to calculate $[A]^{-1}[C]$.

The 3 x 1 solution matrix displays the solution to the system
$$\begin{aligned} 2x - 5y + 3z &= 7 \\ y - 2z &= -3 \\ -4x + 7y + 6z &= 1 \end{aligned}$$

Reading down the matrix, the solution is $x = -0.5$, $y = -1$, and $z = 1$.

TI-83
Graphing Calculator
Reference Manual

TI-83 REFERENCE MANUAL

The TI-83 represents an upgrade to the TI-82 graphics calculator. Many of the procedures, screens, and menus on the two calculators are similar. However, some procedures are located under different menus or have different keystrokes. This part of the manual will describe the TI-83 operations.

Turn the calculator on and off

Press **ON** to turn the calculator on. If nothing appears on the screen, adjust the display contrast (see below). If you still see nothing, check the batteries.

Press **2nd** **ON** to turn the calculator off. The TI-83 is equipped with an automatic power down feature which turns the calculator off if no keys have been pressed for about five minutes.

Keyboard organization

The keys are grouped by both color and location for ease of use.

The keys are color-coordinated.

To choose a command that is in yellow and appearing above a key, press the yellow **2nd** key first. By convention, any key that is in yellow will appear with black, rather than white, letters in this manual. Notice that the cursor changes from a blinking solid black rectangle to a rectangle with a light upward pointing arrow inside when the second key is depressed.

To choose a letter or command which appears in green and above a key, press the green **ALPHA** key first. Notice that the cursor changes from a blinking solid black rectangle to a rectangle with a light A inside when the Alpha key is depressed.

Keys by row beginning at the top of the calculator:

Row 1: Graphing and table keys.

Y =   **WINDOW**   **ZOOM**   **TRACE**   **GRAPH**

Rows 2 and 3: Editing keys.

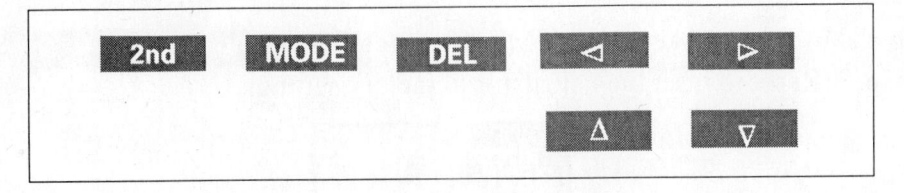

Rows 3 and 4: Advanced functions accessible through pull-down menus.

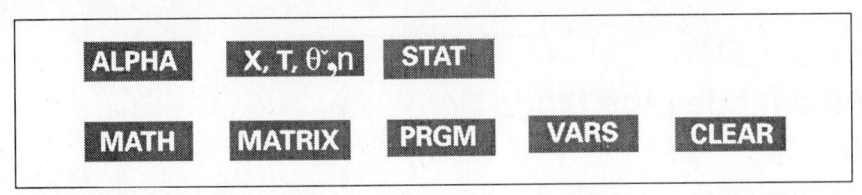

Rows 5–10: Scientific calculator keys.

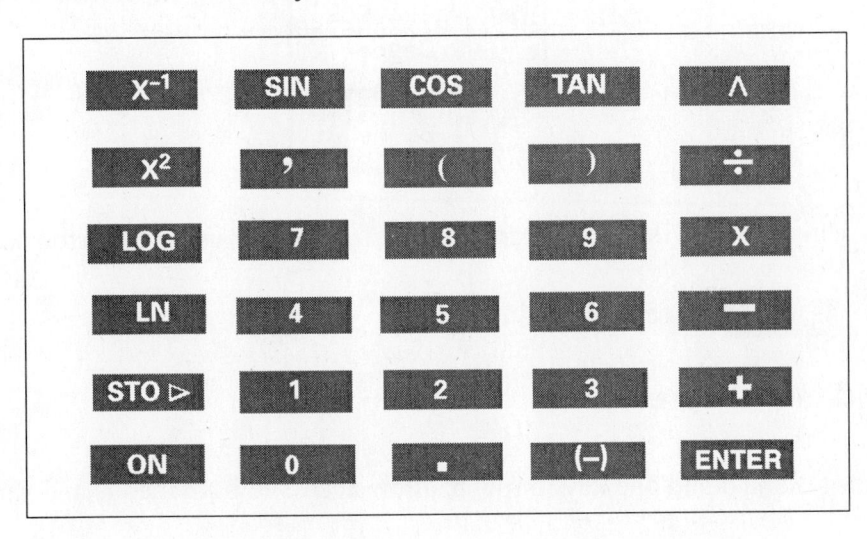

Adjust the display contrast (darken or lighten screen)

Press and release the **2nd** key.

Press and hold **▲** to increase the contrast (darken the screen).

Press and hold **▼** to decrease the contrast (lighten the screen). A black rectangle with a number inside appears in the upper right hand corner of the screen when darkening or lightening the screen. The number indicates the relative darkness of the screen. 9 is corresponds to the darkest setting while 0 indicates that the screen is as light as it will get. If you find it necessary to have the darkness set at 8 or 9 in order to see the screen then it is time to replace the four triple A batteries. In addition, the TI-83 displays a message that the batteries need changing when they are weak.

The calculator can be returned to factory settings by pressing [MEM]. Since MEM is in yellow above the "+" key, access MEM by pressing [2nd] [+]. Select item number **5:Reset** to reset the calculator to factory settings.

```
MEMORY
1:Check RAM…
2:Delete…
3:Clear Entries
4:ClrAllLists
5:Reset…
```

Changing the mode (the Mode Screen)

Press [MODE] to customize the calculator settings. Activated settings are highlighted.

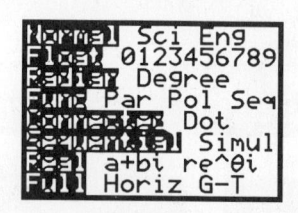

```
Normal Sci Eng
Float 0123456789
Radian Degree
Func Par Pol Seq
Connected Dot
Sequential Simul
Real a+bi re^θi
Full Horiz G-T
```

Change a setting by moving the cursor to the desired setting using the arrow keys.

Press [ENTER] to select a setting.

Menus and menu screens

Keys in rows 3 and 4 and one key in row 5 allow access to both menus and variable values.

Selecting a menu item

Press the key(s) corresponding to the menu to be displayed. For example, pressing the [MATH] opens the first menu shown. The other menus shown are accessed by using the right or left cursor controls to highlight the menu name that appears at the top of the screen.

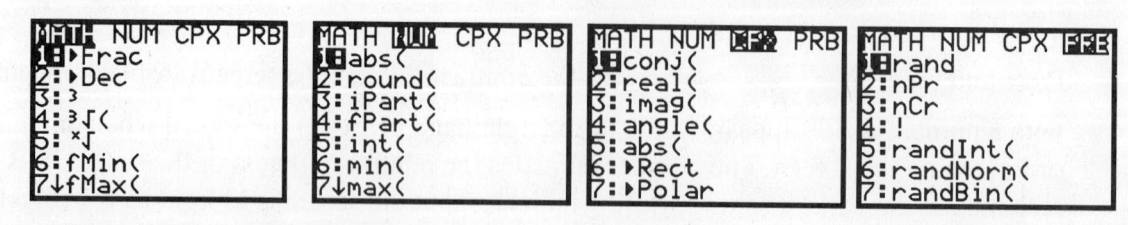

```
MATH NUM CPX PRB
1:▶Frac
2:▶Dec
3:³
4:³√(
5:ˣ√
6:fMin(
7↓fMax(
```
```
MATH NUM CPX PRB
1:abs(
2:round(
3:iPart(
4:fPart(
5:int(
6:min(
7↓max(
```
```
MATH NUM CPX PRB
1:conj(
2:real(
3:imag(
4:angle(
5:abs(
6:▶Rect
7:▶Polar
```
```
MATH NUM CPX PRB
1:rand
2:nPr
3:nCr
4:!
5:randInt(
6:randNorm(
7:randBin(
```

Select a menu item either by using the arrow keys to highlight the desired choice and then pressing [ENTER]; or if a menu item is numbered, press the number of the menu item.

Home screen

The home screen is used for calculations.

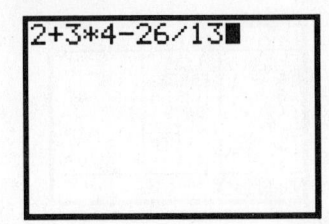

```
2+3*4-26/13█
```

You may return to the home screen at any time by executing the QUIT command. This is done by pressing **2nd** **MODE** . Notice that QUIT appears in yellow above the MODE key.

Graphics screen

The graphics screen is used to display graphs of functions that are entered into the Y= screen or of data which are entered in the list screen and plotted using the **STAT PLOTS** menu.

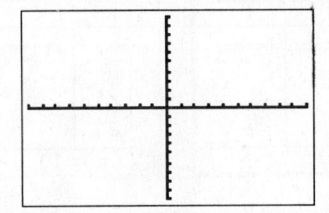

The part of the coordinate plane that is displayed is set using the window screen to be discussed later in the manual. To show the graphics screen press **GRAPH** .

Y= menu screen

The Y= menu screen is used to enter algebraic representations of functions. The graphs of these functions may be displayed on the graphics screen and input output tables generated from these functions may be displayed on the table screen. Ten different functions may be entered in this menu.

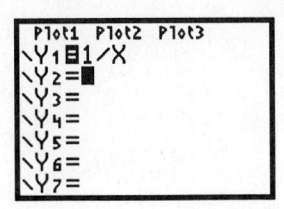

```
Plot1  Plot2  Plot3
\Y₁■1/X
\Y₂=█
\Y₃=
\Y₄=
\Y₅=
\Y₆=
\Y₇=
```

Press **Y=** to display this screen.

Table screen

This screen displays a table of inputs and outputs for each active user-defined function.

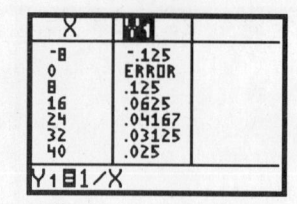

Choose TABLE to display this screen by pressing 2nd   GRAPH . Notice the word TABLE appears above the GRAPH key printed in yellow.

List screen

This screen displays lists where users may enter data. There are six lists in memory but it is possible to create additional list and name the new list with words that are up to five letters long.

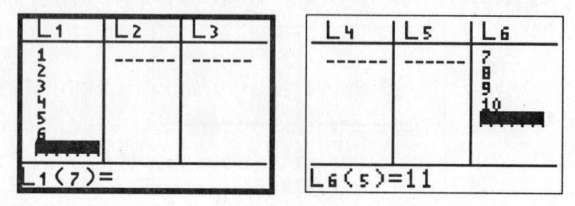

Press STAT and then press 1 to choose the Edit item on this menu. This is the **STAT EDIT** menu.

This shows the list screen and allows you to edit the lists. Press the right cursor control to show more list.

Important keys and their function

Keystroke	Function Description
ON	Turns on calculator; interrupts graph being sketched or other calculations.
CLEAR	Clears the text screen; deletes functions when in the Y= screen.
ENTER	Executes a command; moves cursor to next line.
(−)	The unary operation of opposing (not used for subtraction).
MODE	Displays mode screen to view current operating mode.
DEL	Deletes character at the cursor.
∧	Symbol used for exponentiation
Y=	Accesses screen where up to ten functions may be entered and stored.
X, T, θ, n	Enters the variable x in function mode; enters the variable t in parametric mode; enters Θ in polar mode; enters n in sequence mode.
MATH	Accesses math, number, complex and probability menus and submenus.
2nd	Accesses functions and menus printed in yellow.
ALPHA	Accesses letters and symbols printed in green to the right and above many keys.
STO ▷	Used to assign a numeric value to a variable or store a list.
GRAPH	Displays the graphics screen.
WINDOW	Accesses menus for setting the viewing window.
ZOOM	Accesses ZOOM menu where various adjustments can be made to the viewing window.
TRACE	Activates trace mode and displays graphics screen and cursor. Use the left or right arrow keys to trace along a graph. Use the up and down arrow keys to move the trace cursor from one graph to another.

Accessing functions using the [2nd] key
(commands written in yellow, above and left of center of the keys).

Editing Functions Accessed with the [2nd] key.

Function	Keystrokes	Description
OFF	[2nd] [ON]	Turns the calculator off.
ENTRY	[2nd] [ENTER]	Allows the last line entered to be edited and re-executed.
INS	[2nd] [DEL]	Allows insertion of character(s) to the left of the cursor.
LINK	[2nd] [STAT]	Allows transfer of data or programs between two TI-83 or 82 calculators.
QUIT	[2nd] [MODE]	Exits current screen; returns calculator to home screen.

Table Functions Accessed with the [2nd] key.

Function	Keystrokes	Description
TABLE	[2nd] [GRAPH]	Displays a table of x-values and corresponding output values for functions stored in the **Y=** menu.
TblSet	[2nd] [WINDOW]	Accesses the TABLE SETUP menu where the minimum table value and the table increment are set.
STATPLOT	[2nd] [Y=]	Accesses the STAT PLOT menu where statistical data may be graphically displayed using a scatter plot, line plot, box-and-whisker plot or histogram.

Calculations

Calculations are done in the home screen. If you are not in the home screen, choose `QUIT` which is accessed by pressing `2nd` `MODE`.

Enter expressions as you would write them. Observe the algebraic order of operations.

After you complete typing an expression, press `ENTER` to obtain the answer.

If you make a mistake during entry, use the arrow keys, `INS` (for insert), or `DEL` (for delete) to edit your expressions.

Raising a number to a power `^`

Enter an exponent first entering the base and then pressing the `^` key followed by the exponent. For example, 4^5 should appear on the home screen as shown below. Press `ENTER` to display the value of the expression.

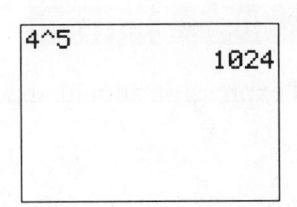

```
4^5
            1024
```

Negative numbers `(-)`

Enter a negative number using the *gray* key `(-)` (located to the left of the `ENTER` key), not the blue key, which is reserved for the operation of subtraction.

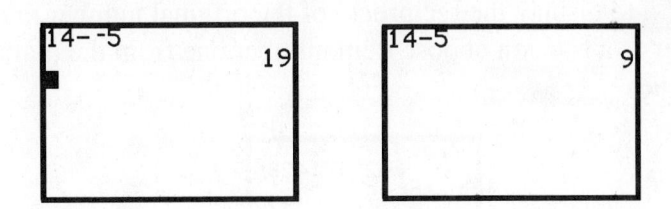

```
14--5
              19
```

```
14-5
               9
```

Scientific notation (3.46E11 means 3.46×10^{11})

This notation uses powers of ten to represent very large and very small numbers. Numbers are written as a decimal times a power of 10.

To enter on the calculator, replace the **X 10** with **EE**. The exponent appears immediately to the right of the E in the display. The display, 3.46E11, actually represents 3.46×10^{11}.

Enter **EE** by pressing [2nd] [,]. Only one **E** appears on the screen to indicate that the next number entered will be accepted as a power of ten.

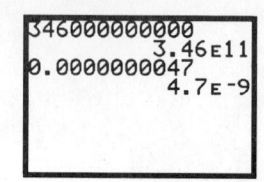

```
346000000000
         3.46E11
0.0000000047
         4.7E-9
```

Editing a previously-entered expression [2nd] [ENTER]

The last expression entered in the calculator can be displayed on the screen and edited. Execute the **Entry** command to display the last expression entered by pressing the following keys.

Choose [ENTRY] by pressing [2nd] [ENTER].

A copy of the previously entered expression should appear on the screen.

This expression may be edited.

The new expression may be executed by pressing [ENTER].

Reciprocal function [x^{-1}]

To find the reciprocal of a number first enter the number whose reciprocal you wish to find on the home screen. If the number is a fraction it must be enclosed in parentheses.

Press [x^{-1}].

Press [ENTER] to display the reciprocal of the original number in decimal form. To display the answer as a fraction choose item number one from the math menu by pressing [MATH] then [1].

```
(7/3)⁻¹
         .4285714286
Ans▶Frac
              3/7
```

INT function

The greatest integer function (**int**) returns the largest integer less than or equal to a number, expression, list or matrix. For nonnegative numbers and negative integers, the **int** value is the same as **iPart**. For negative noninteger numbers **int** returns a value one integer less than **iPart**.

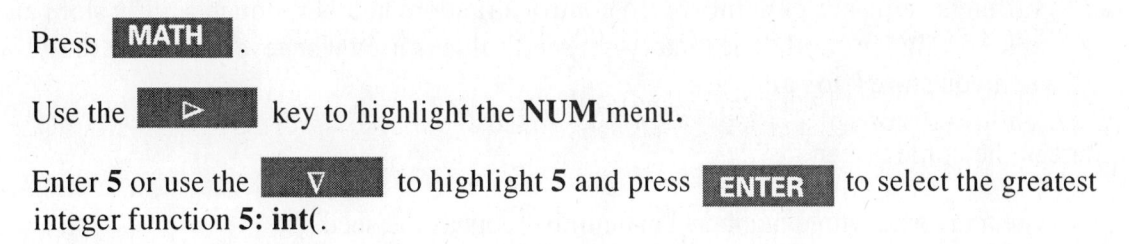

Press **MATH**.

Use the **▷** key to highlight the **NUM** menu.

Enter **5** or use the **▽** to highlight **5** and press **ENTER** to select the greatest integer function **5: int(**.

The calculator returns to previous screen and inserts "**int (**" at cursor. Follow the function name with an input value and close the parentheses. The function returns an output which is the largest integer that does not exceed the input.

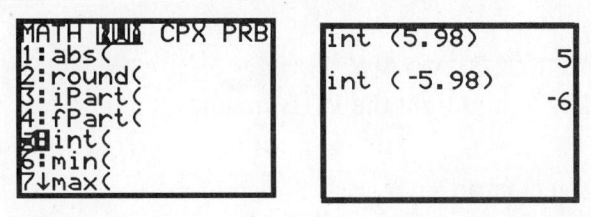

```
MATH NUM CPX PRB
1:abs(
2:round(
3:iPart(
4:fPart(
5:int(
6:min(
7↓max(
```

```
int (5.98)
              5
int (-5.98)
             -6
```

Obtaining fractional results

If a computation results in a decimal output, the result can often be converted to a fraction by using the **Frac** command.

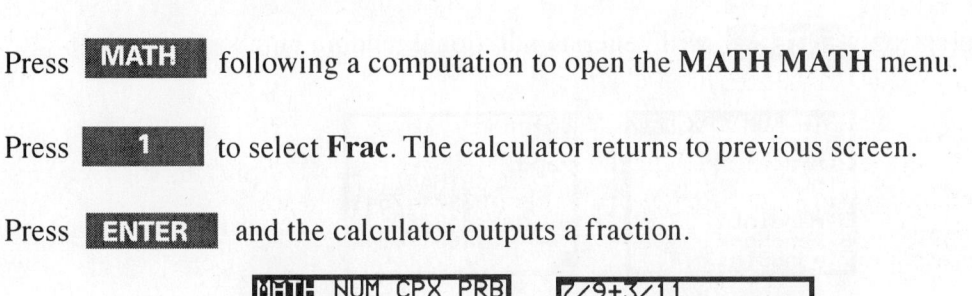

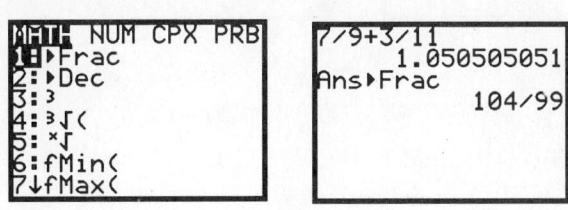

Press **MATH** following a computation to open the **MATH MATH** menu.

Press **1** to select **Frac**. The calculator returns to previous screen.

Press **ENTER** and the calculator outputs a fraction.

```
MATH NUM CPX PRB
1:▶Frac
2:▶Dec
3:3
4:3√(
5:×√
6:fMin(
7↓fMax(
```

```
7/9+3/11
        1.050505051
Ans▶Frac
              104/99
```

405

RAND function

The calculator contains a random number generator that will return random numbers greater than zero and less than one.

The random number generator should be seeded initially so that each calculator will generate a different sequence of numbers. To control a random number sequence, first store an integer seed value in **rand**. The factory-set seed value is used whenever you reset the TI-83 or when you store **0** to **rand**.

Start in the home screen.

> Type a favorite whole number. This number acts as the seed.
>
> Press **STO ▷**. This is the key used for storing a value in the calculator.
>
> Press **MATH**.
>
> Use the **◁** key to highlight the **PRB** menu.
>
> Press **1** to select **rand**.
> The calculator returns to the home screen and inserts the **rand** function.
>
> Press **ENTER** to complete the seeding process.

Recall **rand** by pressing **MATH**, arrowing to **PRB**, and selecting **1**.

> Press **ENTER** to see the first random number.

Continually pressing **ENTER** will generate additional random numbers.

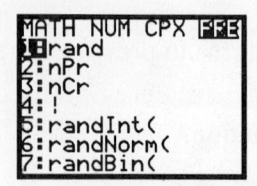

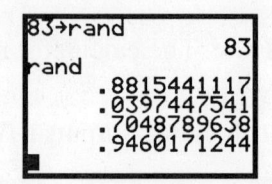

```
MATH NUM CPX PRB
1:rand
2:nPr
3:nCr
4:!
5:randInt(
6:randNorm(
7:randBin(
```

```
83→rand
              83
rand
       .8815441117
       .0397447541
       .7048789638
       .9460171244
```

Factorial function

The factorial of a whole number is the product of all whole numbers less than or equal to the given whole number.

Type a whole number.

Press **MATH**.

Use the ◁ key to highlight the **PRB** menu.

Press **4** to select **!** which represents factorial.

The calculator returns to the home screen and inserts the **!** after the number you typed.

Press **ENTER** to see the result of the factorial.

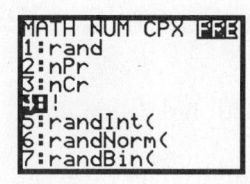

```
MATH NUM CPX PRB
1:rand
2:nPr
3:nCr
4:!
5:randInt(
6:randNorm(
7:randBin(
```

```
13!
        6227020800
```

Entering expressions in the Y = menu **Y =**

When you enter a function the = symbol is highlighted to show the function is selected to be graphed or to display outputs on the table screen. The highlighting may be toggled on and off by using the arrow keys to move the cursor to the = symbol and then pressing **ENTER**.

Define the function:

Press **Y =** to access the function entry screen.

Type the expression for the function after $Y_1 =$. This defines the function as Y_1.

Select the *int* function:

Example: Enter the function $f(x) = x - 8int(x/8)$ as Y_1:

Press **Y =** to access the function entry screen.

Enter the function expression *x − 8* at the cursor.

Press **MATH**.

Use the ▷ key to highlight the **NUM** menu.

Press the **4** key or use the [▽] to highlight **4** and press [ENTER] to select the greatest integer function *4: int*.

Complete the entry of the function by entering (*x* / **8**) then press [ENTER].

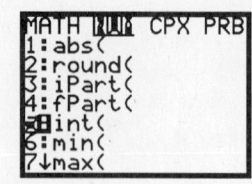

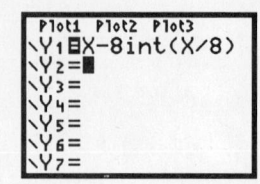

Functions may be removed from this menu by pressing [CLEAR].

Up to ten functions can be defined and accessed simultaneously.

Displaying a table of values from a function in the Y= menu

The TI–83 will display a table of values for a function defined under the Y = menu.

Example: Display a table of values for the function $y(x) = x - 8int\left(\frac{x}{8}\right)$

First Define the function:

Press [Y=] to access the function entry screen and enter *x - 8 int (x/8)* as Y_1.

Press [2nd] [WINDOW] to access [TblSet], the table set up menu.

Set the beginning table value and the increment:

Enter a beginning table value for **TblStart** at the flashing cursor.

Press the [▽] to place the cursor on the number next to ΔTbl = to change the increment value.

ΔTbl represents the increment on the input values. The default value is **1**. If you wish a different increment, enter that value.

Check to see that **Auto** is highlighted for both Indpnt and Depend. If not, use arrow keys to place cursor on **Auto** and press [ENTER] to select it.

408

To view the table:

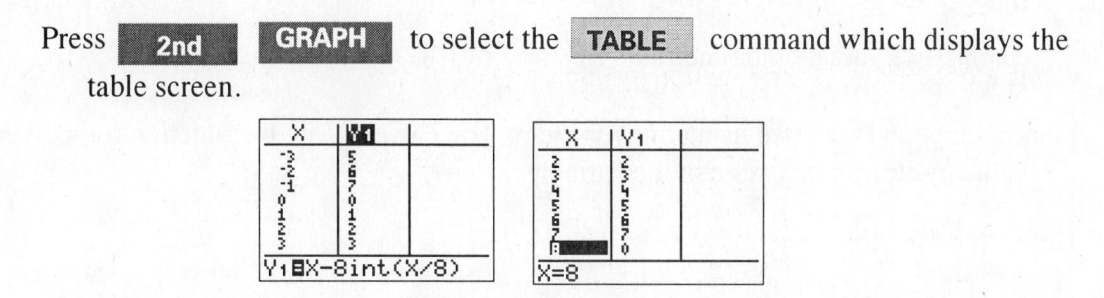

Press 2nd   GRAPH to select the TABLE command which displays the table screen.

By using the up and down arrows, you can scroll through the table, either up or down.

Storing a value for a variable

In the home screen enter the input value and store it in the variable x.

Type the number you wish to store.

Press STO ▷ . This is the key used to store values.

When pressed, an arrow pointing right appears.

Press X, T, Θ, n . This will display the variable x.

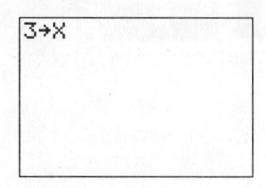

```
3→X
```

Press ENTER . this causes the storage to take place.

A common mistake is to forget to press ENTER .Store the value of 3 in x.

```
3→X
              3
```

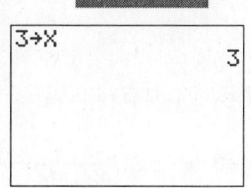

Function evaluation: storing input

Example: Evaluate the function $y(x) = -x^2 + 8x + 19$ for $x = 3$.

Once a function is defined using the **Y =** menu, you can evaluate the function for a given input value by storing the value in the variable x.

Define the function:

　Press **Y=** to access the function entry screen.

　Type the expression $-x^2 + 8x + 19$ for the function after $Y_1 =$.
　This defines the function as Y_1.

Return to the Home Screen:

　Return to the Home Screen by pressing **2nd** **MODE**.

　Enter the **input value** followed by **STO ▷** **X, T, Θ, n** **ENTER**.
　This will have the effect of storing a value for variable x.

Writing more than one command on the same line (chaining commands):

　Enter a colon by pressing **ALPHA** **■**. The colon is used to chain two or more commands together on the same line.

Select the function you wish evaluated:

　Select Y_1 by pressing **VARS** **▷** **1** **1**.

Evaluate the function:

　Press **ENTER**. The function value is displayed.

```
Plot1 Plot2 Plot3
\Y1∎-X²+8X+19
\Y2=
\Y3=
\Y4=
\Y5=
\Y6=
\Y7=
```

```
3→X:Y1
           34
```

Function evaluation using function notation

To evaluate a function using function notation you must first define the function in the Y = menu as one of the Y variables. Then on the Home Screen enter the Y variable with the input value in parentheses next to that Y variable.

Example: Evaluate the function $y(x) = -x^2 + 8x + 19$ at $x = 3$.

First, define the function:

Press **Y=** to access the function entry screen.

Type the expression $-x^2 + 8x + 19$ for the function after $Y_1 =$. This assumes that Y_1 is clear. This defines the function as Y_1.

Second, return to the Home Screen:

Return to the home screen by pressing **2nd** **MODE**.

Third, select the Y variable you wish evaluated:

Type **VARS** **▷** **1** **1** to access Y_1 in the Y variables menu. Y_1 appears in the home screen.

Finally, complete the process by giving the input value in parentheses:

Immediately after Y_1,

Type **(3)** followed by **ENTER**.

This returns the output of the function stored in Y_1 at an input value of 3.

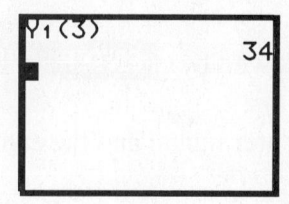

```
Plot1 Plot2 Plot3
\Y1■-X²+8X+19
\Y2=
\Y3=
\Y4=
\Y5=
\Y6=
\Y7=
```

```
Y1(3)
                34
```

Function evaluation using Ask mode with the table screen

You can define a function and, using the table with the independent variable set to Ask mode and the dependent variable set to Auto mode, enter values for the independent variable and by pressing ENTER after each entry, displaying the output of the function for the input entered.

Evaluate the function $y(x) = -x^2 + 8x + 19$ at $x = 3$.

Define the function:

Press **Y =** to access the function entry screen.

Type the expression $-x^2 + 8x + 19$ for the function after $Y_1 =$. This assumes Y_1 is clear. This defines the function as Y_1.

Select the table setup **TblSet** :

Press **2nd** **WINDOW** to access **TblSet**, the table set up screen.

Arrow down to **Indpnt** and over to **Ask**.

Press **ENTER** to select the **Ask** mode for the independent variable. The dependent variable should still be in **Auto** mode.

Press **2nd** **GRAPH** to access **TABLE**, and display the table screen. A blank table appears.

Enter the input value:

Type the input followed by **ENTER**. The output for this input appears.

You may continue to enter inputs and the calculator will return the corresponding outputs.

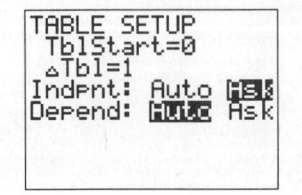

```
TABLE SETUP
 TblStart=0
  ΔTbl=1
 Indpnt: Auto Ask
 Depend: Auto Ask
```

X	Y₁

X=

X	Y₁
3	34

X=

412

Graphing a function

Once a function is defined in the Y = menu, its graph can be displayed on the graphics screen. It may be necessary to adjust the viewing rectangle which controls the portion of the graph shown.

Example: Graph the function $f(x) = -x^2 + 8x + 19$.

Define the function:

Press **Y=** to access the function entry screen.

Type the expression $-x^2 + 8x + 19$ for the function after $Y_1 =$.
This defines the function as Y_1.

Choose a viewing window:

A *viewing window* is a rectangular portion of the coordinate plane.

Xmin puts the left edge of the viewing window at this *x*-coordinate.

Xmax puts the right edge of the viewing window at this *x*-coordinate.

Xscl defines the distance between tick marks on the horizontal axis.

Ymin puts the lower edge of the viewing window at this *y*-coordinate.

Ymax puts the upper edge of the viewing window to this *y*-coordinate.

Yscl defines the distance between tick marks on the vertical axis.

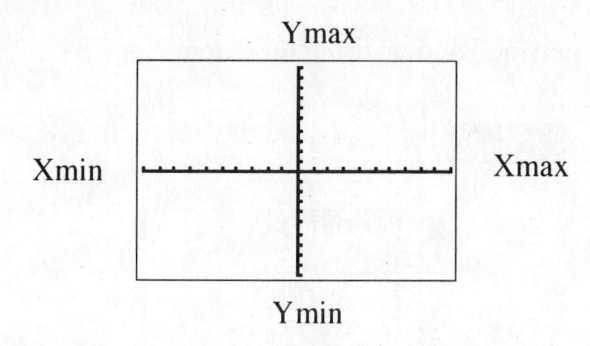

Standard viewing window (factory setting and item 6 in the Zoom menu)

The *standard viewing window* is defined as:

Xmin = –10	Ymin = –10
Xmax = 10	Ymax = 10
Xscl = 1	Yscl = 1

10 (Ymax)

– 10 (Xmin)   10 (Xmax)

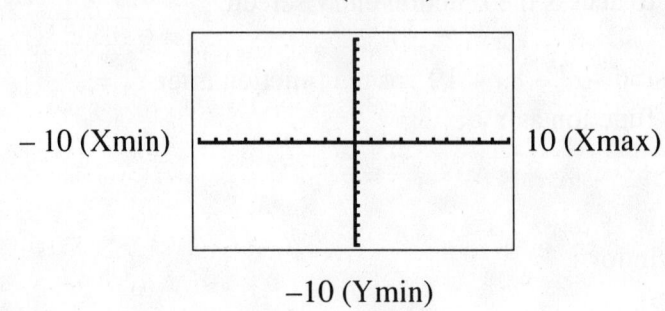

–10 (Ymin)

Set the viewing window:

Press **WINDOW** and enter the viewing window values.

Press the **GRAPH** key to display the graphics screen with a portion of the graph shown.

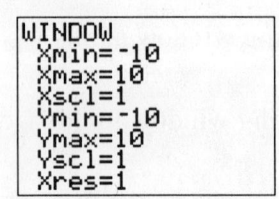

```
WINDOW
 Xmin=-10
 Xmax=10
 Xscl=1
 Ymin=-10
 Ymax=10
 Yscl=1
 Xres=1
```

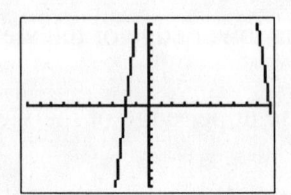

To select the standard viewing window using the Zoom menu:

Press **ZOOM** **6** .

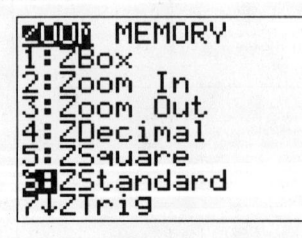

```
ZOOM MEMORY
1:ZBox
2:Zoom In
3:Zoom Out
4:ZDecimal
5:ZSquare
6:ZStandard
7↓ZTrig
```

414

Changing the viewing window

The previous graph is only a portion of the complete graph. To see more of the graph, change the viewing window. This is done by pressing the **WINDOW** key which displays the window screen. First decide how to change the window. The table feature helps determine a better viewing window by displaying output values of the function.

Press **2nd** **GRAPH** to display the table.

Determine the smallest and largest output value between Xmin (–10) and Xmax (10).

Examine the table values for x = –10 and x = 10.

X	Y1
-10	-161
-9	-134
-8	-109
-7	-86
-6	-65
-5	-46
-4	-29

Y1■-X²+8X+19

X	Y1
-3	-14
-2	-1
-1	10
0	19
1	26
2	31
3	34

Y1■-X²+8X+19

X	Y1
4	35
5	34
6	31
7	26
8	19
9	10
10	-1

Y1■-X²+8X+19

The smallest output value shown is –161. The largest output value shown is 35.

Press **WINDOW**.

Enter a value for Ymin which is smaller than the smallest output –161. –170 would be a good choice.

Enter a value for Ymax which is larger than the largest output value 35. Say, 40.

For Yscl, choose a reasonable value for the distance between tick marks. Approximately one-tenth of the distance between Ymin and Ymax is a reasonable choice. Set Yscl = 20.

Graph the function:

Press **GRAPH**.

```
WINDOW
Xmin=-10
Xmax=10
Xscl=1
Ymin=-170
Ymax=40
Yscl=20
Xres=1
```

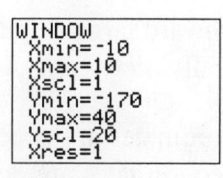

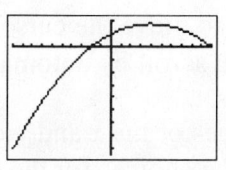

Adjusting the viewing window:

Change Xmin and Xmax to see more of the graph to the left or to the right. For instance one might increase the Xmax value to 20. It would then be wise to increase the Xscl value so there won't be an excessive number of tick marks on the x-axis.

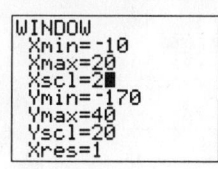

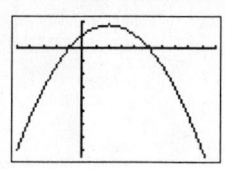

```
WINDOW
Xmin=-10
Xmax=20
Xscl=2
Ymin=-170
Ymax=40
Yscl=20
Xres=1
```

Trace: displaying coordinates of points on a graph

The cursor can be moved from one plotted point to the next along the graph of a function. The coordinates of the point located by the cursor are displayed at the bottom of the screen.

Example: Trace along the graph of $f(x) = -x^2 + 8x + 19$.

Define the function:

Enter the expression $f(x) = -x^2 + 8x + 19$ as Y_1 in the Y = menu.

Verify the viewing window settings:

If the viewing window is not in standard setting, enter the following:

Xmin = –10   Ymin = –170

Xmax = 10   Ymax = 40

Xscl = 1   Yscl = 20

Graph the function:

Press **GRAPH**. The graph of the function is displayed.

Press **TRACE**.

A flashing cursor appears on the graph along with the x and y-coordinates of the point at the bottom of the screen.

Press the left arrow key to move the cursor toward smaller inputs. Moving to the left of Xmin causes the graph to scroll by automatically decreasing Xmin.

Press the right arrow key to move the cursor toward larger inputs. Moving to the right of Xmax causes the graph to scroll by automatically increasing Xmax.

As the cursor moves, the values of the *x* and *y*-coordinates at the cursor appear at the bottom of the screen. Note that the choices for the *x*-coordinates are determined by the programming of the graphing utility. The coordinates which are displayed can be controlled as is explained in the next section of this manual.

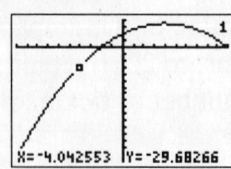

Friendly viewing windows to control the Trace Display

A "friendly" viewing window displays values for x with at most one decimal position. There are 95 pixels across the screen. The distance between Xmin and Xmax is divided by 94 to determine the size of the increment in x when moving the cursor in the graphics window.

To obtain a "friendly" window, Xmin and Xmax are selected so that when the difference Xmax − Xmin is divided by 94, the quotient terminates by the tenths position.

Tracing by x increments of 0.1: Zoom Menu Item Number 4:ZDecimal

Press **ZOOM** **4** to automatically set the "friendly" viewing window to

Xmin = −4.7   Ymin = −3.1

Xmax = 4.7   Ymax = 3.1

Xscl = 1   Yscl = 10

Notice that Xmax − Xmin = 9.4.

Divide this difference by 94, to determine the x-increment displayed. The quotient is 0.1.

The x values displayed by the Trace feature will increment by 0.1 in this window.

Using the ZOOM options to adjust the viewing window

Use the **ZOOM** key to open the following menu of options for altering the viewing window.

1:ZBox: creates a new viewing window based on a rectangle that you draw.

Press **ZOOM** **1** to return to the graphics window.

Use the arrow keys to move the cursor to one corner of the desired viewing window.

Press **ENTER** to set one corner.

Move the cursor to the opposite corner of the desired viewing window.

Press **ENTER** and the box (rectangle) becomes the new viewing window.

2:Zoom In: Use this item to zoom in using a preset factor to decrease the size of the viewing window. The factors can be set by selecting the Memory submenu from the Zoom screen and selecting item number **4:SetFactors** in that submenu.

Press `ZOOM` `2` which returns you to the graphics window.

Use the arrow keys to position the cursor at the desired center of the new window.

Press `ENTER` .

You can now repeatedly zoom in by moving the cursor to the center of the desired window and pressing `ENTER` .

3:Zoom Out: Use this item to zoom out using a preset factor to increase the size of the viewing window. The factors can be set by selecting the Memory submenu from the Zoom screen and selecting item number **4:SetFactors** in that submenu.

Press `ZOOM` `3` which returns you to the graphics window.

Use arrow keys to position the cursor at the desired center of the new window.

Press `ENTER` .

You can now repeatedly zoom out by moving the cursor to the center of the desired window and pressing `ENTER` .

4:Decimal: This item sets a viewing window so that the x values will increment by 0.1 as the cursor moves across the window or when using the trace feature. These are "friendly" values for analyzing the graph.

Press `ZOOM` `4` to change the window so that Xmin = –4.7, Xmax = 4.7, Ymin = –3.1, and Ymax = 3.1. Both Xscl and Yscl are 1.

5:Square: sets the viewing window so that the unit used on the x and y axes are the same distance. Graphs, such as circles, will appear distorted if this aspect ratio is not one.

Press `ZOOM` `5` .

6:Standard: returns graph to the standard viewing window.

Press `ZOOM` `6` to graph in the standard viewing window.

7:Trig: sets a "friendly" viewing window for graphing trigonometric functions.

Press `ZOOM` `7` to select this window.

8:Integer: sets the viewing window so that the x values will increment by one as the cursor moves across the window. These are "friendly" values for analyzing graphs.

   Press `ZOOM` `8` to change the window.

9:ZoomStat: This item resets the viewing window so that all the points in the active StatPlots are displayed.

   Press `ZOOM` `9` to change the window.

0:ZoomFit: This item resets the Ymin and Ymax to include the minimum and maximum values of the selected functions between the current Xmin and Xmax and graphs the selected functions in this new viewing window. Xmin and Xmax are not changed.

   Press `ZOOM` `0` to change the window.

Connected and Dot Modes for graphs

The TI-83 has a graphics mode setting icon for each function in the Y= menu. The icon appears to the left of the y variable and may be changed by placing the cursor on the icon and pressing the enter key to cycle through the seven choices. For **Dot** mode the icon should appear as three separate dots. Connected mode is the factory setting and is four dots close together.

As an example the graph of the function $y(x) = x^2 - 1$ is shown with both mode settings using the "friendly" viewing window set by choosing item number **4:ZDecimal** in the **Zoom** menu.

Dot mode

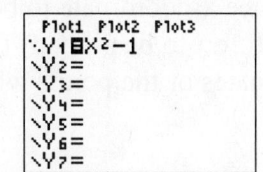

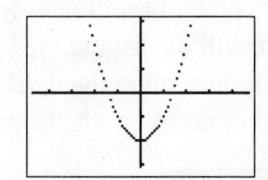

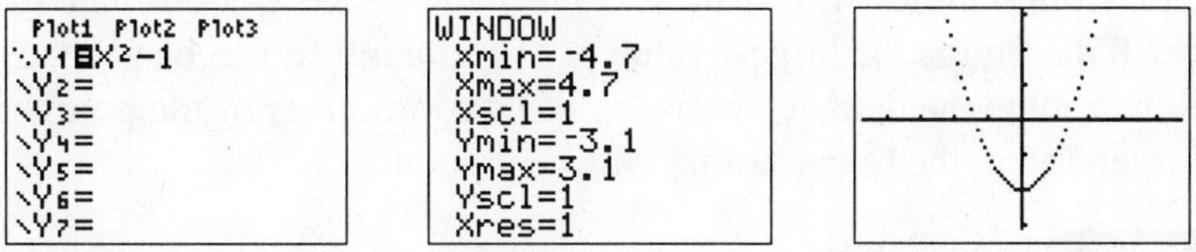

Connected mode

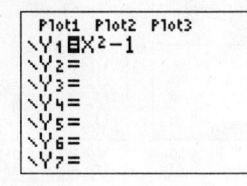

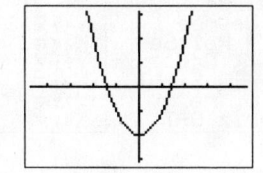

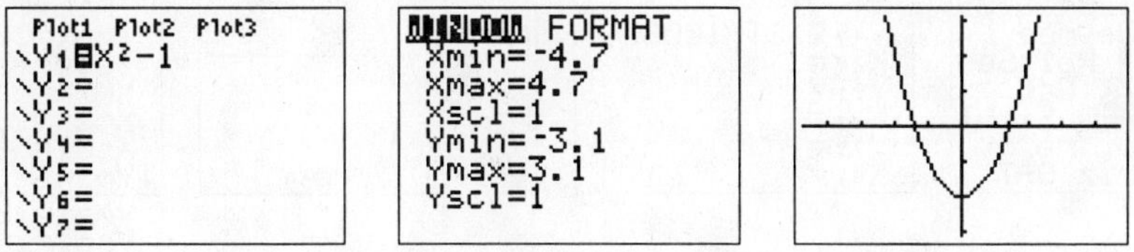

Parametric graphing

Parametric equations consist of two equations, representing the x-component and the y-component of a graph, each expressed in terms of a third independent variable, T. Parametric equation may be used to control the points which are plotted on the graph of a function when **Dot** mode is selected.

Before entering the two parametric equations, you must change from function mode to parametric mode and adjust your viewing window.

Press **MODE** and use the ▽ key to highlight **Par** on the mode screen.

Press **ENTER**.

Example: Suppose we wish to plot the function $y(x) = 10x - 1000$ but only show the points on the graph whose x-coordinates are 0, 5, 10,... 200. First we select Parametric in the Mode window.

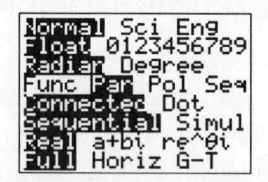

Now we open the Y= menu and define $x = t$ and $y = 10t - 1000$ and select dot mode using the mode icon next to the y variable.

Press **Y=** and enter t for X_{1T} and $10t - 1000$ for Y_{1T} and select the dot icon next to X_{1T}.

Press **WINDOW** and set the viewing window appropriately. (Shown below.)

The value of Tmin will determine the smallest t and hence x-coordinate to be plotted and the Tmax will determine the largest value of t and therefore x to be plotted. The size of the Tstep will determine the distance between the x-coordinates of the points which are plotted. The larger Tstep, the fewer points will be plotted.

Press **GRAPH** when all this is set up.

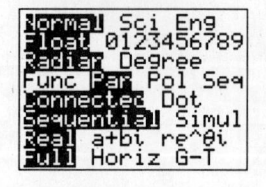

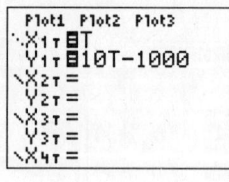

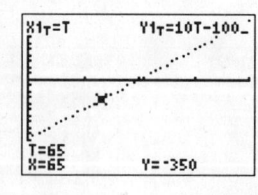

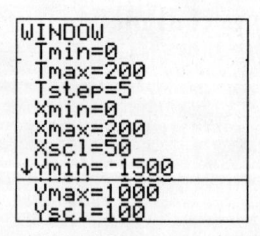

When you use the trace feature you will see the points on the graph highlighted by the cursor and the t, x and y- coordinates will be displayed at the bottom of the screen. As you move the cursor only those points whose x-coordinates are multiples of 5 starting with 0 are displayed.

Graphing the solution interval to inequalities in one variable

Inequalities in one variable, such as $2x + 3 < 5$, can be entered directly in the Y= menu. The calculator will graph the solution set by drawing a horizontal line at $y = 1$ for all x values that make the inequality true. The "1" is the calculator's way of saying "true" whereas "0" is what the calculator uses for "false".

Set the calculator to **Func** mode:

Press **MODE**, arrow down to **Func**, and press **ENTER**.

Press **Y=**.

Type the left hand side of the inequality.

Open the **TEST** menu by pressing **2nd** **MATH**.

Select the desired inequality relation by typing the number next to the relation, in this case **5**. The calculator returns to the Y= screen and prints the inequality symbol.

Type the right hand side of the inequality.

Set an appropriate viewing window. It is important that $y = 0$ and $y = 1$ be clearly visible on the graphics screen. It would be appropriate to use Ymin = –2 and Ymax = 2.

Press **GRAPH**. A horizontal line segment at $y = 1$ appears for x values that satisfy the inequality.

You can trace on the graph. An output of 1 indicates the corresponding x is in the solution set to the inequality. An output of 0 indicates the corresponding x makes the inequality false.

Example: Graph the solution set to $2x + 3 < 5$.

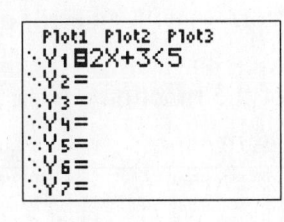

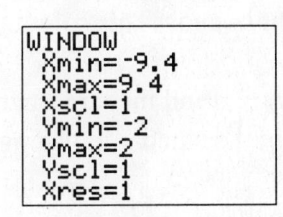

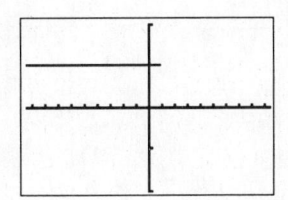

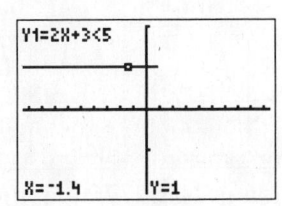

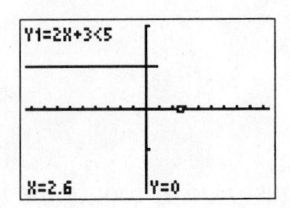

Finding zeros using the CALCULATE menu item 2: zero

Enter the function in the Y= menu (the mode setting should be **Func**) and graph the function in a window so that the points where the graph crosses the horizontal axis are visible in the graphics screen.

Press `2nd` `TRACE` to display the **CALCULATE** menu, abbreviated as CALC in yellow above the Trace key.

Type `2` to select **2:zero** from the menu.

The calculator returns to the graphics window and prompts you to select a **Left Bound** for one of the roots.

Use the **left arrow key** to move the cursor until the value for x at the cursor is smaller than the desired zero.

Press `ENTER` and the x value of the cursor is selected as the left bound.

The calculator prompts you to select a **Right Bound** for the zero.

Press the **right arrow key** at least twice to move the cursor until the value for x at the cursor is larger than the desired zero.

Press `ENTER` and the x value of the cursor is selected as the right bound.

The calculator prompts you to select a **Guess** for the zero.

Press the **left arrow key** to move the cursor close to the zero, which is where the graph intersects the x-axis. Be careful not to move the cursor past the left bound. The guess must be between the left and right bounds. When the cursor is close to the zero, press `ENTER`.

The calculator begins approximating the zero. A moving vertical line segment is displayed in the upper right corner of the screen to indicate that the calculator is computing. When finished, the calculator displays the approximate zero within the interval defined by the left and right bounds. This solution is correct with an error of no more than 0.00001.

Example: Find the largest zero of the function $y = x^2 + x - 3$. The function is graphed in the standard viewing window.

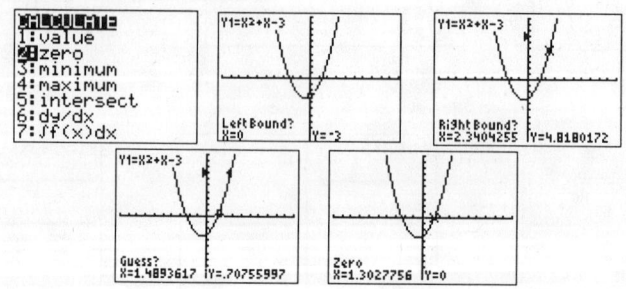

The largest zero or root of the function is 1.3027756, approximately.

Finding intersection points using the CALCULATE menu

Enter the two functions in the Y= menu as Y_1 and Y_2. (You may use any of the other y variables if you do not wish to erase functions you may have entered for Y_1 and Y_2.)

Graph both functions in a window so that the desired intersection point is visible on the graphics screen.

Press **2nd** **TRACE** to display the **CALCULATE** menu.

Press **5** to select **5:intersect.**

The calculator returns to the graphics screen. The calculator prompts you to select the **First curve**, meaning the graph of one of the functions. The cursor is on the graph of Y_1.

Press **ENTER** to select Y_1 as the first curve.

Now the calculator prompts you to select the **Second curve**. The cursor is on the graph of Y_2. (If there were more than two graphs on the graphics screen you could move the cursor to one of the other graphs by using the up or down cursor control.)

Press **ENTER** to select Y_2 as the second curve.

The calculator prompts you to supply a **Guess** for the intersection point.

Do this by using the right or left arrow keys to move the cursor near the intersection point.

Press **ENTER** when the cursor is close to the intersection point.

The calculator begins approximating the intersection point. A moving vertical line segment is displayed in the upper right corner of the screen to indicate that the calculator is computing. When finished, the calculator displays the approximate ordered pair for the intersection point. The solution is correct with an error of no more than 0.00001.

Example: Find the intersection point of the functions $y = 2x - 1$ and $y = -3x + 7$. The graphs appears in the standard viewing window.

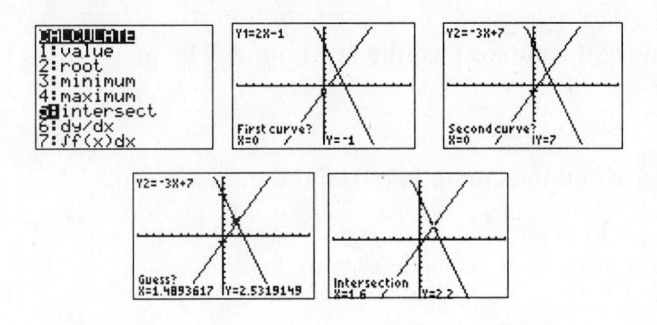

The intersection point is $(1.6, 2.2)$.

Finding a function's maximum using the CALCULATE menu

Enter the function in the Y= menu.

Graph the function in a window so that the graph displays a highest point.

Press **2nd** **TRACE** to open the **CALCULATE** menu.

Type **4** to select **4:maximum**.

The calculator returns to the graphics window. The calculator prompts the user to select a **Left Bound** for the interval containing the function's maximum.

Use the **left arrow key.** Move the cursor until it is to the left of the function's maximum.

Press **ENTER** to select this x value as the left bound.

Now the calculator prompts the user to select an **Right Bound** for the interval containing the function's maximum.

Press the **right arrow key** at least twice to move the cursor until the cursor is to the right of the function's maximum.

Press **ENTER** to select this x value as the right bound.

The calculator prompts the user to select a **Guess** for the function's maximum.

Press the **left arrow key** to move the cursor close to the function's maximum.

Be careful not to move the cursor past the left bound. The guess must be between the left and right bounds.

Press **ENTER** when the cursor is close to the maximum.

The calculator begins approximating the function's maximum. A moving vertical line segment is displayed in the upper right corner of the screen to indicate that the calculator is computing. When finished, the calculator displays the approximate maximum of the function within the interval defined by the left and right bounds. This solution is correct with an error of no more than 0.00001.

Example: Find the vertex of $y = -x^2 + x + 4$. For this function the vertex is the highest point on the graph. The graph appears in the standard viewing window.

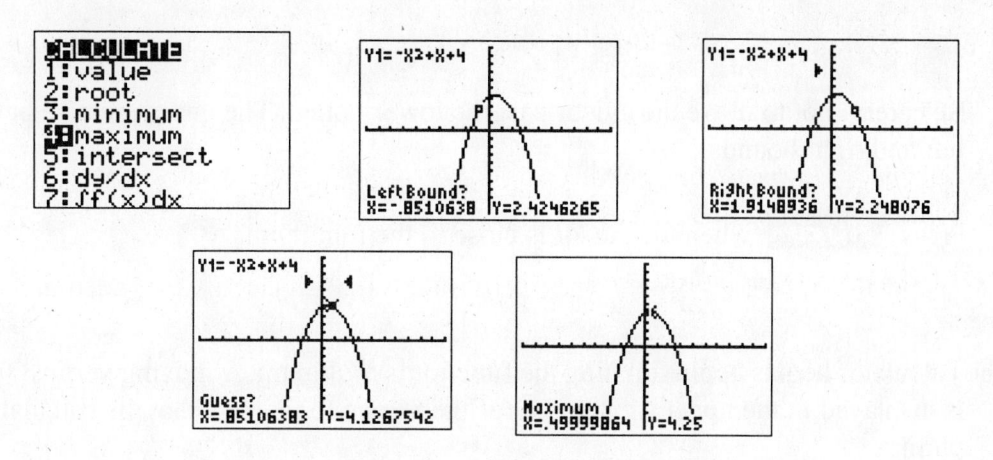

The vertex is actually the point (0.5, 4.25). The x-coordinate shown is an approximation.

Finding a function's minimum using CALC

Enter the function in the Y= menu.

Graph the function in a window so that the graph displays a lowest point.

Press **2nd** **TRACE** to open the **CALCULATE** menu.

Type **3** to select **3:minimum**.

The calculator returns to the graphics window. The calculator waits for you to select a **Left Bound** for the interval containing the function's minimum.

Use the **left arrow key** to move the cursor until it is to the left of the function's minimum.

Press **ENTER** and this *x* value becomes the left bound.

The calculator waits for you to select a **Right Bound** for the interval containing the function's minimum.

Press the **right arrow key** at least twice to move the cursor until it is to the right of the function's minimum.

Press ENTER and this *x* value becomes the right bound.

The calculator waits for you to select a **Guess** for the function's minimum.

Press the **left arrow key** to move the cursor close to the function's minimum.

Be careful not to move the cursor past the lower bound. The guess must be between the left and right bounds.

Press ENTER when the cursor is close to the minimum.

The calculator begins approximating the function's minimum. A moving vertical line segment is displayed in the upper right corner of the screen to indicate that the calculator is computing.

When finished, the calculator displays the approximate minimum of the function within the interval defined by the left and right bounds.

This solution is correct with an error of no more than 0.00001.

Example: Find the vertex of $y = x^2 - 3x - 2$. The vertex of the graph of this function is the lowest point on the graph. The graph appears in the standard viewing window.

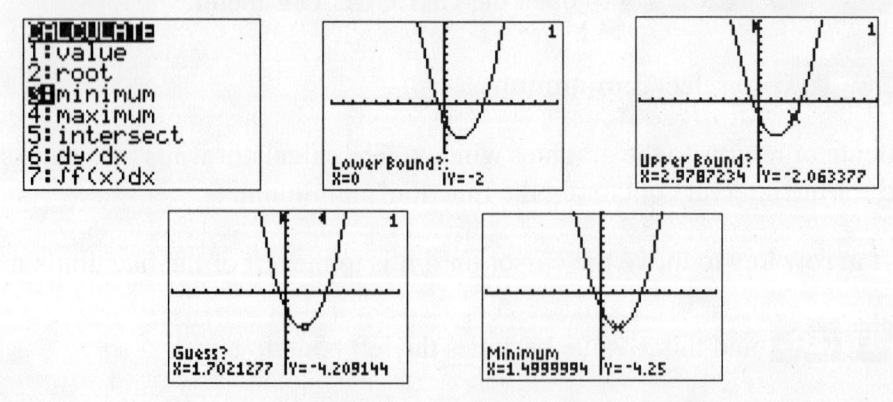

The vertex is actually (1.5, −4.25). The x-coordinate shown is only an approximation.

Working With Lists

Entering data into a list

A list of data can be entered, displayed, copied to another list, stored, sorted, used to graph families of curves, and used in mathematical expressions. Data that has been entered in lists can be displayed graphically using the **Stat Plots** menu. The types of plots available include scatter plots, line plots, box-and-whisker plots, and histograms.

From the Home Screen:

Press **STAT** **ENTER** to select **1:Edit** form the **STATEDIT** menu and to display the list screen.

With the lists displayed:

Position the cursor in the column with heading L_1 using the arrow keys. At the bottom of the screen you should see $L_1(1) =$. The 1 in the parentheses indicates that the cursor is at the first position in the list.

Enter each of the data elements.

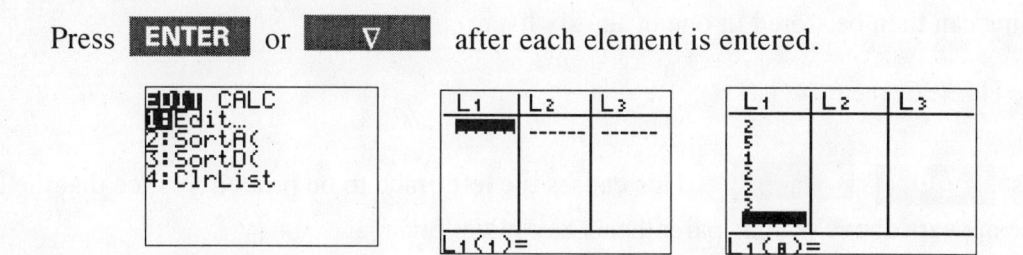

Press **ENTER** or **▽** after each element is entered.

Clearing data from a list

If data is already entered in a list you wish to use and you would like to erase the entire list:

Use the **△** , **▷** or **◁** keys and position the cursor in the column heading over the list name.

Press **CLEAR** **ENTER** to remove the existing data elements and clear the list.

The empty list is ready for entry of new data elements.

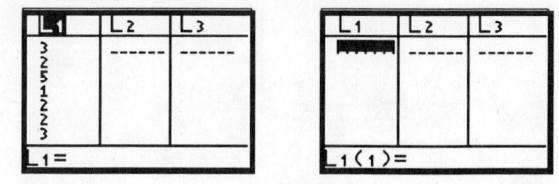

Changing numbers in an existing list

A number in a list can also be changed by positioning the cursor over that number and typing in the desired new number, followed by **ENTER** .

Inserting new numbers into an existing list

If you wish to insert a new number into an existing list without removing any of the numbers from the list first position the cursor where you wish to place the new number. Then put the calculator in **insert** mode by pressing **2nd** **DEL** . The abbreviation for **insert** appears above the **DEL** key in yellow. A zero appears in the list where you wish to insert a new number. Type the new number and press the enter key.

Entering data into a list from the Home Screen

Data entered on the Home Screen using list notation and with each element followed by a comma can then be stored in one of the six lists.

From the Home Screen:

Press **2nd** **(** . This causes the left brace to be printed. Notice that the left brace appears above the left parenthesis key in yellow.

Enter the data elements separating each element with a comma.

Press **2nd** **)** after the last data element is entered. This prints the right brace which appears above the right parenthesis key in yellow.

Storing and viewing the newly-generated list:

After entering the list in braces on the home screen press **STO ▷** **2nd** **1**, then press **ENTER** to store the list in L_1. Notice that L_1 appears above the key with 1 on it, printed in yellow.

Press **STAT** **ENTER** to select **1:Edit** from the **STATEDIT** menu to view the newly-generated list on the list screen.

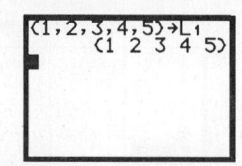

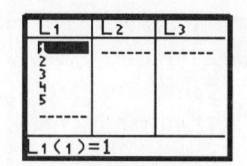

Pressing **STO** **2nd** **2**, and then **ENTER** would store the list in L_2.

Press **STAT** **ENTER** to select **1:Edit** from the **STATEDIT** menu and view the newly-generated list on the list screen.

Generating new data on the list screen:

Press **STAT** **ENTER** to select **1:Edit** from the **STATEDIT** menu and to display L_1, L_2, and L_3.

Position the cursor on the column heading L_2 using the arrow keys.

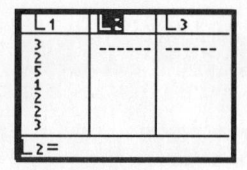

Create a new list L_2, by adding three to each element of L_1:

Type $L_1 + 3$ and press **ENTER**. The calculator adds 3 to each element of L_1 and stores this new list in L_2.

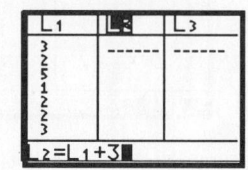

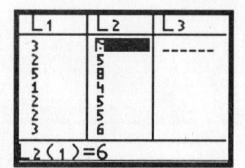

Applying arithmetic operations to an existing list to create new list

A new list can be created either on the home screen and then stored in a blank list or on the list screen.

Generating new data from the Home Screen

A new list of data can be generated from the home screen using an existing list name. The data generated can be stored as a new list and viewed on the list screen.

From the Home Screen:

Press **STAT** **ENTER** to select the **EDIT** menu and to display L_1, L_2, and L_3.

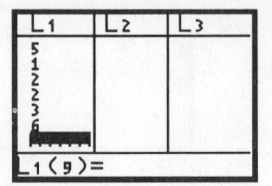

Create a new list L2, adding three to each element of an existing list L1:

Press **2nd** **QUIT** to return to the Home Screen.

Type $L_1 + 3$

Store and view the newly-generated list into L_2.

Press **STO** **2nd** **2**, then press **ENTER** to store the list in L_2.

Press **STAT** **ENTER** to select **1:Edit** from the **STATEDIT** menu and view the newly-generated list.

Use right and left arrows to view additional elements of the list on the Home Screen.

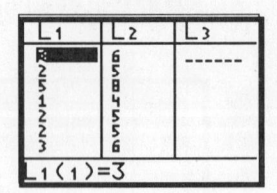

Generating a list of data using the sequence function in the LIST OPS menu

From the Home Screen access the **LIST OPS** menu:

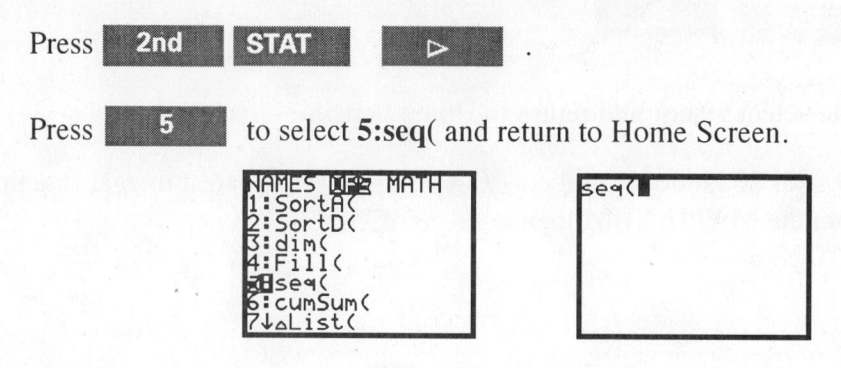

Press `2nd` `STAT` `▷` .

Press `5` to select **5:seq(** and return to Home Screen.

```
NAMES OPS MATH
1:SortA(
2:SortD(
3:dim(
4:Fill(
5:seq(
6:cumSum(
7↓ᴀList(
```

```
seq(
```

Type $x^2, x, 1, 10, 1)$ to generate the squares of the numbers 1– 10.

Press `ENTER` to display the list of elements on the Home Screen. Use left and right arrows to view additional list elements on the Home Screen.

```
seq(X²,X,1,10,1)
```

```
seq(X²,X,1,10,1)
{1 4 9 16 25 36…
```

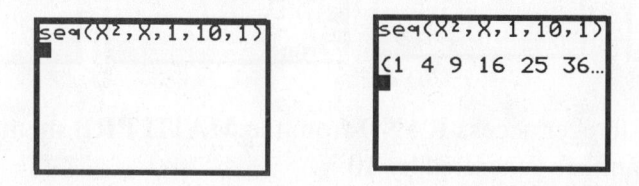

Note: The general format for entry of a sequence is:

seq (generating rule, variable used in generating rule, initial value of the variable, ending value of the variable, increment of the variable)

The increment of the function may be left out on the TI-83 if you wish the increment to be the default value of 1. Commas are necessary and must be included after each input for the sequence function.

To store and view the newly-generated list in L₁:

Press `STO` `2nd` `1` , then press `ENTER` to store the list in L_1.

Press `STAT` `ENTER` to return to the list screen.

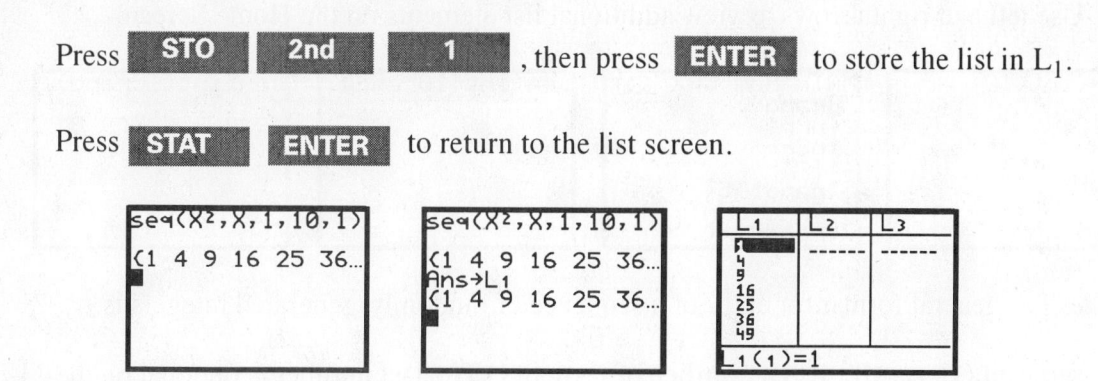

```
seq(X²,X,1,10,1)
{1 4 9 16 25 36…
```

```
seq(X²,X,1,10,1)
{1 4 9 16 25 36…
Ans→L1
{1 4 9 16 25 36…
```

```
L1   L2   L3
1
4
9
16
25
36
49
L1(1)=1
```

431

Generating a list of random numbers using sequences

From the Home Screen access the **LIST OPS** menu:

Press `2nd` `STAT` `▷` .

Press `5` to select **5:seq(** and return to Home Screen.

Generate and store a list of 20 random numbers (0 – 9) using the greatest integer function, **int**, accessed from the **MATH NUM** menu.

Press `MATH` .

Use the right arrow to highlight **NUM**. Enter `5` to select the greatest integer function *5: int*.

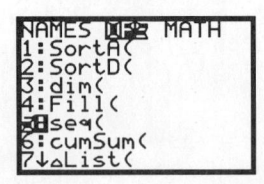

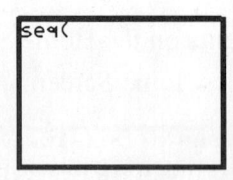

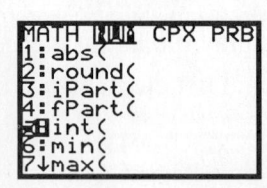

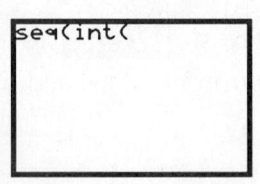

To generate a random integer access **RAND** from the **MATH PRB** menu and multiply the random decimal number generated by 10:

Enter **(10** and Press `MATH` .

Use the right or left arrow to highlight **PRB**.

Press `ENTER` to select the random number generator.

Complete the sequence by typing **), *x*, 1, 20, 1)**. The screen should look like the one below.

Press `ENTER` to display the list of elements on the Home Screen.

Use left and right arrows to view additional list elements on the Home Screen.

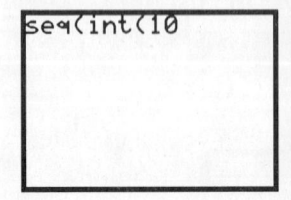

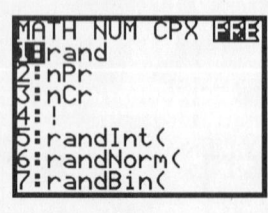

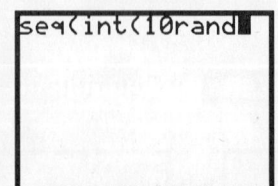

Note: The general format for entry of a sequence of randomly-generated integers is:

seq (int (10 rand), *x*, 1, number of terms, 1) The last number is optional on the TI-83.

Storing the newly-generated list in L_1:

Press STO   2nd   1 , then press ENTER to store the list in L_1.

Press STAT   ENTER to view the list on the list screen.

Use up and down arrows to view additional list elements on the list screen.

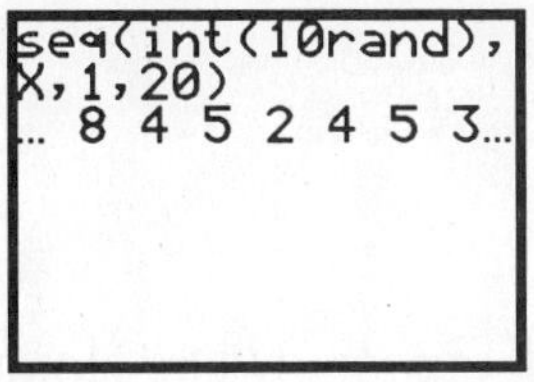

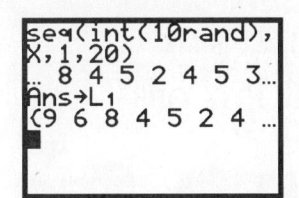

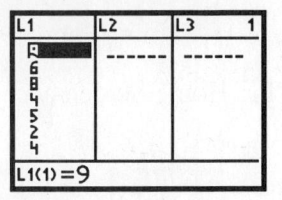

Adding a random number to each element of a list:

With the lists screen displayed:

Position the cursor on column heading L_2. At the bottom of the screen is displayed $L_2 =$

Complete the right-hand side by entering L_1 + int(10 RAND).

Press ENTER to display the list.

Use up and down arrows to view additional list elements.

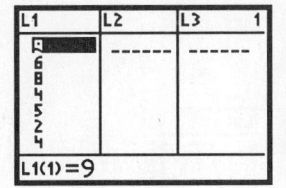

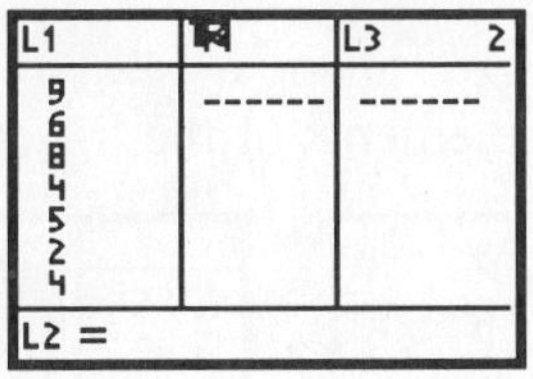

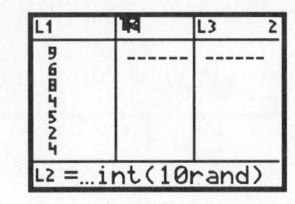

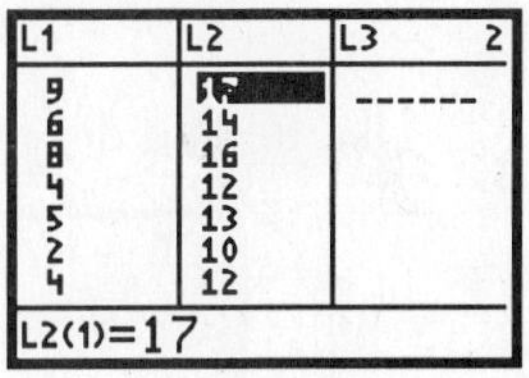

NOTES:

- **RAND** generates and returns a random number between zero and 1 (0 < rand < 1).

- To generate more random number sequences, have individuals each store a different **integer** seed value in **RAND** first.

- If using a new set of TI-82s or TI-83s, have each person store an integer (social security numbers are a good choice) to generate random numbers.

- When you reset a TI-82 or TI-83 or store **0** to **RAND**, the factory-set seed value is restored.

Generating data using a sequence and an existing list (from the Home Screen)

A new list of data can be generated using an existing list with a sequence command from the Home Screen or with the lists displayed.

Using the list of twenty numbers randomly generated, return to the Home Screen.

Press [2nd] [QUIT].

From the Home Screen access the **LIST OPS** menu:

Press [2nd] [STAT] [▷] [5] to select **5:seq(**. The Home Screen displays **seq(**.

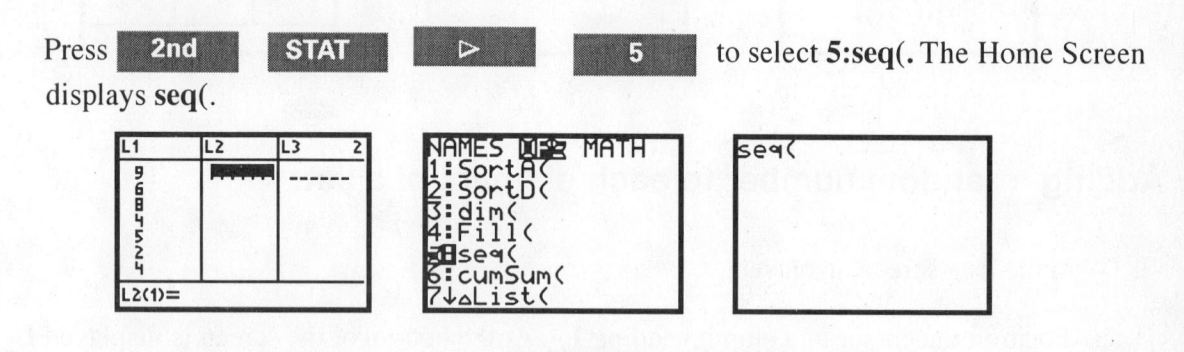

Type $L_1(x) + 3, x, 1, 20, 1)$ then press [ENTER] to display the list.

Press [STO] [2nd] [2], then press [ENTER] to store the list in L_2.

Press [STAT] [ENTER] to return to the list screen.

Use up and down arrows to view additional elements of the list.

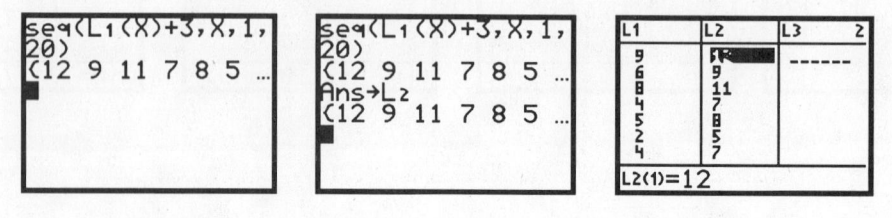

Note: The general format for entry of a sequence using an existing list is:

seq (L_{old} (x) with operation(s), x, 1, number of elements in the list,1) The last one is optional on the TI-83. It need only be included when you wish the step to be something other than 1.

Constructing a scatter plot of data

A scatter plot is a two dimensional graph of ordered pairs and requires two lists of data values, one for the first coordinates and one for the second coordinates. Scatter plots plot the data points from two lists called the Xlist and Ylist in the **SCATTER PLOTS** menu.

You can select one of three options for highlighting the points: (a) a box (□), (b) a cross (+) or (c) a dot (•). The number of data points in the Xlist and in the Ylist must be the same. The frequency of occurrence of each data value does not apply to scatter plots.

From the Home Screen:

Press **Y =** and turn off all selected Y= equations.

Press **STAT** **ENTER** to view the list screen and display L_1, L_2, and L_3.

If data has been previously entered in L_1, L_2, and L_3, clear existing data.

Enter the first set of data in L_1 and the second set of data in L_2.

Press **2nd** **Y =** to access the **STAT PLOTS** menu.

Select **1** for **Plot1.** Highlight **On** and press **ENTER**.

Use the **▽** and **▷** to highlight the first icon under the **Type:** options.

Press **ENTER**. This darkens the first icon which is the **scatter plot** icon.

Use **▽** the list under **Xlist.** Type **L_1** for the **Xlist,** and type **L_2** for the **Ylist.** Choose the square in the **Mark:** options.

Press **ENTER** after each selection.

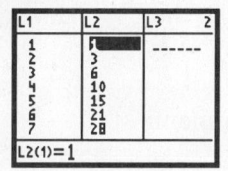

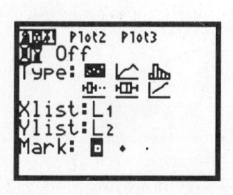

Set a viewing window that contains all the data points and then display the scatter plot on the graphics screen.

Press **ZOOM** **9** to select **9:ZoomStat** from the **ZOOM** menu, set a viewing window and display the scatter plot.

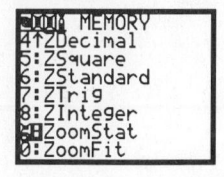

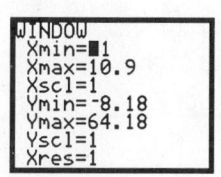

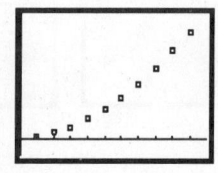

Constructing a line plot of data

Enter the data in lists L_1 and L_2.

Press **2nd** **Y=** to access the **STAT PLOTS** menu.

Select **2** for **Plot2.** Highlight **On** and press **ENTER**.

Use the **▽** and **▷** to highlight the second icon in the **Type:** options.

Press **ENTER**. This selects the second type which is the **line plot** option.

Use the **▽** and type L_1 in **Xlist,** and L_2 in **Ylist** if it isn't already selected.

Press **ENTER** after each selection.

Select **ZoomStat** by pressing **ZOOM** **9**. This command sets a viewing window that contains all the data points and then displays the scatter plot. Turn off **Plot1**.

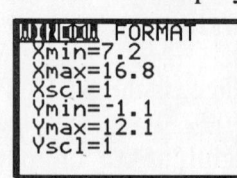

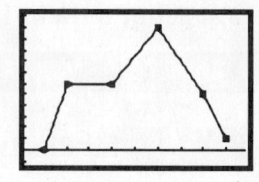

Turning off Stat Plots

When finished with them, it is a good idea to turn off the statistical plots. Not doing so often results in confusing graphs and error messages at a later time when you may be graphing functions entered in the Y= menu.

The TI-83 displays the Plot names at the top of the Y= menu. Any Plot which is darkened is turned on. To turn a plot off simply move the cursor to the Plot name and press the enter key.

Press **2nd** **Y=** to access the **STAT PLOTS** menu.

Select **4** for **PlotsOff**. The calculator returns to the home screen and displays the message **PlotsOff**.

Press **ENTER**. The message **Done** appears indicating that the statistical plots are now off.

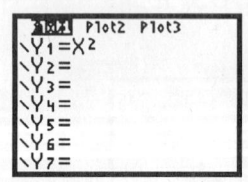

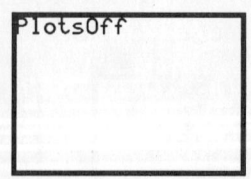

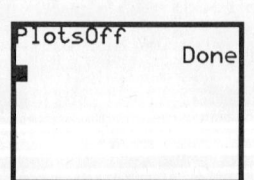

Finite differences and ratios using the sequence function

Finite differences can be calculated using the sequence function for data entered in a list or by using the ΔList item in the **List Ops** menu. If the data consists of ordered pairs, enter the x-coordinates in the first list and the y-coordinates in the second list.

Example: Enter the following table into two list on the list screen and calculate the finite differences for the values in the second column:

X	S(x)
1	35,000.00
2	37,000.00
3	39,000.00
4	41,000.00
5	43,000.00
6	45,000.00
7	47,000.00
8	49,000.00
9	51,000.00
10	53,000.00

If the function is known and entered in the Y= menu, then the ordered pairs can be entered into two lists element by element or by using the function rule with the sequence command. In the table above, the left column is a list of sequential numbers (years) from 1 to 10; the right column is a list of the first ten years of salaries for a fictitious company named CD-R-Us, given an initial starting salary of $35,000, with annual pay raises of $2000 each subsequent year.

Enter the column of sequential years, X, as the first list, L_1, using the sequence function:

Press **STAT** **∘ENTER** to view the list screen and to display L_1, L_2, and L_3.

Position the cursor on the column heading L_1 using the arrow keys.

Press **2nd** **STAT** **▷** **5** to select **5:seq(**.

Type x, x, 1, 10) to generate the list of numbers 1 – 10. Press ENTER.

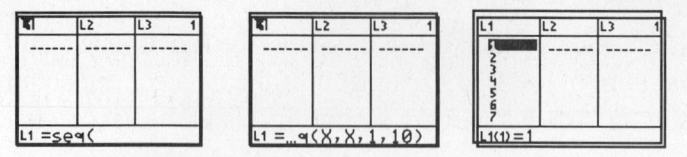

Enter the column of salaries, S(X), as the second list, L_2:

Position the cursor on the column heading L_2 using the arrow keys.

Press 2nd STAT ▷ 5 to select 5:seq(

Type: $35000 + 2000(x-1)$, x, 1, 10) to generate the list of salaries. Press ENTER.

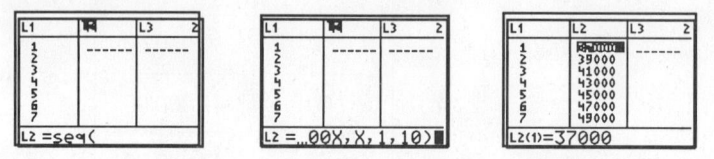

Generate the first finite differences of L_2 and store in L_3 using the sequence function:

Position the cursor on the column heading L_3 using the arrow keys.

Press 2nd STAT ▷ 7 to select 7:ΔList.

Type L_2) to generate the list of first finite differences.

Press ENTER.

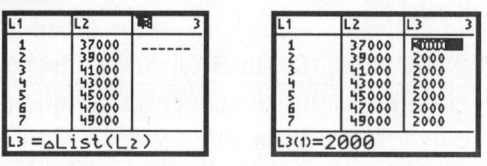

Generate the finite ratios of L_2 and store in L_4:

Position the cursor on the column heading L_4 using the arrow keys.

Press 2nd STAT ▷ 5 to select 5:seq(.

Type L_2 $(x + 1) / L_1 (x)$, x, 1, 9) to generate the list of finite ratios.

Press ENTER.

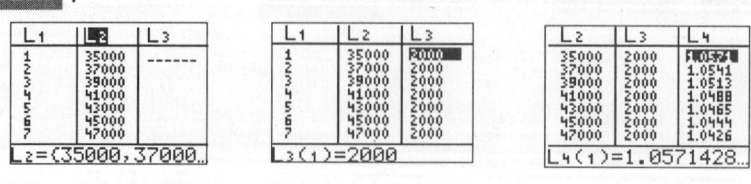

Note: The first finite difference is constant, an indication that the function is linear.

438

Using finite differences to select an appropriate regression equation

Data generated by an unknown sequence can be investigated using finite differences in order to select an appropriate regression equation and determine the function.

Example: Use finite differences to investigate the sequence

$$1, 3, 6, 10, 15, 21, 28, 36, 45, 55, \ldots$$

The first list consists of the positions of the known terms in the sequence; the second list consists of the terms that are given of the unknown sequence.

Position: 1, 2, 3, 4, 5, 6, 7, 8, 9, 10, ...
Seq term: 1, 3, 6, 10, 15, 21, 28, 36, 45, 55, ...

Enter the position numbers as the first list, using a sequence.

Press **STAT** **ENTER** to select the **STAT EDIT** menu and to display L_1, L_2, and L_3.

Position the cursor on the column heading L_1 using the arrow keys.

Press **2nd** **STAT** **▷** **5** to select **5:seq(**.

Type x , x, 1, 10) to generate the list of numbers 1 – 10.

Press **ENTER**.

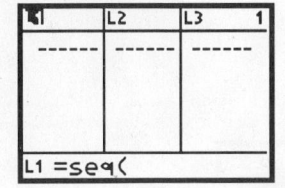

Enter the terms of the sequence as the second list, L_2, element by element.

Position the cursor in L_2 for the first entry.

Enter the terms of the sequence, element by element. Press **ENTER** after each term.

439

Generate the first finite different as L₃:

Position the cursor on the column heading **L₃** using the arrow keys.

Press 2nd   STAT   7   to select **7:ΔList(**.

Type **L₂)** to generate the list of first finite differences.

Press ENTER .

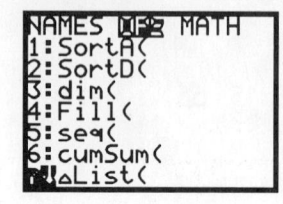

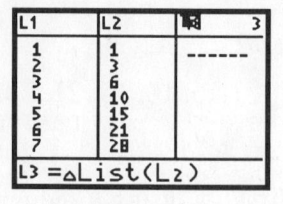

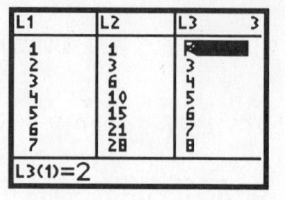

A linear model is indicated by the first finite difference being constant. If, however, the first finite differences are not constant, the function is not linear and you should continue to take finite differences.

Generate the second finite difference as L₄:

Position the cursor on the column heading **L₄** using the arrow keys.

Press 2nd   STAT   ▷   7   to select **7:ΔList(**.

Type **L₃)** to generate the list of second finite differences.

Press ENTER .

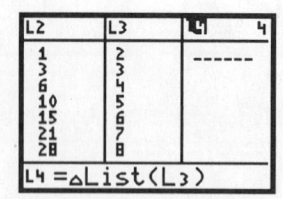

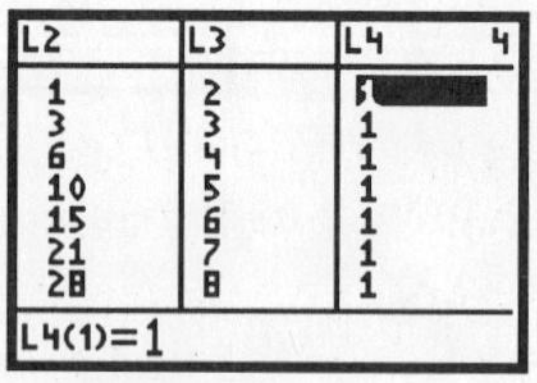

Note: The second finite difference is constant, an indication that the function is quadratic. The function can be determined using the quadratic regression model.

Creating a linear regression model

The TI–83 has the capability of creating the equation of a line that best fits a set of paired data. Such a line is called a linear regression model for the data.

Example: Enter the following paired data as lists L_1 and L_2.

x	1	2	7	9	12
y	4	11	23	32	45

Enter the paired data in two lists, L_1 and L_2.

Construct a scatter plot of the data (see index of procedures).

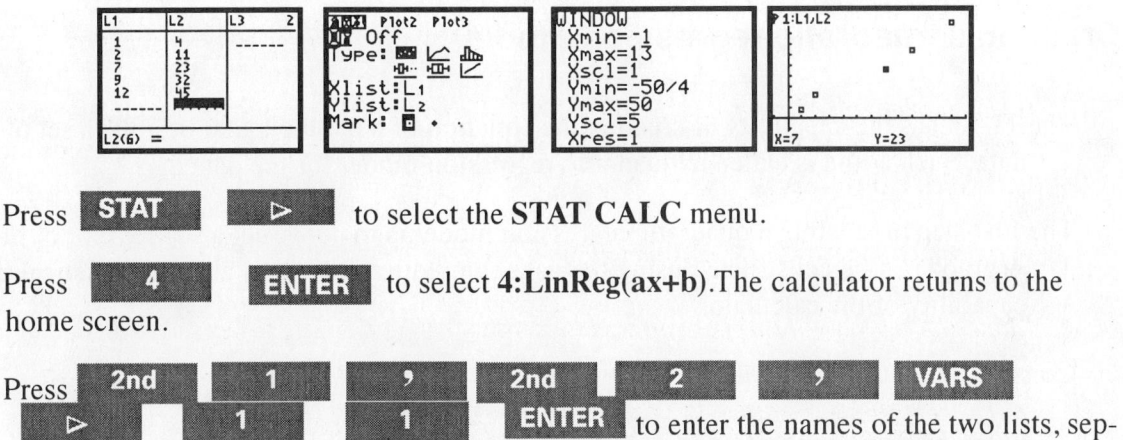

Press **STAT** ▷ to select the **STAT CALC** menu.

Press **4** **ENTER** to select **4:LinReg(ax+b)**. The calculator returns to the home screen.

Press **2nd** **1** **,** **2nd** **2** **,** **VARS** ▷ **1** **1** **ENTER** to enter the names of the two lists, separated by a comma and the Y variable to which the function will be automatically pasted. The best–fit linear model is displayed.

The value of *a* is the slope of the line.

The value of *b* is the vertical intercept of the line.

The value of *r* measures how well the line fits the data. The closer the absolute value of *r* is to one, the better the fit. If your calculator does not display the values of r and r^2 choose

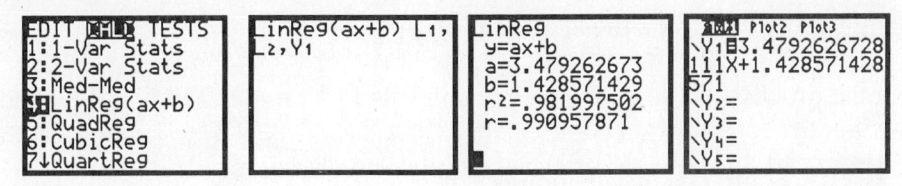

the CATALOG above the zero key and scroll down to DiagnosticOn and press the enter key.

The function has been pasted into the Y= menu for Y_1.

Graph the regression model

The regression model is stored in the **Y =** menu.

Adjust the viewing window if necessary.

Press **GRAPH** to display the regression model.

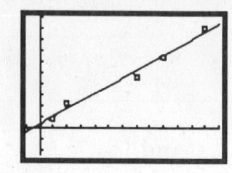

Creating a quadratic regression model

The TI–83 has the capability of creating the equation of a parabola that best fits a set of paired data. Such a line is called a quadratic regression model for the data.

The first step in creating a quadratic regression model is to construct a scatter plot of the data. A graph can be constructed using the data entered into two lists and the statistical plot capability of the calculator.

Construct a scatter plot of data

Enter the data in lists L_1 and L_2 as indicated below.

Press **2nd** **Y =** to access **STAT PLOT**.

Select **1** for **Plot1**.

Highlight **On** and press **ENTER**.

Use the arrow keys to move down to **Type** and highlight the first figure (scatter plot).

Press **ENTER** to turn on the **scatter plot.**

Repeat the process of using arrow keys and typing L_1 under **Xlist**, and L_2 under **Ylist.**

Press **ENTER** after each selection is typed.

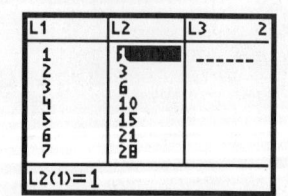

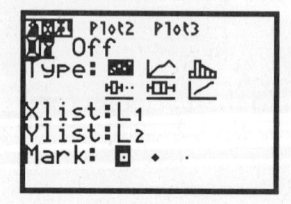

Set a viewing window that contains all the data points and then display the scatter plot.

> Set an appropriate viewing window and graph the data.

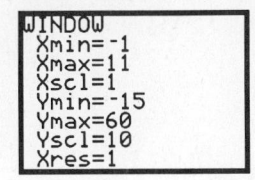

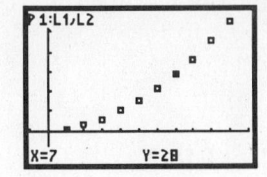

Select an appropriate regression model.

The data points displayed on the scatter plot give the appearance of being quadratic. A good first choice would be the quadratic regression model.

> Press **STAT** **▷** to select the **STAT CALC** menu.
>
> Press **5** **ENTER** to select **QuadReg**. The Home Screen is displayed.
>
> Press **2nd** **1** **,** **2nd** **2** **,** **VARS** **▷** **1** **ENTER** to enter the names of the two lists, separated by a comma and the Y variable to which the function is to be pasted. The best-fit quadratic model is displayed on the home screen and pasted in the Y= menu.

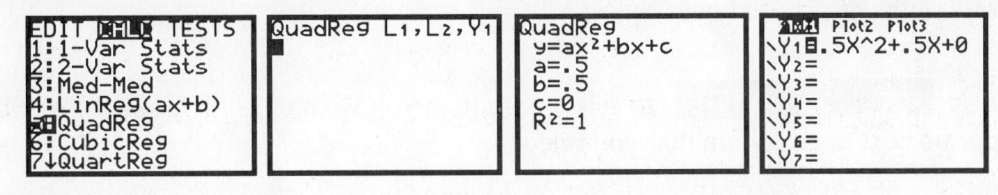

The value of a is the coefficient of the quadratic term. The value of b is the coefficient of the linear term and the value of c is the vertical intercept and constant term.

Graph the regression model.

> Press **GRAPH** to display the graph of the regression model.
>
> Adjust the viewing window if necessary. Use the table generated by the equation entered in the Y = menu to view additional values of the sequence.

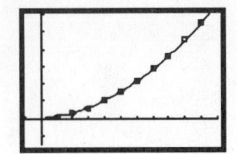

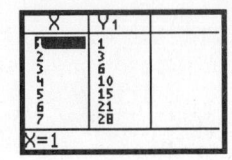

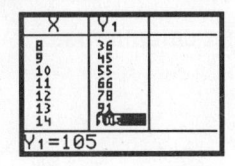

Transferring data between calculators: using LINK

Connect the two calculators using the connection cable.

Person receiving the data:

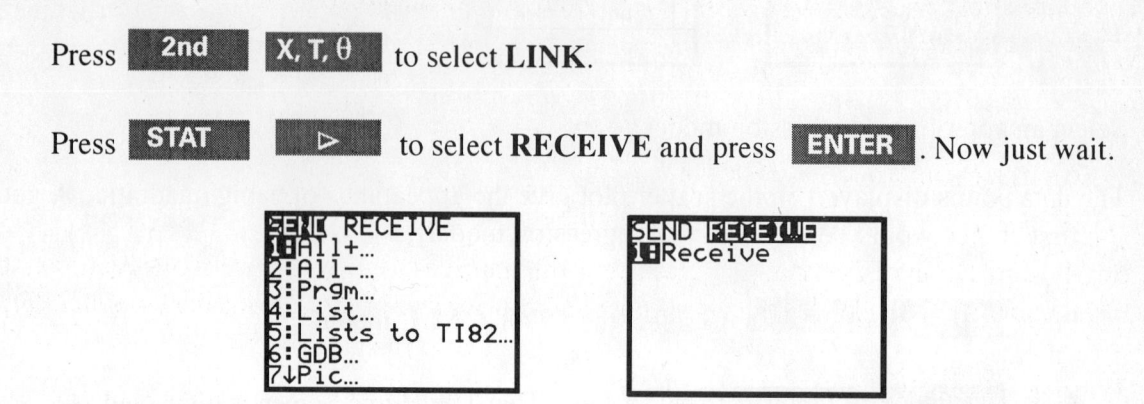

Press `2nd` `X, T, θ` to select **LINK**.

Press `STAT` `▷` to select **RECEIVE** and press `ENTER`. Now just wait.

```
SEND RECEIVE
1:All+...
2:All-...
3:Prgm...
4:List...
5:Lists to TI82...
6:GDB...
7↓Pic...
```

```
SEND RECEIVE
1:Receive
```

Person sending the data:

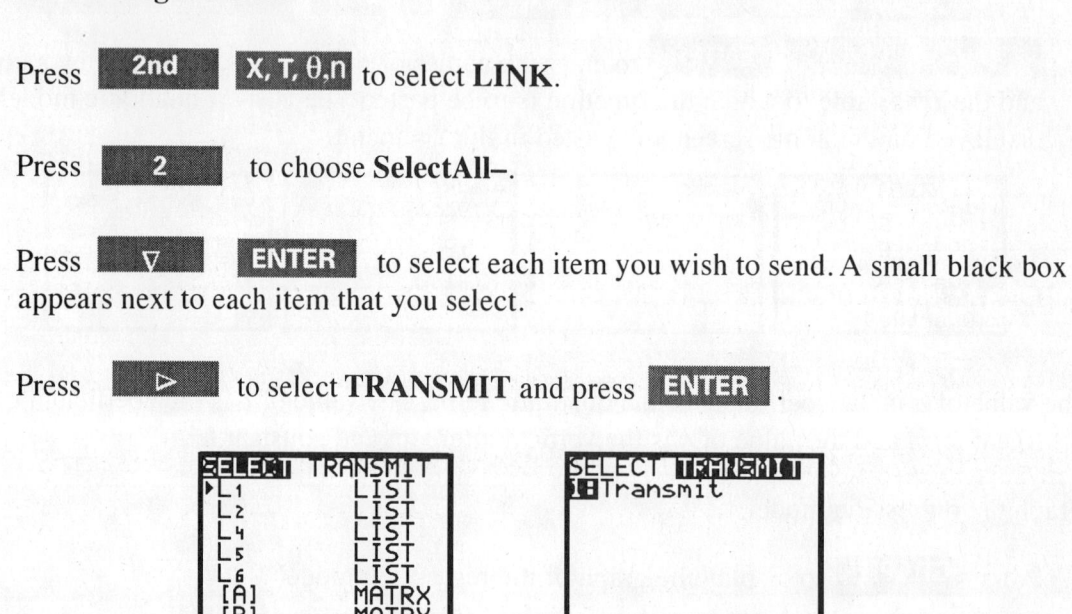

Press `2nd` `X, T, θ, n` to select **LINK**.

Press `2` to choose **SelectAll–**.

Press `▽` `ENTER` to select each item you wish to send. A small black box appears next to each item that you select.

Press `▷` to select **TRANSMIT** and press `ENTER`.

```
SELECT TRANSMIT
▶L1      LIST
 L2      LIST
 L4      LIST
 L5      LIST
 L6      LIST
 [A]     MATRX
 [B]     MATRX
```

```
SELECT TRANSMIT
1:Transmit
```

Disconnect calculators.

Entering and displaying a matrix

A matrix is a two-dimensional array of rows and columns. The dimensions of a matrix, the number of rows and the number of columns, is designated with the row number preceding the column number. The TI-83 stores as many as ten matrices using the matrix variables [A], [B], [C], [D], [E], [F], [G], [H], [I] and [J]. You can display, enter, or edit a matrix in the matrix editor.

Example: Enter the 3 x 3 matrix $\begin{bmatrix} 2 & -5 & 3 \\ 0 & 1 & -2 \\ -4 & 7 & 6 \end{bmatrix}$.

Set the dimension of the matrix. If you are using matrices to solve a system of equations, the number of equations is the row number and the number of variables is the column number.

Press **MATRIX** and arrow to **EDIT**.

Type **1** to select matrix [A].

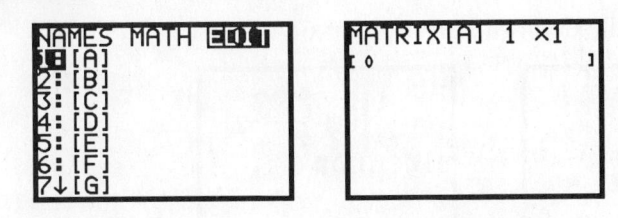

Enter the size of the matrix (in this case, 3 x 3) followed by the matrix elements.

Press **2nd** **MODE** to return to the Home Screen.

Press **MATRIX** **1** **ENTER** to display matrix [A] in the Home Screen.

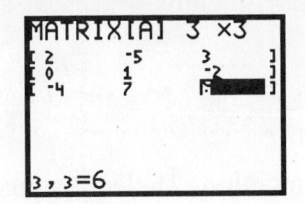

Calculating the determinant of a matrix

Example: Find the determinant of the 3 x 3 matrix $\begin{bmatrix} 2 & -5 & 3 \\ 0 & 1 & -2 \\ -4 & 7 & 6 \end{bmatrix}$.

Enter the matrix, if you haven't done so, as matrix [A].

Return to the Home Screen.

Press **MATRIX** and arrow to **MATH**.

Type **1** to select **det**, which represents a determinant.

The calculator returns to the Home Screen and displays **det**.

Press **MATRIX** **1** to display [A] after **det**.

Press **ENTER** to see the determinant of matrix [A].

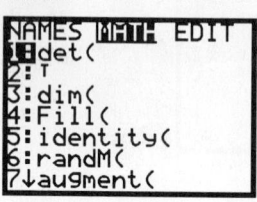

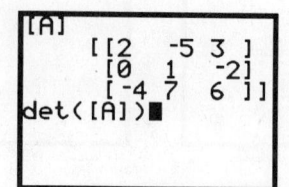

Deleting a matrix from memory

Press **2nd** **+** to access the **MEM** menu.

Press **2** to display the 2: Delete options.

Press **5** to display the matrix deletion options. Select the matrix you wish to delete.

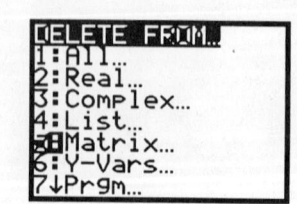

Press **ENTER** to delete the selected matrix.

446

Finding the inverse of a square matrix

Example: Find the inverse of the 3 x 3 matrix $\begin{bmatrix} 2 & -5 & 3 \\ 0 & 1 & -2 \\ -4 & 7 & 6 \end{bmatrix}$.

Enter the matrix, if you haven't done so, as matrix [A].

Return to the Home Screen.

Press **MATRIX** **1** **x⁻¹** **ENTER** to display the inverse of matrix [A] in decimal form.

The left and right arrow keys can be used to see all elements of the 3 x 3 inverse matrix.

```
[A]⁻¹
[[1.666666667 4...
 [.6666666667 2...
 [.3333333333  ...
```

```
[A]⁻¹
...5  .5833333333]
...   .3333333333]
...   .1666666667]]
```

Press **MATH** **1** **ENTER** to display the inverse matrix in fractional form.

```
...5  .5833333333]
...   .3333333333]
...   .1666666667]]
Ans▶Frac
[[5/3  17/4  7/12...
 [2/3  2     1/3  ...
 [1/3  1/2   1/6  ...
```

Solving linear systems using matrix inverses

A system of linear equations can be solved by entering the coefficients as elements in an augmented matrix, then finding the solution using matrix row operations. If the coefficient matrix is square (same number of rows and columns) and the determinant is not equal to zero, you may solve the system by finding the product of the inverse of the coefficient matrix and the constant matrix.

Example: Solve the system
$$\begin{aligned} 2x - 5y + 3z &= 7 \\ y - 2z &= -3 \\ -4x + 7y + 6z &= 1 \end{aligned}$$

The coefficient matrix is $[A] = \begin{bmatrix} 2 & -5 & 3 \\ 0 & 1 & -2 \\ -4 & 7 & 6 \end{bmatrix}$ and the constant matrix is $[C] = \begin{bmatrix} 7 \\ -3 \\ 1 \end{bmatrix}$.

Enter the coefficient matrix as matrix[A], if you haven't done so.

Press **MATRIX** and arrow to **EDIT**.

Type **3** to select matrix [C].

Enter the size of the matrix (in this case, 3 x 1) followed by the matrix elements.

Press **2nd** **MODE** to return to the Home Screen.

The determinant of [A] is 12. The system has one solution since the determinant of the coefficient matrix is not zero.

If the linear system has a unique solution, the computation $[A]^{-1}[C]$ will return a column matrix that contains the solution to the system.

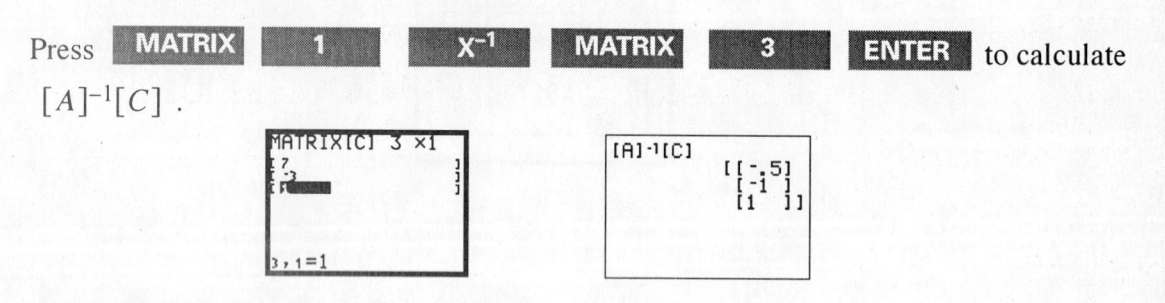

Press **MATRIX** **1** **X⁻¹** **MATRIX** **3** **ENTER** to calculate $[A]^{-1}[C]$.

The 3 x 1 solution matrix displays the solution to the system
$$\begin{aligned} 2x - 5y + 3z &= 7 \\ y - 2z &= -3 \\ -4x + 7y + 6z &= 1 \end{aligned}$$

Reading down the matrix, the solution is $x = -0.5$, $y = -1$, and $z = 1$.

TI-82
Index of Procedures

and

TI-83
Index of Procedures

Index of TI-82 Procedures

A

Access functions using the 2nd key 349
Add a random number to each element of a list 379
Adjust the display contrast 345
Adjust view window 364
Apply arithmetic operations to an existing list to create new list 375
Arithmetic computations 350
ASK mode 359

C

CALC 369
Calculate the determinant of a matrix 392
Calculations 350
Calculator organization 344
Change the mode 345
Change numbers in an existing list 374
Change the view window 364
Clear data from a list 373
Connected Mode for graphs 366
Construct a scatter plot of data 381
Construct a line plot of data 382
Control the TRACE option 364
Create a linear regression model 387
Create a quadratic regression model 389

D

Darken or lighten screen 345
Default view window 361
Delete a matrix from memory 393
Determinant of a matrix 392

Discrete graph 366
Display coordinates of points on a graph 363
Display a discrete graph 366
Display a matrix 392
Display a table of values 355
Dot Mode for graphs 366

E

Edit a previously-entered expression 351
Edit an existing list 374
Enter values in a matrix 392
Enter data into a list 373
Enter data into a list from the Home Screen 374
Enter expressions in the Y = menu 354
Evaluate a function 357, 358
Exponentiation 350

F

Factorial function 355
Find zeros of a function 369
Find intersection points 370
Find a function's minimum 372
Find a function's maximum 371
Finite differences 383
Finite ratios 384
Friendly view windows 364
Function evaluation: storing input 357
Function evaluation: using Ask mode 359
Function evaluation: using function notation 358

G

Generate data using an existing list 380
Generate data from the Home Screen 376
Generate new data on the list screen 375
Generate a list of data using the sequence function 377
Generate a list of random numbers 378
Graphics screen 347
Graph a function 360
Graph the regression model 388
Graph the solution interval to inequalities in one variable 368

H

Home screen 346, 380

I

Important keys and their function 348
Inequalities: graphing a solution interval 368
Insert new numbers into an existing list 374
INT function 351
Intersection points 370
Inverse of a Square Matrix 393

K

Keyboard organization 344

L

Linear regression models 387, 388
Linear systems graphing solution 394
List screen 347
Lists of data 374

M

Maximum of a function 371
Matrix entry and display 392
Matrix inverses 393, 394
Menus 346
Menu screens 347
Minimum of a function 372

N

Negative numbers 350

O

Obtain fractional results 352

P

Parametric graphing 367

Q

Quadratic Regression models 389

R

Raise a number to a power 340
RAND function 352, 379
Reciprocal function 351
Regression equation 385

S

Scientific notation (3.46E11means) 350
Select a menu item 346
Setting a view window 362
Solve linear systems graphically 394
Square matrix 393
Standard viewing window 361
STAT PLOT 382

Store a function input 357
Store the regression model 388
Store a value for a variable 356
Store a newly-generated list 375, 377

T

Table screen 347
TRACE 363
Transfer data between calculators 391
Turn on/off Stat Plots 382
Turn the calculator on/off 344

V

View a newly-generated list 375
View a table of values 355

Z

Zeros of a function 369
ZOOM options to adjust the view window 364

Index of TI-83 Procedures

A

Access functions using the 2nd key 402

Add a random number to each element of a list 433

Adjust:
    the display contrast 397
    the view window 417

Apply arithmetic operations to an existing list to create new list 429

Arithmetic computations 403

ASK mode 412

C

CALC 422, 423, 424

Calculate the determinant of a matrix 446

Calculations 403

Change:
    the mode 398
    the view window 415

Clear data from a list 427

Connected Mode for graphs 419

Construct:
    a scatter plot of data 435
    a line plot of data 436

Continuous graph 419

Control the TRACE option 417

Coordinates of points on a graph 416

Create:
    a linear regression model 441
    a quadratic regression model 442

D

Darken or lighten screen 397

Default view window settings 414, 417

Delete:
    an equation from the Y= menu 407
    a list of data 428
    a matrix from memory 446

Determinant of a matrix 446

Discrete graphs 419

Display:
    coordinates of points on a graph 416
    a discrete graph 419
    a matrix 445
    a table of values 408
    view window settings 417

Dot Mode for graphs 419

E

Edit:
    an existing list 428
    a previously-entered expression 404

Enter:
    data into a list 427
    data into a list from the Home Screen 428
    expressions in the Y = menu 407
    a matrix 445

Evaluate a function 411, 412

Exponentiation 403

F

Factorial function 407

Find:
    a function's maximum 424
    a function's minimum 425
    intersection points 423
    zeros of a function 422

Finite differences 437, 439

Finite ratios 438
Friendly view windows 417
Function evaluation:
    storing input 410
    using Ask mode 412
    using function notation 411

G

Generate:
    data using an existing list 433, 434
    data from the Home Screen 430
    new data on the list screen 429
    a list of data using the sequence function 431
    a list of random numbers 432
Graph:
    a continuous function 419
    a discrete function 419
    a function 413
    the regression model 442
    the solution interval to inequalities in one variable 421
Graphics screen 399

H

Home screen 399, 428, 430

I

Important keys and their function 401
Inequalities: graphing a solution interval 421
Insert new numbers into an existing list 428
INT function 405
Intersection points: finding 423
Inverse of a Square Matrix 447

K

Keyboard organization 396

L

Line plot of data 436
Linear regression models 441
Linear systems solutions 423, 448
LINK function 444
LIST OPS menu 431
List screen 400, 429
Lists of data 429

M

Maximum of a function 424
Matrix
    entry 445
    determinant of 446
    display 445
    inverses of a square 447, 448
    soltuions of linear systems 448
Menus 398
Menu screens 398
Minimum of a function 425
MODE menu 398

N

Negative numbers 403

O

Obtain fractional results 405

P

Parametric graphing 420

Q

Quadratic Regression models 442

R

Raise a number to a power 403
Random number generation 432
RAND function 406, 432
Reciprocal function 404
Regression equation 439
Regression models:
    graphing 442
    linear 441
    quadratic 442

S

Scatter plot of data 435
Scientific notation (3.46E11means) 404
Select a menu item 398
Sequence function 431, 432, 434, 437
Setting a view window 398, 414, 417
Standard viewing window 414
STAT PLOT 436
Store:
    a function input 411
    a newly-generated list 429
    a regression model 442
    a value for a variable 409

T

Table screen 400
TRACE 416, 417
Transfer data between calculators 444
Turn on/off Stat Plots 436
Turn the calculator on/off 396

V

Variable substitution in an expression 409
View:
    a graph 413, 419
    a matrix 445
    a newly-generated list 429
    a previously-entered expression 404
    a table of function values 408
    a line plot of data 436
    a scatter plot of data 435
    an existing list 428
    an expression 407
View window:
    adjustments 417
    a friendly 417
    standard (default) settings 414, 417

Y

Y = menu 399, 407

Z

Zeros of a function 422
ZOOM options: adjust the view window 417